ATLAS OF ZEBRAFISH DEVELOPMENT

ATLAS OF ZEBRAFISH DEVELOPMENT

Robert Bryson-Richardson
School of Biological Sciences, Monash University

Silke Berger
Australian Regenerative Medicine Institute, Monash University

Peter Currie
Australian Regenerative Medicine Institute, Monash University

ELSEVIER

AMSTERDAM • BOSTON • HEIDELBERG • LONDON • NEW YORK • OXFORD
PARIS • SAN DIEGO • SAN FRANCISCO • SINGAPORE • SYDNEY • TOKYO
Academic Press is an imprint of Elsevier

Academic Press is an imprint of Elsevier
32 Jamestown Road, London NW1 7BY, UK
225 Wyman Street, Waltham, MA 02451, USA
525 B Street, Suite 1800, San Diego, CA 92101-4495, USA

First edition 2012

British Library Cataloguing-in-Publication Data
A catalogue record for this book is available from the British Library

Library of Congress Cataloging-in-Publication Data
A catalog record for this book is available from the Library of Congress

ISBN : 978-0-12-374016-8

For information on all Academic Press publications
visit our website at www.elsevierdirect.com

Typeset by MPS Limited, a Macmillan Company, Chennai, India
www.macmillansolutions.com

Printed and bound in China

12 13 14 15 10 9 8 7 6 5 4 3 2 1

CONTENTS

CHAPTER 1

INTRODUCTION

THE NEED FOR AN ATLAS OF ZEBRAFISH DEVELOPMENT

Over the past two decades zebrafish have been established as a genetically tractable model system to investigate many different aspects of vertebrate development. The strengths of this model system stem from a unique combination of embryological manipulability and the optical clarity of the early embryo and larvae, which allows, by simple optical inspection, the visualization of cell biological events within an in vivo context. This is further enhanced by the ability to apply invertebrate-style forward genetics to questions of vertebrate development. While the credentials of the system as a developmental model have long been recognized, it is the application of the zebrafish to a different aspect of biology that has generated a surge in popularity and greatly increased zebrafish laboratory use. Many investigators have now turned to directly modeling human disease states in zebrafish, realizing that many of the same strengths that made it a compelling model for the study of development also allow it to complement existing mammalian disease models (Lieschke and Currie 2007).

Numerous human disease conditions are now being modeled in zebrafish, from addiction syndromes to carcinoma. Remarkably, in the vast majority of instances the results of these studies indicate that the zebrafish models share a similar disease pathogenesis with humans (Darland and Dowling 2001; Langenau et al. 2003; Amsterdam et al. 2004; Carvan et al. 2004; Lockwood et al. 2004; Berghmans et al. 2005; Shepard et al. 2005; Patton et al. 2005). Zebrafish have also come of age as a reverse genetic model system. The ability to specifically knock down and mutate individual loci by a number of different techniques has allowed function of human disease gene orthologues to be directly examined in zebrafish. Antisense oligonucleotides or morpholinos have been utilized very effectively to knock down specific gene function during the embryonic and early larval period. Targeted lesion detection (Wienholds et al. 2003) has produced mutant models of a number of human diseases. This strategy has identified p53 mutant zebrafish that have a predisposition to cancer, rag1 deficient zebrafish with immunodeficiency (Wienholds et al. 2002, Berghmans et al. 2005), and has also been utilized to generate mutations in the APC gene that predispose to colorectal cancer (Hurlstone et al. 2003).

These approaches possess the ability to revolutionize the way that zebrafish biologists approach disease modeling in this system. The commitment of the Wellcome Trust's Sanger Centre to producing a fully sequenced and integrated high-quality genome sequence for the zebrafish has enabled the development of many of the new reverse genetic techniques available to zebrafish researchers as well as facilitating the cloning of randomly generated mutations. Analysis of the zebrafish genome has demonstrated a considerable level of synteny with the human genome, and direct orthologues of human genes can nearly always routinely be located within the zebrafish genome. An ambitious new project that aims to mutate every gene within the zebrafish genome using a combination of the approaches outlined above has also been recently proposed. There is a possibility that a percentage of mutant phenotypes produced by the zebrafish knockout consortium will be phenotyped in a systematic way and catalogued in a database.

Another recent development has raised the stakes considerably for zebrafish as a human disease model. The application of chemical genetics to zebrafish biology has allowed for the first time the simultaneous analysis of a mutant phenotype representative of a human disease and the screening for drugs that will influence this phenotype in vivo (Peterson et al. 2004). The ability to screen in vivo using zebrafish allows the drug discovery process to leapfrog a number of the initial hurdles of toxicity and specificity that plague drug discovery employing in vitro models.

The unique features of zebrafish biology, such as external fertilization, high fecundity, ease of adding putative drugs to larval fish, make this the only fully integrated vertebrate model system for drug discovery. Collectively, these recent advances have heralded a watershed period for researchers utilizing zebrafish as a tool for modeling human disease and indicate a further expansionary phase in its use as a model system given that many of its strengths complement existing mammalian models so fully. However, despite the advances in many areas of zebrafish biology and the generation of sophisticated models of human disease, one area has substantially lagged behind. There is a lack at this current point in time of a detailed anatomical reference for zebrafish. While the first 24 hours of zebrafish development have been well studied, comparatively little is known about the development of larval, juvenile and adult anatomy. Such knowledge is critical for the analysis of developmental processes extending later in development and the accurate modeling of human disease in zebrafish, as these periods are the most likely to generate information that is relevant to human disease pathogenesis. Therefore the aim of this book is to provide

Atlas of Zebrafish Development. 10.1016/B978-0-12-374016-8.00001-7

an anatomical account of the development of the zebrafish from embryo to adult. Our hope is that this resource, in conjunction with the online three-dimensional (3D) anatomical resource FishNet (http://www.FishNet.org.au, discussed in Chapter 2) from which the images in this book are derived, can be utilized as a standard reference text on which to base further analyses that will investigate aspects of normal and perturbed zebrafish development.

EXISTING ANATOMICAL RESOURCES FOR ZEBRAFISH

The major online resource for zebrafish is the zebrafish information network ZFIN (http://www.zfin.org). ZFIN curates information on tools, publications and laboratories involved in zebrafish research, as well as information about the zebrafish strains and DNA stocks held by the Zebrafish International Resource Center (ZIRC). Importantly, ZFIN also contains the first online collection of anatomical images comprising of nearly 80 sections covering the first 5 days of development. Other anatomical resources are scarce for zebrafish and piecemeal in their representation of anatomy. Several specialized zebrafish anatomy atlases exist in book form, including atlases of adult and embryonic brain anatomy. Necessarily these types of atlases are limited by showing only two-dimensional (2D) images in restrictive views. For instance the adult brain atlas (Wullimann et al. 1996), despite its high degree of anatomical complexity, shows 2D brain structure in complete isolation from any other tissues. The embryonic brain atlas is also limited in scope, showing only transverse views of a very limited number of stages (Mueller and Wulliman 2005). Other embryonic atlases and staging series have been published, which collectively approximate the resources currently online at ZFIN (Kimmel et al. 1995; Schilling 2002). All of these resources rely on 2D imaging.

Several online organ-specific 3D atlases of development have now been developed for zebrafish. A 3D atlas of zebrafish vasculature created by microangiography in the Weinstein laboratory shows the formation of the vasculature from 1 to 7 days in a series of 3D renderings (http://uvo.nichd.nih.gov/atlas.html; Isogai et al. 2001; Kamei et al. 2004). This site allows the user to view a series of images at each stage showing the vasculature in 3D utilizing confocal microscopy of specimens labeled using microangiography to chart the forming vasculature. This series of images represents the first example of a 3D developmental model of any organ system in zebrafish. However, the lack of ability to simultaneously visualize both the vasculature and the tissue through which it forms limits the usefulness of the atlas. Furthermore, the use of confocal microscopy to generate 3D models is limited to stages at which the laser light can suitably penetrate the tissue, and necessarily this is restricted to the earliest stages of zebrafish development.

A high resolution 3D anatomical atlas of the zebrafish developing brain is also being developed within the group of Steve Wilson at UCL in London (http://www.zebrafishbrain.org). This atlas relies on the optical clarity of the zebrafish embryo, coupled with a host of transgenic and antibody labelings to highlight discrete neuroanatomical structures within the developing brain. It uses confocal microscopy exclusively to generate highly detailed 3D interactive models of embryonic and larval brain development. Although this will be a unique and important neuroanatomical resource this approach will necessarily be restricted to relatively early stages of zebrafish brain development and will not be applicable to juvenile and adult zebrafish.

A growing repository of 2D section information is being deposited at the Penn State Atlas project (http://zfatlas.psu.edu), which, when completed, will be a significant data set. Although currently mostly unannotated, the high resolution nature of the material will suit those researchers looking at the cellular anatomical level across a wider range of developmental stages. A significant online zebrafish anatomical resource already exists in the form of a curated anatomical nomenclature. Standardized terms for anatomical structures are deposited and curated at ZFIN by the zebrafish research community, and represent a significant framework on which any annotated developmental atlas can now be built. It also represents a mechanism by which names of new or uncertain anatomical structures can be reviewed and approved relatively rapidly. It ensures that any anatomical atlas that uses the nomenclature represents the experience of the wider zebrafish research community, rather than the more limited experience of the individuals creating the atlas. The committee has also used standardized nomenclature to facilitate comparison between species. This book has endeavored to use the standard anatomical terms curated with ZFIN to describe the structures we have annotated in this book.

In summary, the FishNet website and the images from it that we present within this book represent the first full 3D annotated life history anatomical reference, from embryo to adult, of zebrafish. Indeed it represents the first such anatomical reference for any animal model system, covering the complete range of development from early embryo until adult. It fills a current need for a general gross anatomical reference for zebrafish researchers and we hope it will be an important reference text for all individuals who utilize zebrafish as research and teaching tools.

CHAPTER 2

METHODOLOGY AND INTERPRETATION OF THE ATLAS

This atlas aims to provide a reference at the gross anatomical level for the zebrafish, identifying the major anatomical structures in samples covering the range of development from the early embryo until adulthood. The atlas has been generated using the technique of Optical Projection Tomography (OPT, described below). This technique has a number of advantages for the generation of an anatomical atlas. One of the clear advantages is that it is possible to generate a complete three-dimensional (3D) model from a single sample. As a result the sections in each chapter are images from the same sample for all three orientations. Using the OPT approach, however, we are not able to provide cellular resolution for the range of sample sizes encompassing the development of zebrafish from embryo to adulthood. To generate the images we have used in this atlas, we have relied on the autofluorescence of the tissue following paraformaldehyde fixation to generate the contrast of the images. As a result of this approach, structures such as blood, cartilage, and bone appear much brighter than other tissues. Similarly, reflection of light from the eye results in a much brighter appearance. Differences in the level and pattern of fluorescence are necessary to provide the tissue definition in the images but, as a result, users of the atlas may not be immediately familiar with this type of image. Therefore we will provide here not only the contents of the atlas and the methodology but also a brief guide to the interpretation of the OPT images.

STRUCTURE OF THE ATLAS

There are nine chapters of images containing samples at 24, 48, 72, 96, and 120 hours post-fertilization and at 6 mm, 9 mm, 14 mm, and 16 mm total length. Each sample is displayed in the classical anatomical section views of transverse, sagittal, and coronal. Transverse sections are listed anterior to posterior, sagittal right to left, and coronal dorsal to ventral, and are evenly spaced along the axis of sectioning (Figure 2.1).

METHODOLOGY

We produced a series of 18 3D models of zebrafish development from 24 hours post-fertilization until adulthood using the method of OPT as described in Bryson-Richardson et al. (2007). Zebrafish were obtained from either wild-type or golden mutant strains and staged by age for the first 5 days of development and by total length (most anterior point until the tip of the tail) for the remainder of the staging series, due to variability of size with age after free feeding. Samples were fixed in 4% paraformaldehyde overnight at 4°C, and then washed with PBS containing 0.1% Tween 20. Samples were used immediately or progressed through a methanol series for storage in 100% methanol at −20°C. Pigmented embryos were bleached in several changes of 5% hydrogen peroxide in methanol on a lightbox. Samples were hydrated through a methanol series if necessary, washed in water, and then embedded in a 1.5% low melting point agarose gel. An agarose block containing the sample was then attached to a metal mount. The mounted samples were then cleared through a series of 25%, 50%, 75%, and 100% methanol prior to immersion in a 2:1 benzyl alcohol:benyzl benzoate solution. The samples were then scanned on a prototype OPT scanner as described in Sharpe et al. (2002).

OPTICAL PROJECTION TOMOGRAPHY (OPT)

Optical Projection Tomography (OPT) allows the generation of 3D models from cleared biological samples using conventional brightfield and fluorescent staining procedures or, as in this atlas, imaging without labeling utilizing the intrinsic fluorescence of the sample. A series of images is taken of the sample as it is rotated through 360°. These images are projections through the sample that can be used to generate a section image using back projection (Figure 2.2). In transmitted illumination OPT the projection images record the loss of light as it passes through the sample. In the epi-fluorescent OPT imaging used in the atlas the images record the total amount of fluorescence emitted by the sample along the axis of imaging. The 3D models used to generate the atlas were generated from a series of

Atlas of Zebrafish Development. 10.1016/B978-0-12-374016-8.00002-9

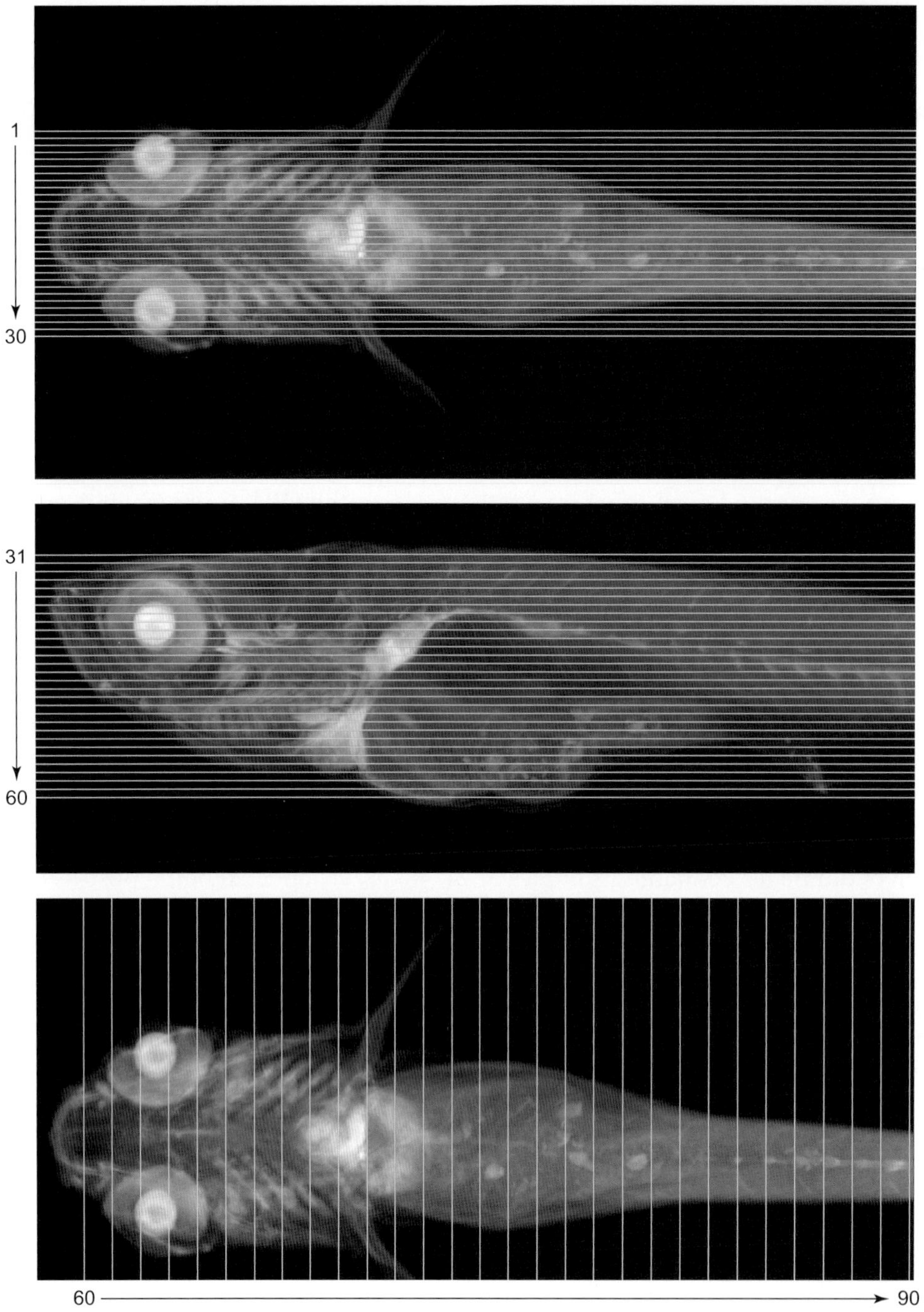

Figure 2.1 Representation of section positions for Chapter 10. In each chapter, sections are evenly spaced and numbered from right to left, dorsal to ventral, and anterior to posterior.

400 images captured over a 360° rotation. Image reconstruction was carried out using filtered back projection (as described by Sharpe et al. 2002). This is very similar in principle to other forms of projection tomography, including magnetic resonance imaging and x-ray computer tomography, the key difference being the use of visible wavelengths to generate the information, as a result of which the sample must be optically clear and have a consistent refractive index throughout, hence the bleaching and clearing described above. In contrast to x-ray and other imaging techniques, however, the resolution is limited by the image-forming optics. The tomographic approach used requires the sample to be in focus for at least 180° of the rotation and therefore the depth of focus should extend at least half-way into the sample. Image resolution is proportional to the numerical aperture of the objective whilst the depth of focus is inversely proportional. Therefore, there is a trade-off between resolution and the thickness of the sample; as a result we are unable to image at a constant resolution throughout the wide range of sample sizes

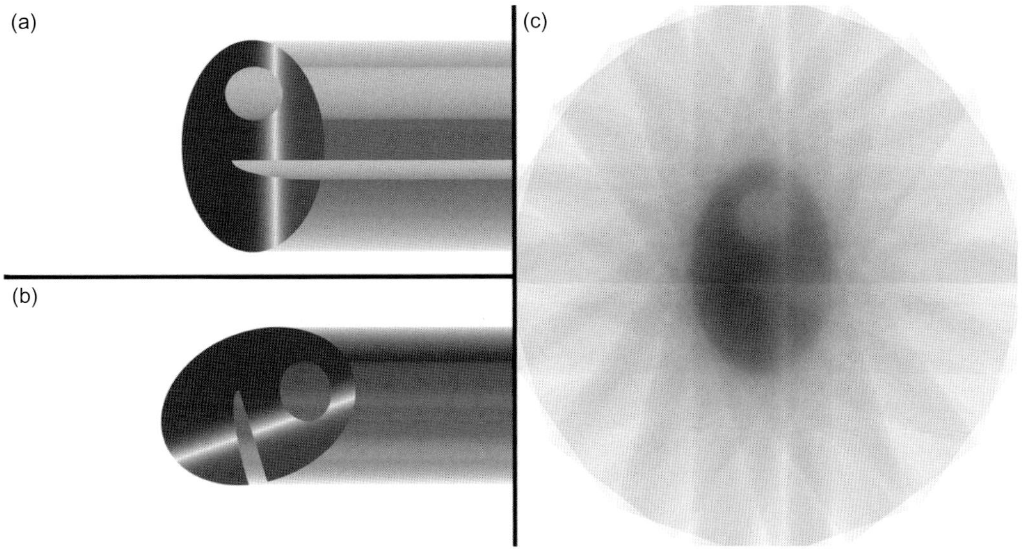

Figure 2.2 Reconstruction of 3D models from projection images. (a) The sample (blue) is illuminated from the left and a projection image formed. Where the sample is thickest less light is transmitted through the sample. (b) The sample is rotated over 360°. (c) Overlapping the projection images reconstructs the shape of the sample, demonstrating the principle of projection tomography.

presented here. A significant advantage of this imaging technique over conventional physical sectioning approaches is that the sample remains intact. Therefore distortion of the sample as a result of the cutting and mounting of sections is eliminated.

INTERPRETATION OF THE ATLAS

As mentioned above, the technique of OPT can generate 3D models of embryonic development; however, the images that are produced have a different appearance to those produced through conventional sectioning and histological staining approaches. In order to assist in the interpretation of the atlas we will present a brief guide to OPT images. The observed differences between tissues are due to varying levels of fluorescence resulting from both natural fluorescence of the tissue and that introduced by fixation. There are various cell components, structures, and chemicals, within tissue samples that are known to cause autofluorescence. Flavins and flavoproteins, for example, emit fluorescence in the wavelengths used for the OPT imaging of the samples presented here but are greatly reduced by fixation and dehydration. Collagen and elastin show high levels of autofluorescence, resulting in the bright appearance of extracellular matrix, blood vessels, and bone in the images. Aldehyde fixatives can react with proteins within the sample to generate fluorescent products. Samples fixed with paraformaldehyde, as used in the atlas, have a much lower level of fluorescence compared to samples after fixation with glutaraldehyde. However, the extremely high level of autofluorescence resulting from gluteraldehyde fixation prevents its use in OPT imaging as the sample appears solid and tissues below the surface cannot be visualized.

In the earliest developmental stages, where contrast between forming tissues is the least, OPT is more informative in relation to shape rather than individual tissue boundaries. However, the earliest stages of development are those that are represented by the largest number of existing anatomical resources, and as the embryo rapidly develops the contrast evident in the images also rapidly increases. In the older samples fluorescent OPT imaging provides sufficient contrast to distinguish the major anatomical structures. The brightest structures in the images are the blood, bone, and lens tissue, which appear saturated in the images (Figure 2.3). The use of fluorescence imaging also establishes very defined boundaries for the air-filled swim bladder and also the ventricles of the brain. Fluid and air-filled structures such as these frequently lose their morphology following physical sectioning but the imaging of the intact specimen using OPT ensures that these structures remain intact and distortion is minimized. More subtle differences in signal between neighbouring tissues allow the distinction of boundaries as illustrated in Figure 2.4. In an individual section, generated by either conventional sectioning techniques or by OPT, it can sometimes be difficult to unambiguously identify structures. A great advantage of techniques such as OPT where we can generate a complete 3D model is that an individual point or structure can be viewed from multiple angles within the same sample. This approach was frequently used in the annotation of the images contained in the atlas and the complete models are available through the accompanying website described in the next chapter.

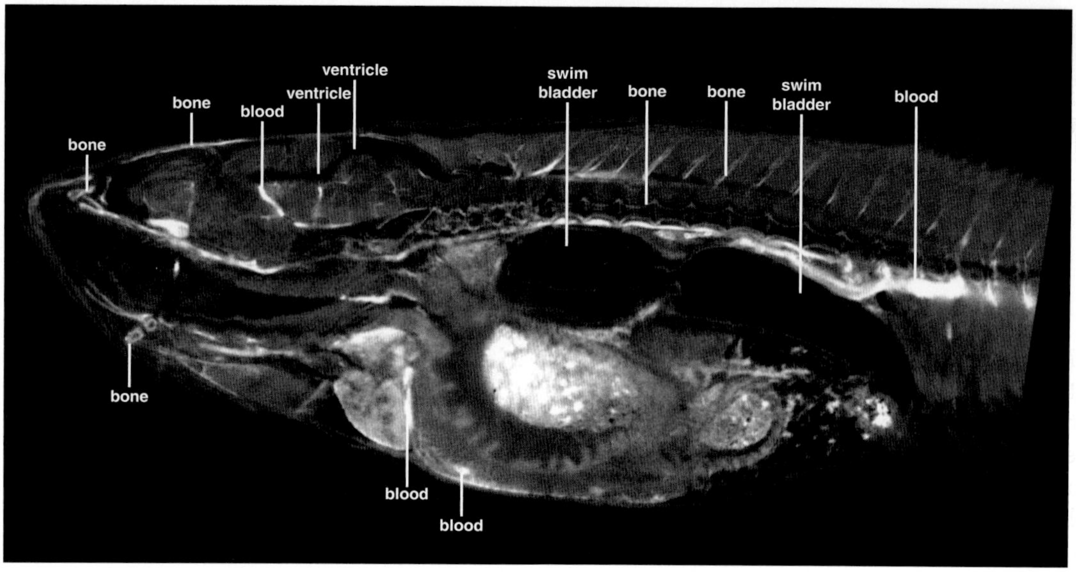

Figure 2.3 Section through a 16 mm fish. The highest levels of autofluorescence are seen in blood vessels and bone.

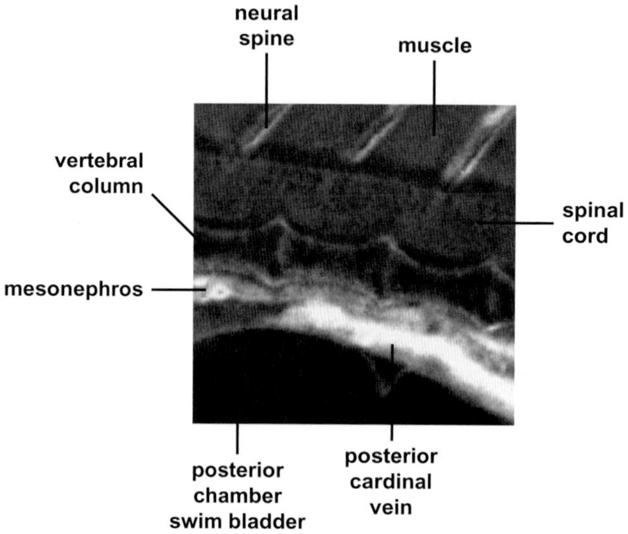

Figure 2.4 Examples of tissue boundaries. Changes in the level of autofluorescence between neighbouring tissue allows the boundaries to be determined. Region of a sagittal section from a 16 mm zebrafish.

CHAPTER 3

FISHNET: AN INTERACTIVE DATABASE OF ZEBRAFISH DEVELOPMENT

The information presented in this book is a condensed version of the total information available at FishNet (http://www.FishNet.org.au). We have previously described the publication of the internet-based resource in detail (Bryson-Richardson et al. 2007) and will therefore not repeat the information here. We will instead describe some of the major features of the website and how it will continue to develop. Additionally, as it is our intention that the two resources will complement each other to provide a complete anatomical reference for the zebrafish research community, we will discuss the pros and cons of each of the digital and print versions both now and into the future, with expected advances in internet publishing technologies.

ADVANTAGES OF THE PRINTED BOOK OVER THE INTERNET RESOURCE

The delivery of an anatomical resource such as this via the internet holds a number of advantages over a traditional, printed version, as explained below. The print reference, however, still has many significant benefits.

Despite improvements in internet access, the time taken for delivery of high-resolution images can still be long, and in order to facilitate browsing and interaction with large datasets image compression, with a corresponding reduction in quality, is still required. Whilst we have made every effort to ensure the image quality at the website remains high we have no such limitations in the delivery of images for the print version. Similarly, whilst display technology is ever improving, it still remains the case that it is often easier to evaluate images in high-quality prints. Alongside the improvements in delivery and display the quality and resolution of our imaging techniques are similarly improving and therefore the print version of the atlas will continue to have benefits. Perhaps the greatest advantage of this print version is its portable nature, allowing its use in the laboratory or aquarium where access to the internet resource may not be possible or appropriate.

ADVANTAGES OF THE INTERNET RESOURCE

The accompanying website to this book adds a number of features not possible in print form. The generation of the 3D models and subsequent section images, described in the previous chapter, has resulted in a wealth of information that cannot be represented in its entirety in print/paper format due to publication costs. In the book we are constrained to presenting a set of selected key stages in the development of the zebrafish and from within each of those stages, several representative section images that will allow the visualization of the gross anatomy. As such, significant temporal and spatial gaps exist in the information presented.

The complete stage series and all sections within each sample embryo, however, have been captured and may be accessed through the website. The increased temporal resolution available online is a significant advantage to the developmental biologist who requires information pertaining to subtle morphological changes throughout development. The interactive nature of the website also allows the virtual sectioning of the 3D models at any of the represented stages in any of the standard orthogonal sectioning planes, and to zoom in and out to different image magnifications. It is also possible to view volume or surface renderings of the complete sample that may be manipulated on screen, which are particularly useful for mental conceptualization of anatomical morphology. The provision of the in-built tools on the website allows the data to be accessed in a user-friendly manner and without the need for specialist software or IT knowledge. Furthermore, the complete 3D models may be downloaded for examination and manipulation locally by the user.

The above describes the existing capabilities of the website, which is not only interactive in its use but also dynamic in its development. We have only begun to explore the possibilities for data presentation in the first release of the website and this dynamic nature will allow the continued improvement of functionality and image quality. There have indeed been several improvements since its original release, including adoption of technology solutions that have significantly improved the speed of image delivery and reduced the effort required for site maintenance. More significant enhancements are in progress, such as development of tools to allow sectioning of the virtual 3D

Atlas of Zebrafish Development. 10.1016/B978-0-12-374016-8.00003-0

embryos in any arbitrary section plane and viewing of multiple angles simultaneously. We are also developing search tools to allow the resource to be searched directly by the use of anatomical names (as opposed to section browsing), again making use of the advantages of internet delivery, and we expect all of these to be made available to users shortly. There are also, as discussed in Chapter 1, significant resources in the form of existing websites for the zebrafish research community. These collective resources house an incredible amount of information and as we develop the site we aim to make it easier for other resources to reference the information contained within FishNet and also improve the connections from FishNet back to these resources. This will allow users to more easily access the collective resources available.

Another benefit of the dynamic nature of the website is the possibility for information to be easily updated, which allows both for addition of further annotations and any modification where necessary. This also allows for the exciting possibility of community annotation by expert users across the wider zebrafish community. It is our hope that in the long term it will be possible for users of the site to contribute their expertise directly into the database, thereby ensuring that individual structures are labeled by the most relevant experts. In the lead up to this community annotation scenario (which will require infrastructure changes to allow for remote editing by researchers across the world), we welcome comments from the community regarding any possible errors, omissions, and changes in current nomenclature which can be updated on the website in a short timeframe. Community annotation efforts would also benefit future versions of the print atlas.

Greater scope still exists in the visualization of 3D images. Currently the manipulation and visualization of large 3D datasets requires specialist software and can require substantial computer processing time. As software and hardware develops it will ultimately be possible to complete many of these tasks remotely through the website, interacting directly with the 3D models, making arbitrary sections, virtual dissections, and adjusting rendering parameters, ensuring that all users can more fully utilize the datasets without the requirements of specialist software, hardware, and expertise.

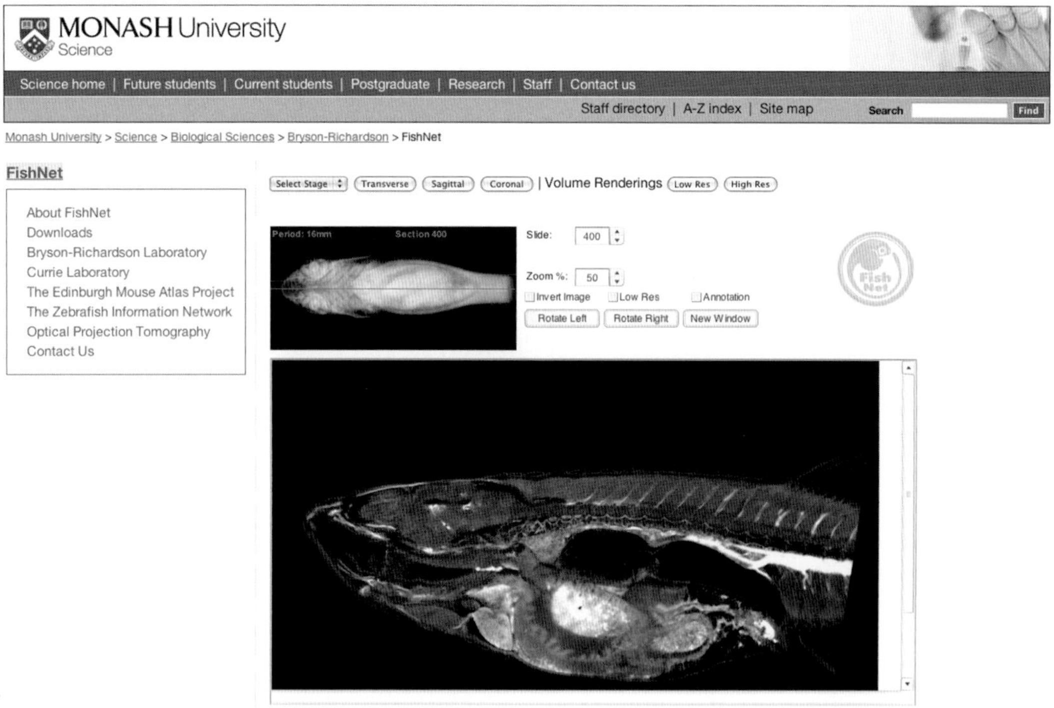

Figure 3.1 The FishNet website. The current interface for the website presents a rendered image of the 3D data set for navigation and a virtual section image. Virtual sections can be examined in any of the three standard section views: transverse, sagittal, or coronal. The section viewed can be adjusted by clicking on the navigation image, typing a number into the slide value box, or by clicking on the up and down arrows. The "Low Res" option allows smaller versions of the images to be viewed, facilitating use of the resource with slow internet connections. Annotated images are evenly spaced throughout the 3D models and selecting annotation will show the closest annotated section to the currently viewed image. In addition to the section images, volume renderings may be viewed. By selecting the "Low Res" or "High Res" buttons in the top right of the screen another window will open containing the rendered model. By clicking on these models and depressing the mouse button the angle at which the models are viewed can be altered.

PART I

THE EARLY EMBRYO

CHAPTER 4

24 HOURS

Sagittal Sections

Figure 4.1

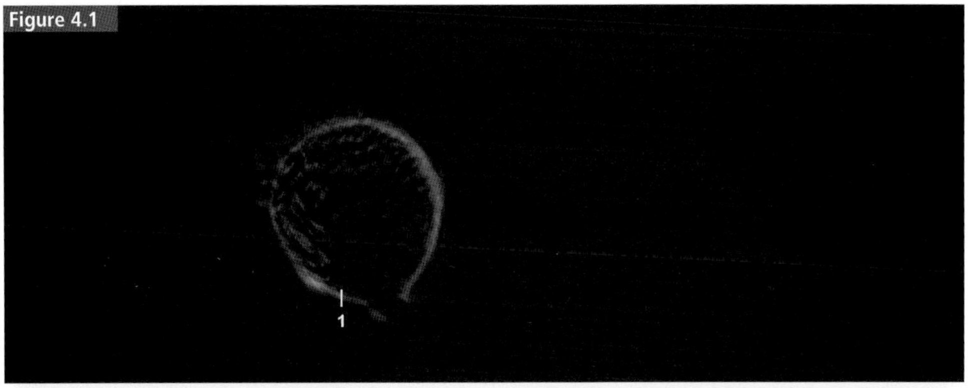

Figure 4.2

Figure 4.3

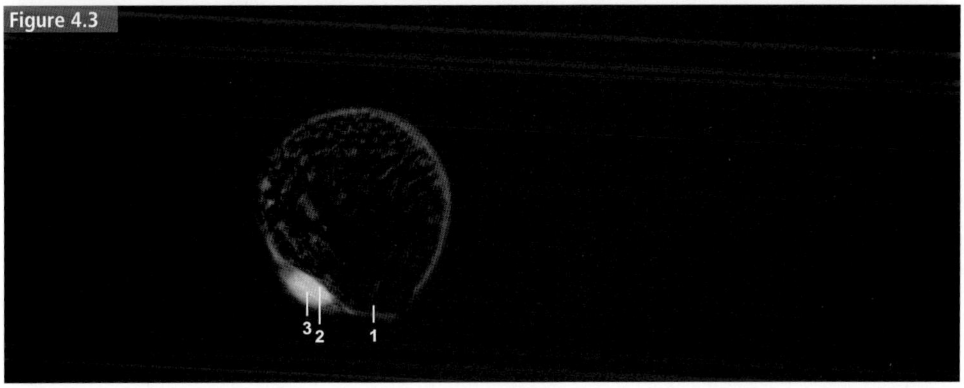

1. yolk	**2.** optic cup	**3.** lens

Atlas of Zebrafish Development. 10.1016/B978-0-12-374016-8.00004-2

Figure 4.4

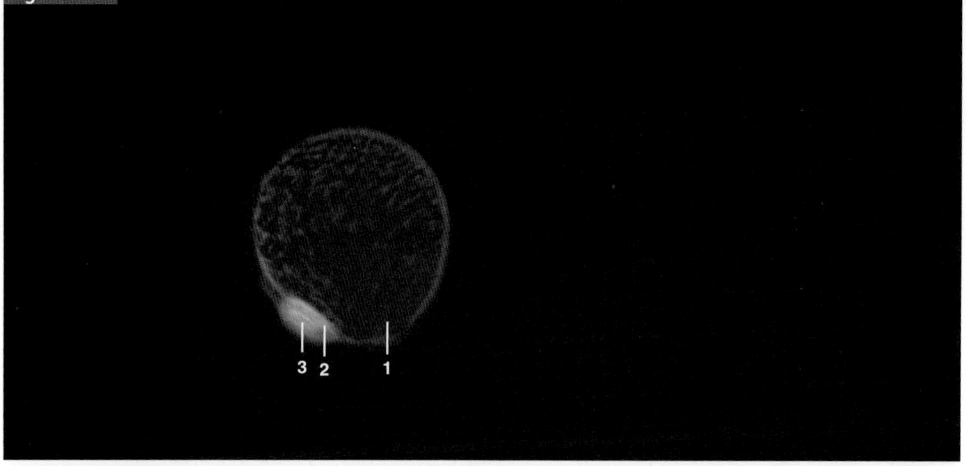

Figure 4.5

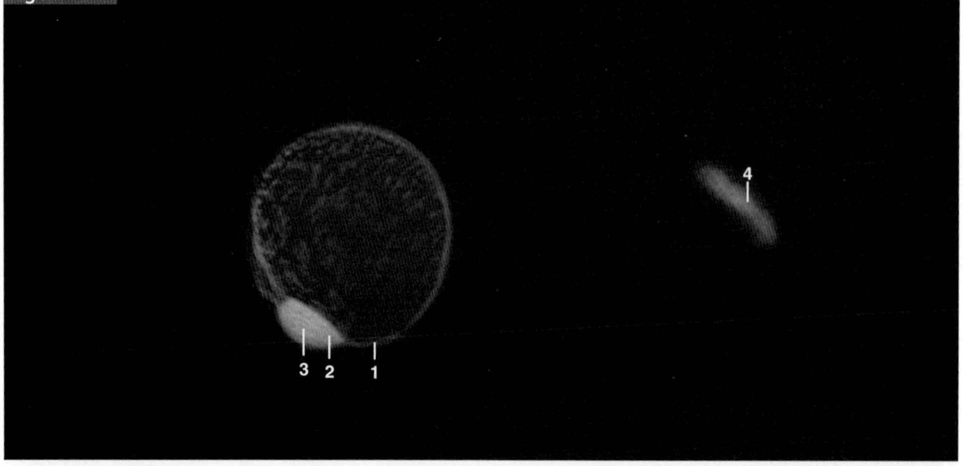

Figure 4.6

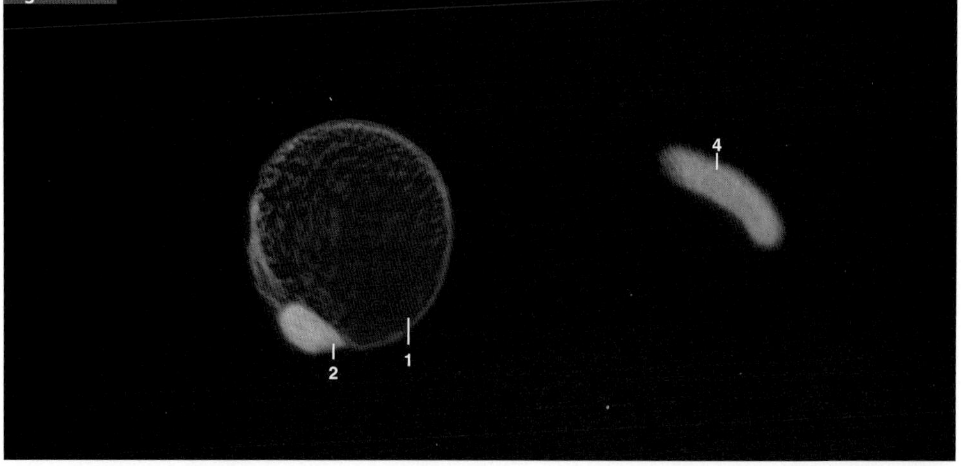

1. yolk	**2.** optic cup	**3.** lens	**4.** myotome

Figure 4.7

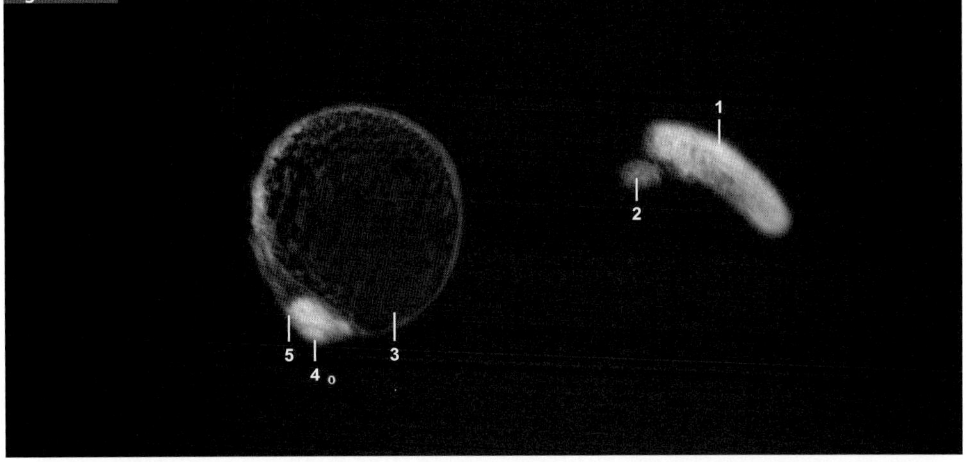

Figure 4.8

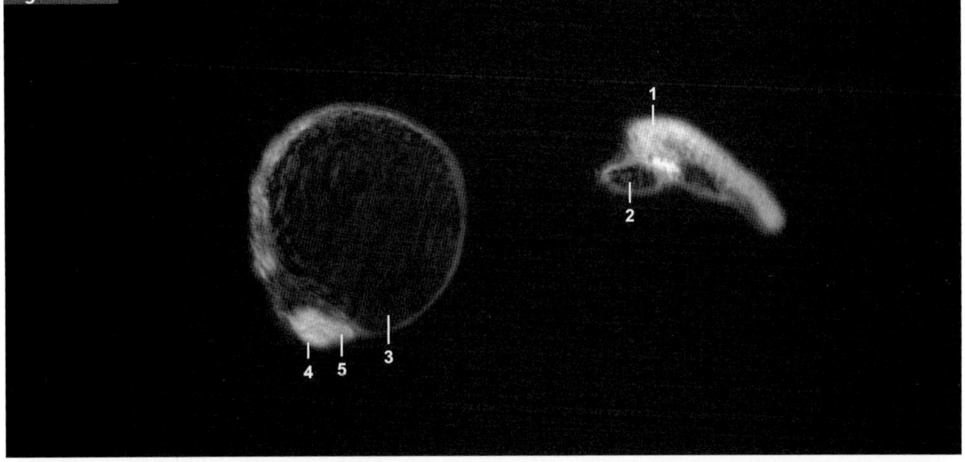

Figure 4.9

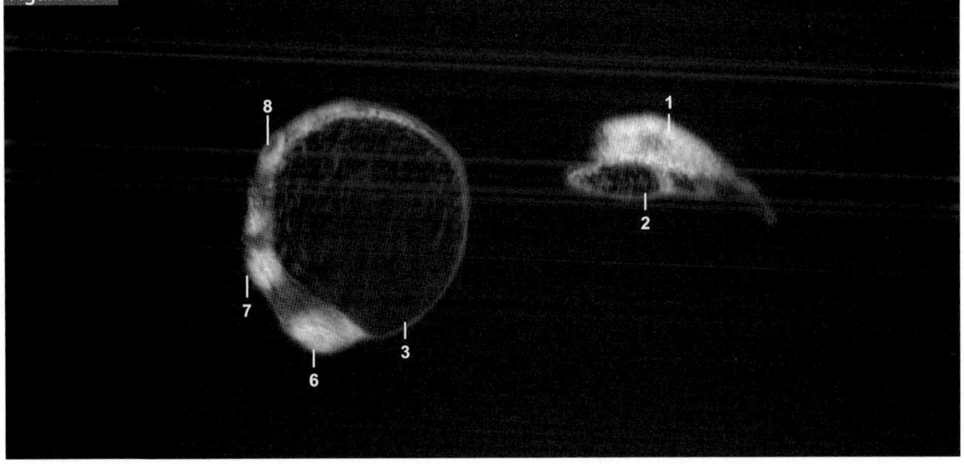

1. myotome	**3.** yolk	**5.** optic cup	**7.** optic tectum
2. yolk extension	**4.** forebrain	**6.** telencephalon	**8.** otic vesicle

Figure 4.10

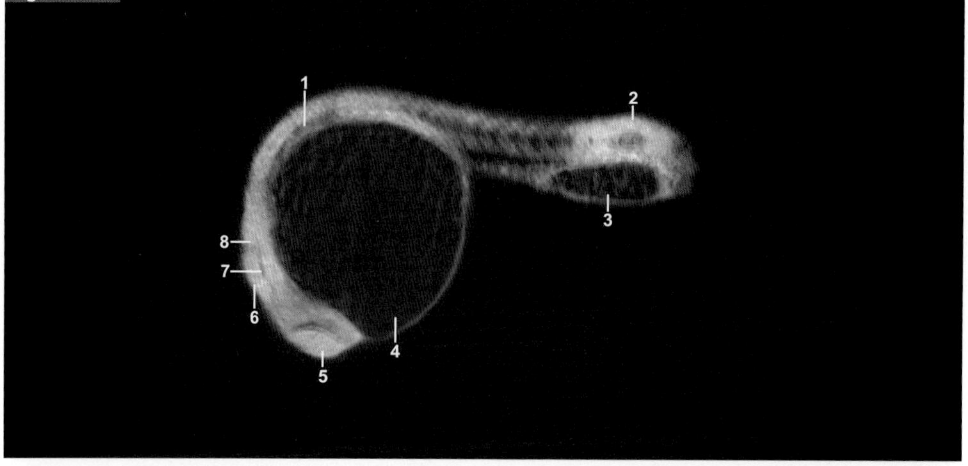

Figure 4.11

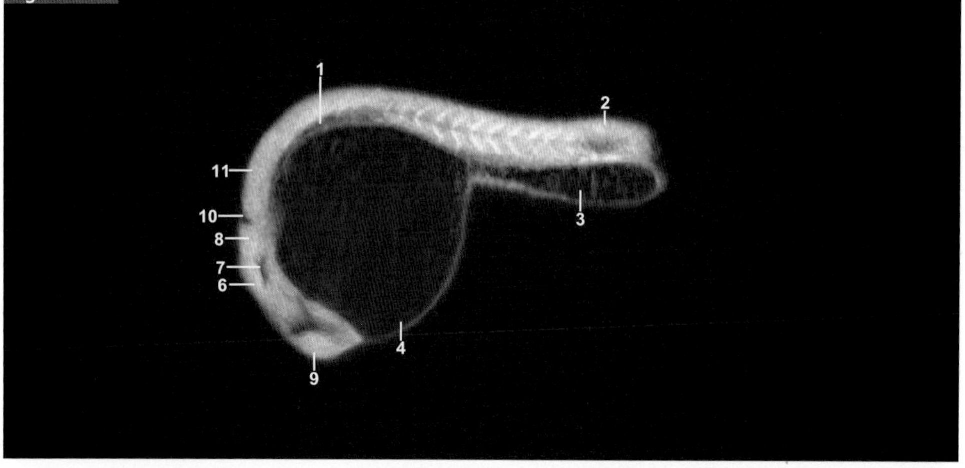

Figure 4.12

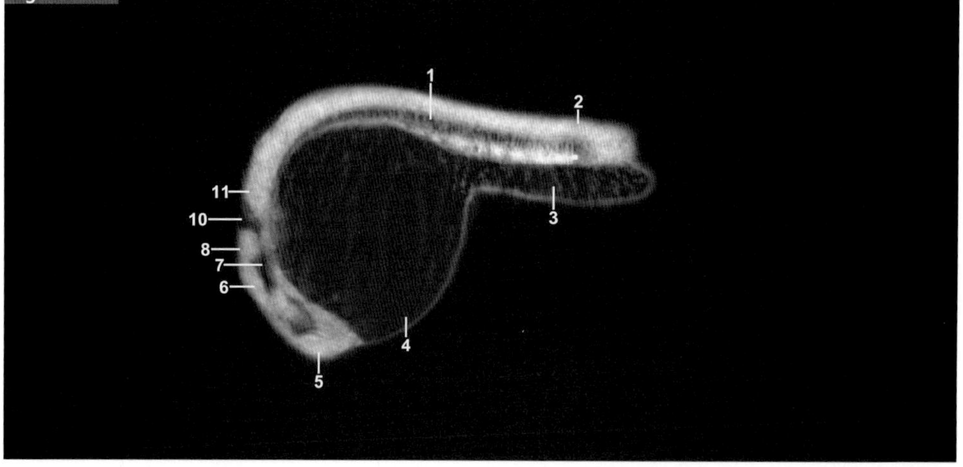

1. notochord	**4.** yolk	**7.** tectal ventricle	**10.** fourth ventricle
2. myotome	**5.** forebrain	**8.** cerebellum	**11.** hindbrain
3. yolk extension	**6.** optic tectum	**9.** telencephalon	

Figure 4.13

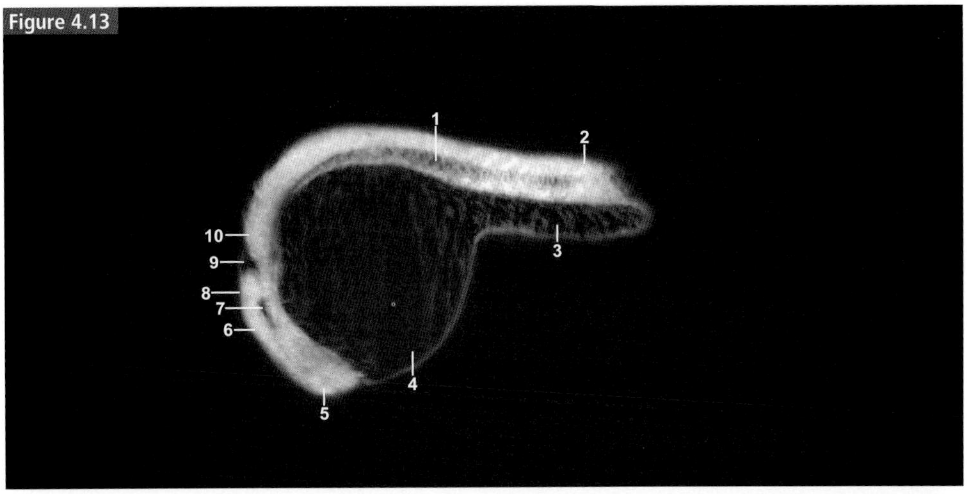

Figure 4.14

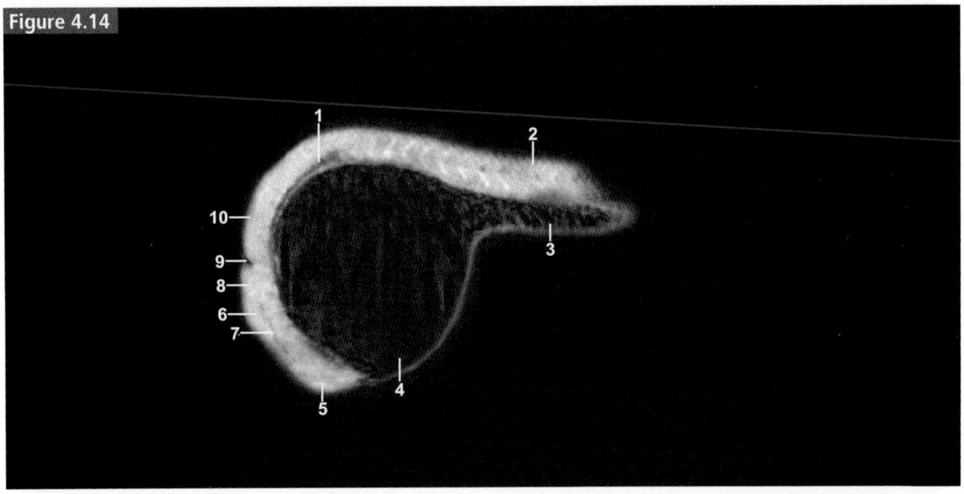

Figure 4.15

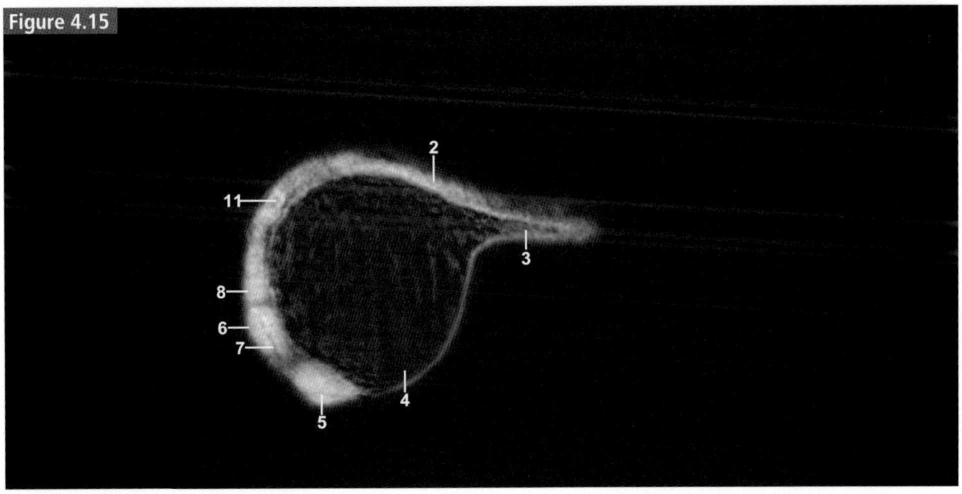

1. notochord	**4.** yolk	**7.** tectal ventricle	**10.** hindbrain
2. myotome	**5.** forebrain	**8.** cerebellum	**11.** otic vesicle
3. yolk extension	**6.** optic tectum	**9.** fourth ventricle	

Figure 4.16

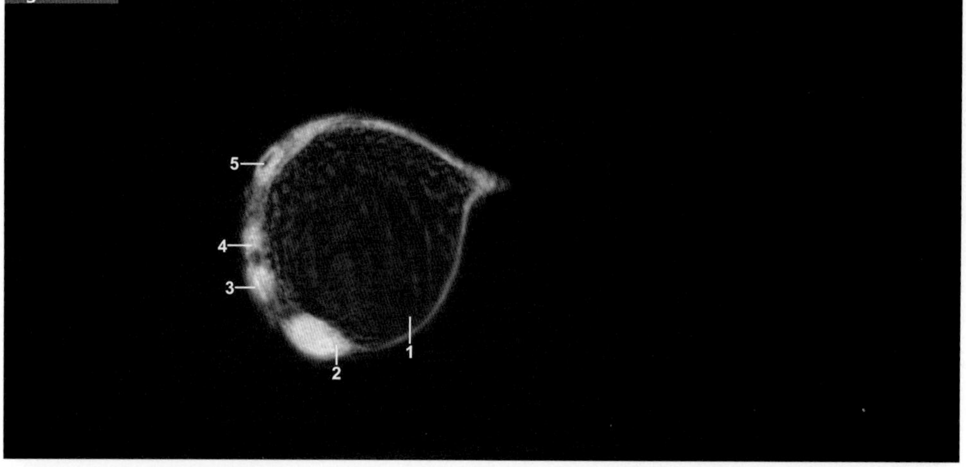

Figure 4.17

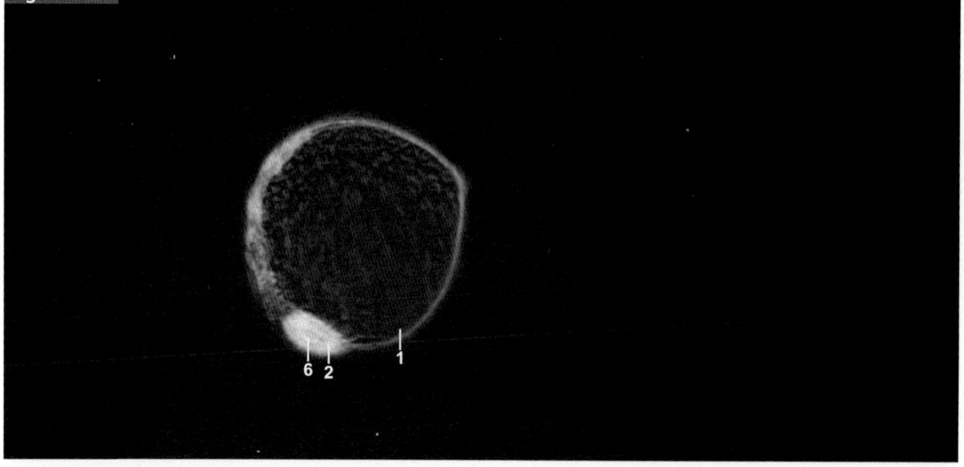

Figure 4.18

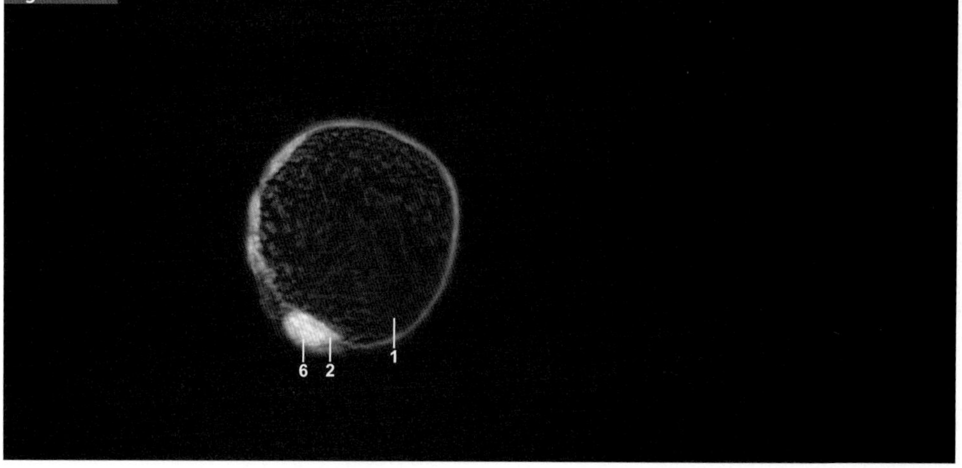

1. yolk	**2.** optic cup	**3.** optic tectum	**4.** cerebellum	**5.** otic vesicle	**6.** lens

Figure 4.19

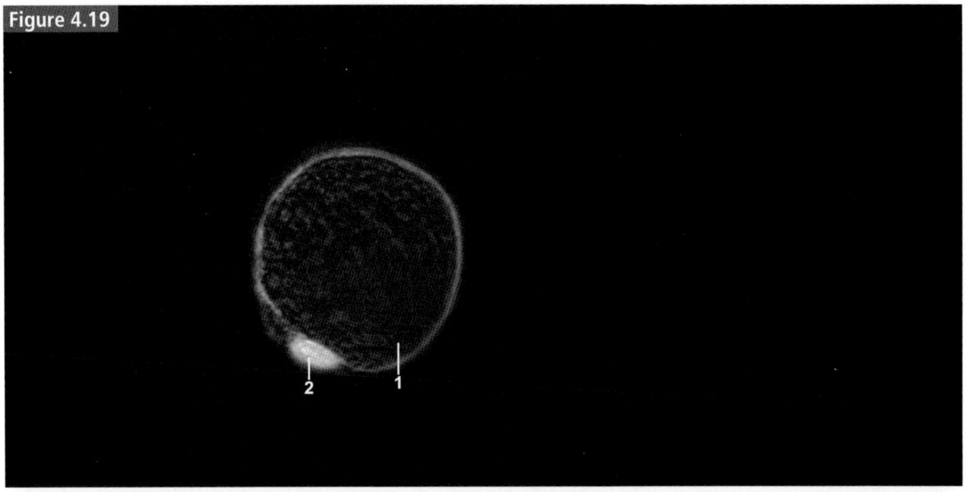

Figure 4.20

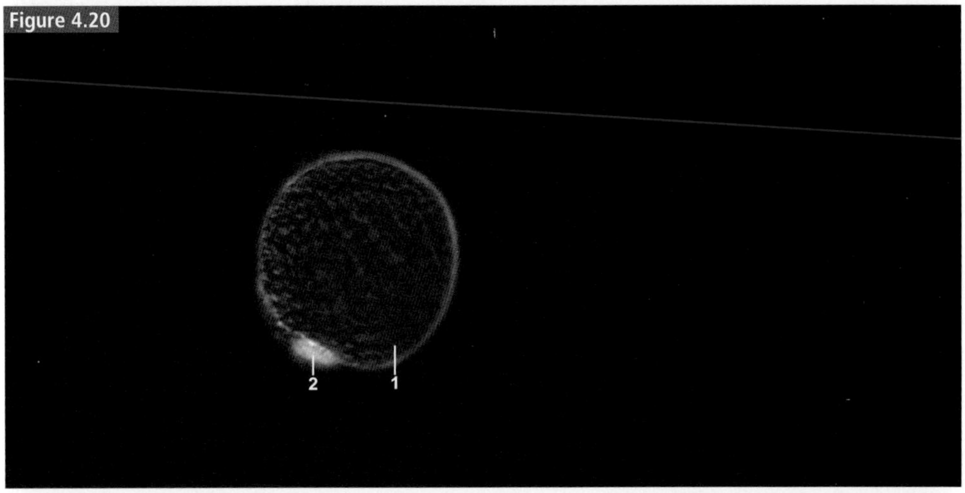

Figure 4.21

1. yolk	**2.** optic cup

Coronal Sections

Figure 4.22

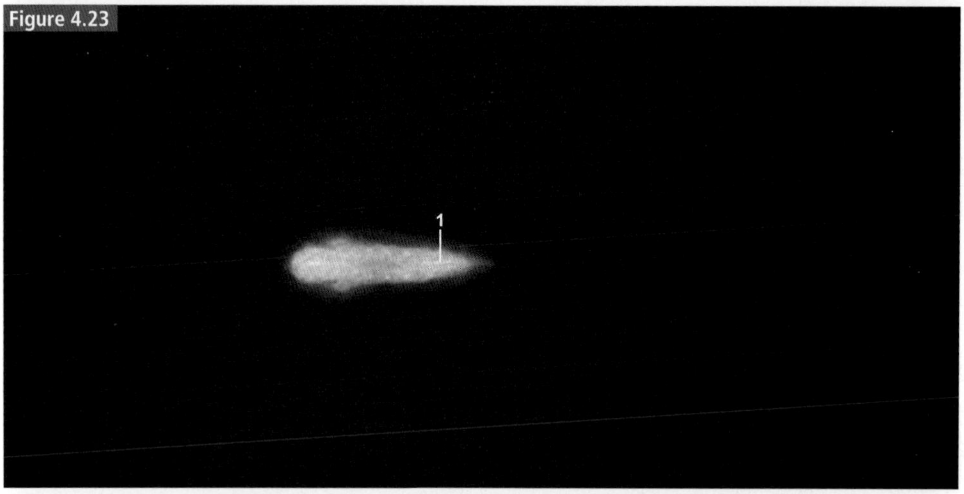

Figure 4.23

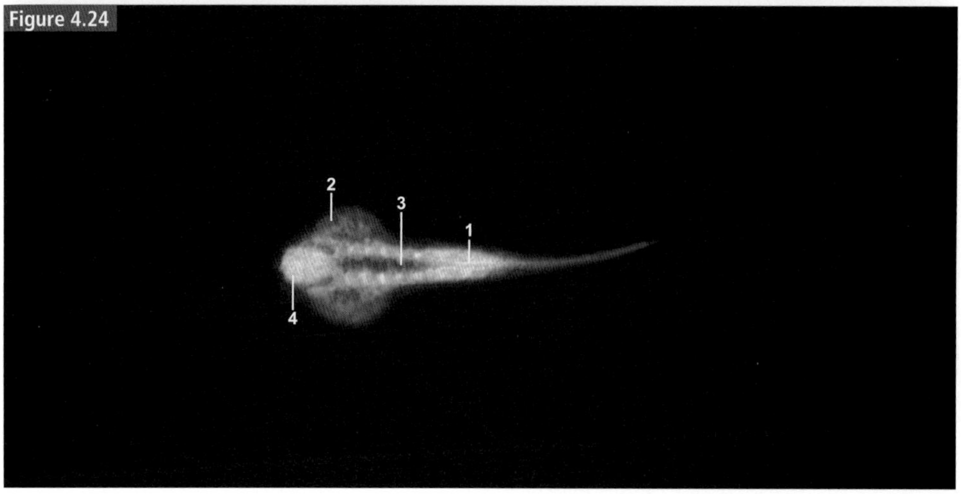

Figure 4.24

1. neural tube	**2.** yolk	**3.** notochord	**4.** hindbrain

Figure 4.25

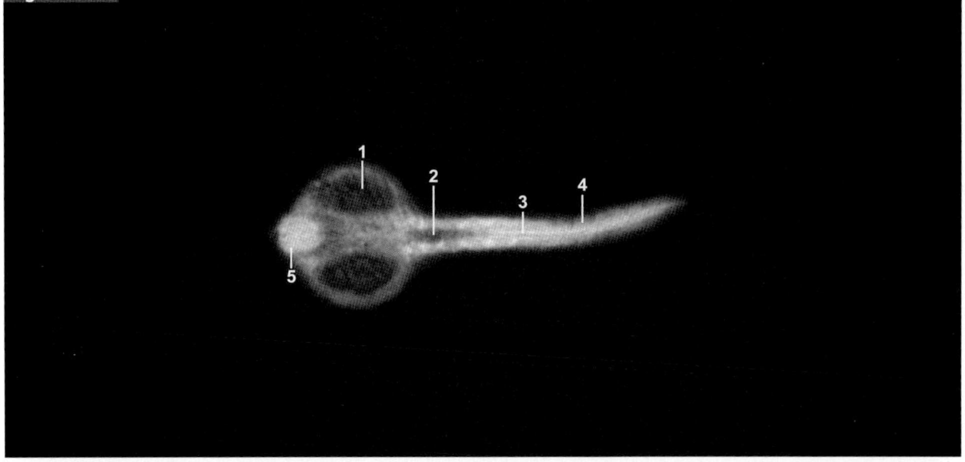

Figure 4.26

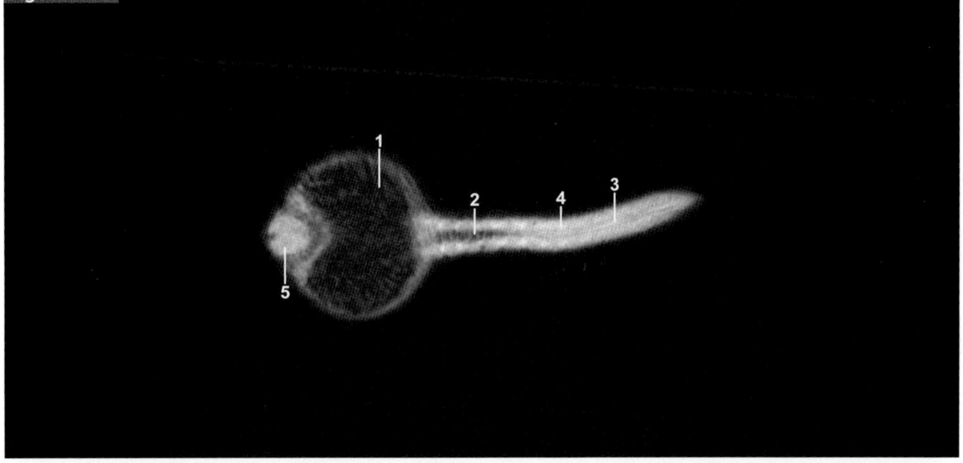

Figure 4.27

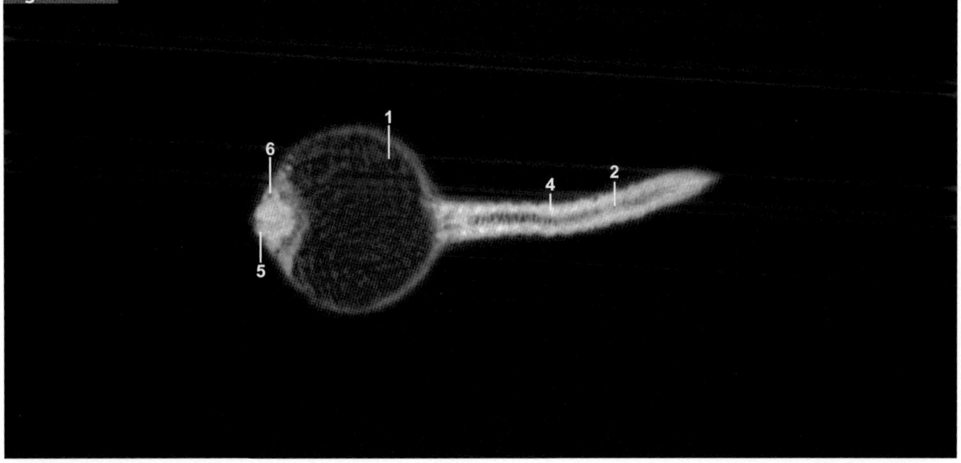

1. yolk	**2.** notochord	**3.** neural tube	**4.** myotome	**5.** hindbrain	**6.** otic vesicle

Figure 4.28

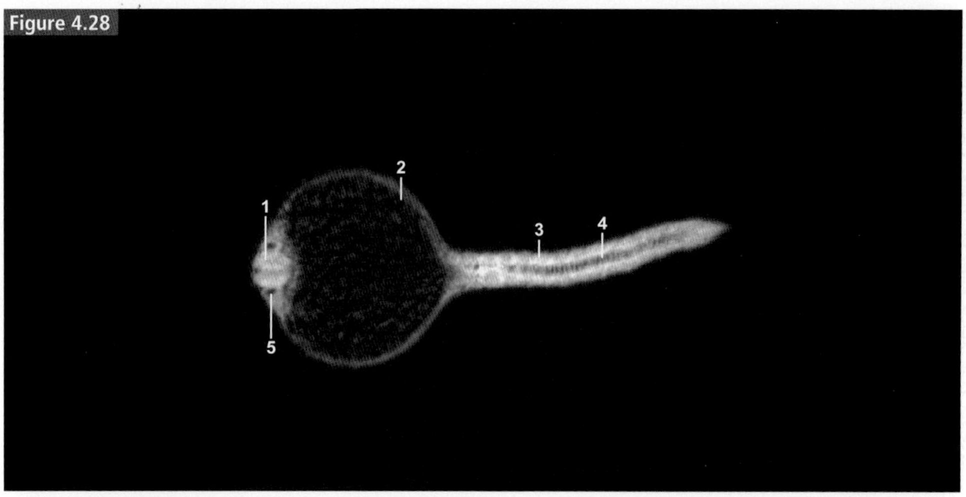

Figure 4.29

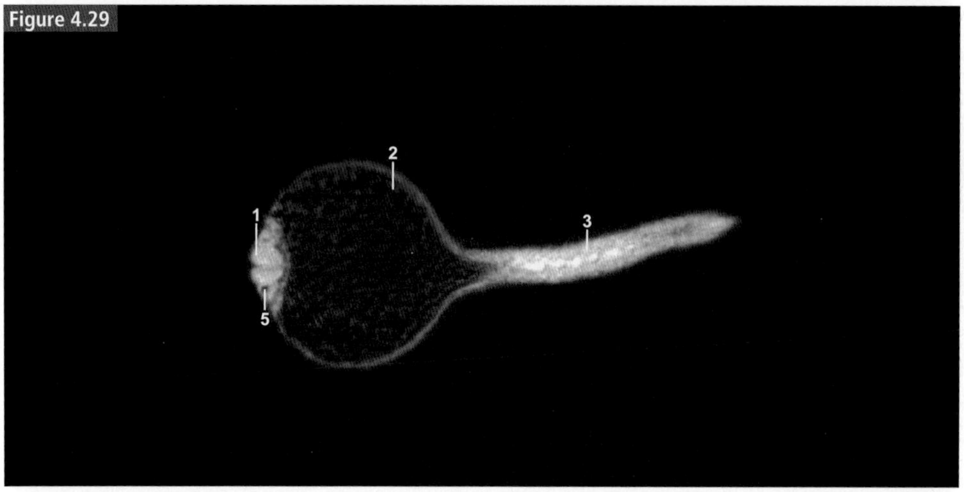

Figure 4.30

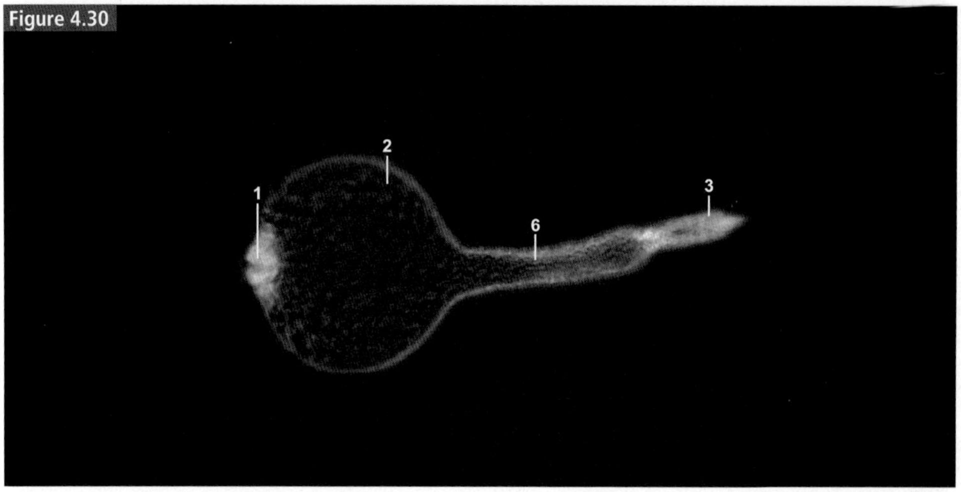

1. hindbrain	**2.** yolk	**3.** myotome	**4.** notochord	**5.** otic vesicle	**6.** yolk extension

Figure 4.31

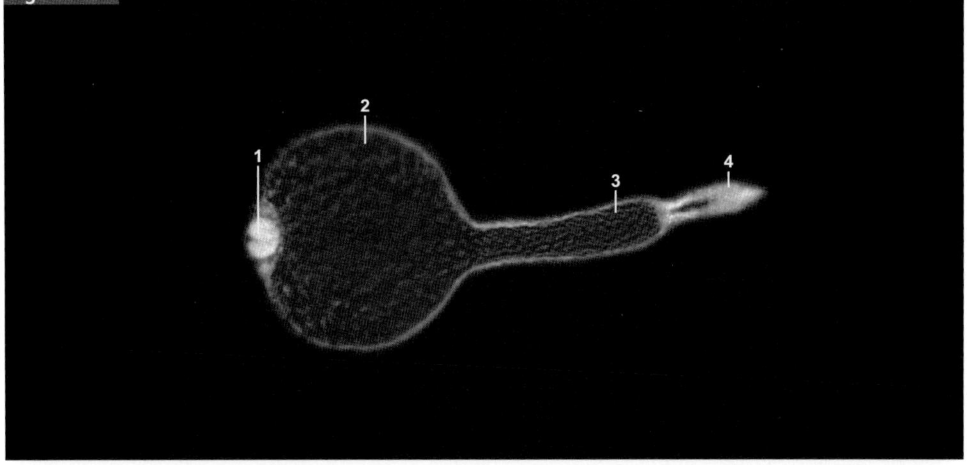

Figure 4.32

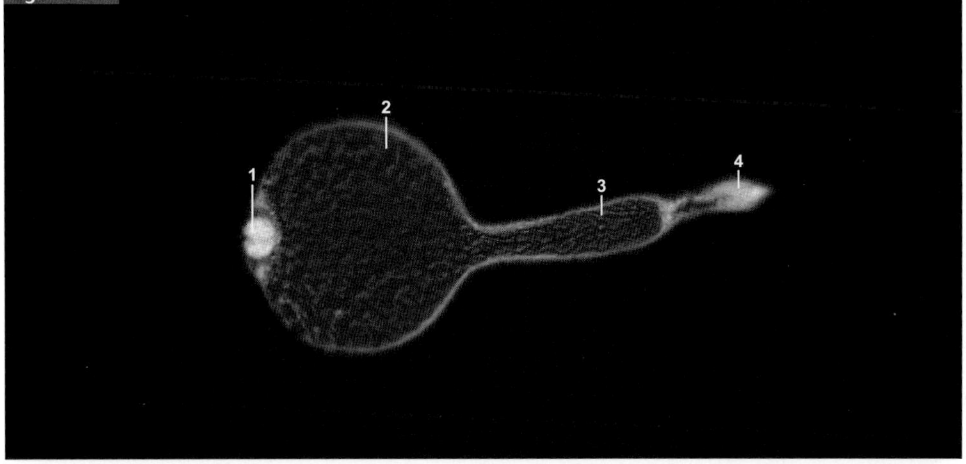

Figure 4.33

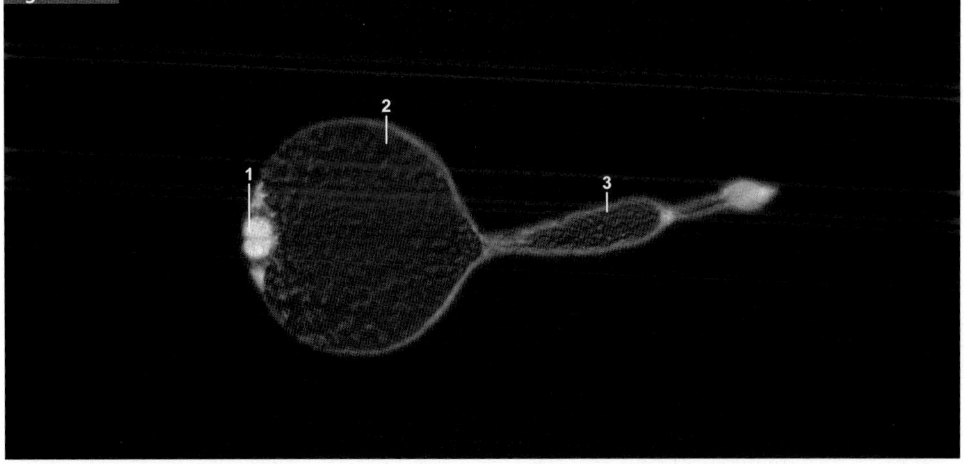

1. hindbrain	**2.** yolk	**3.** yolk extension	**4.** myotome

Figure 4.34

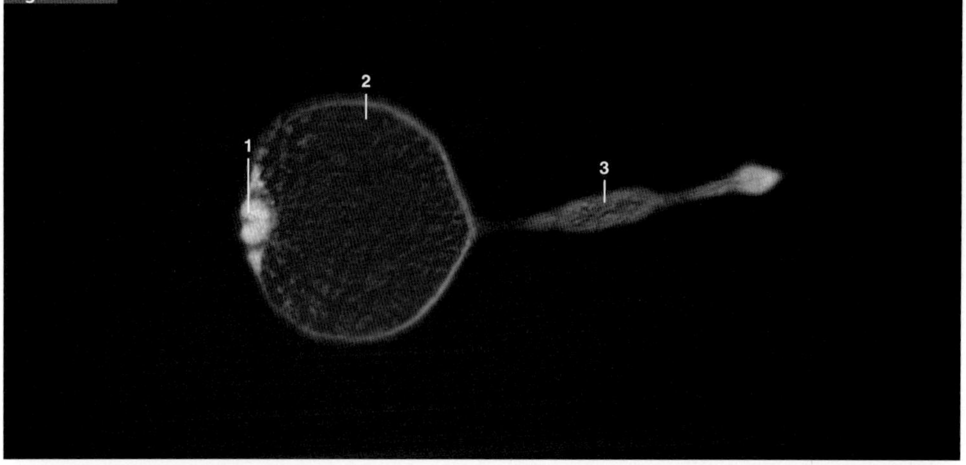

Figure 4.35

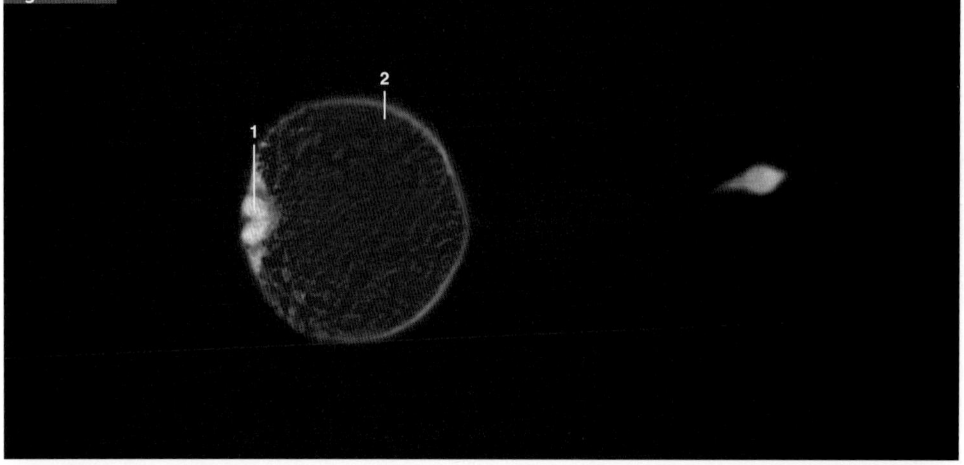

Figure 4.36

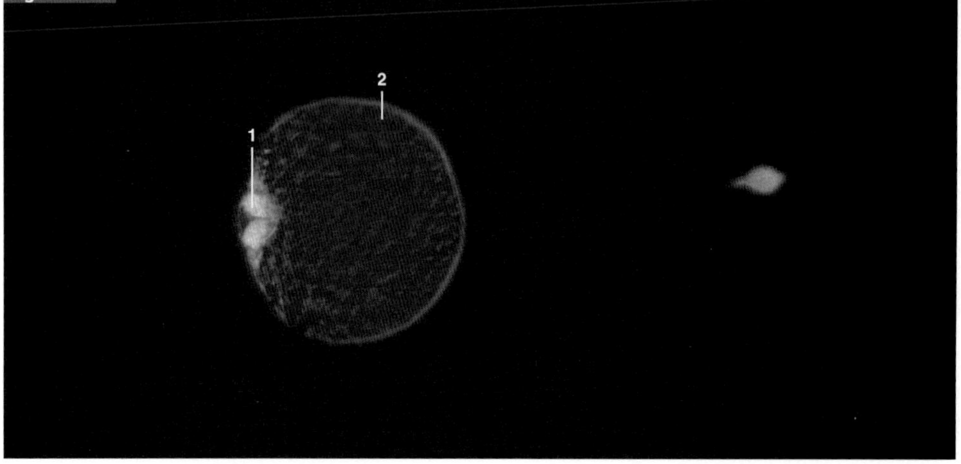

1. hindbrain	**2.** yolk	**3.** yolk extension

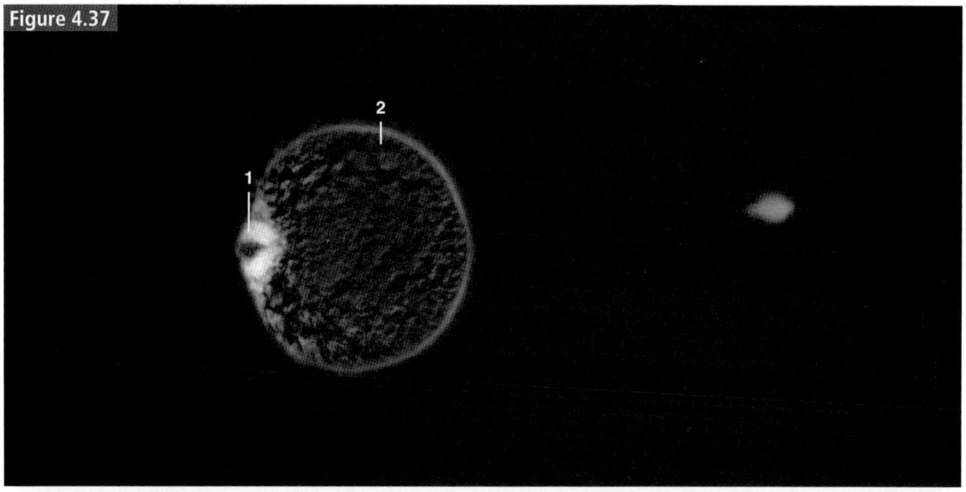

Figure 4.37

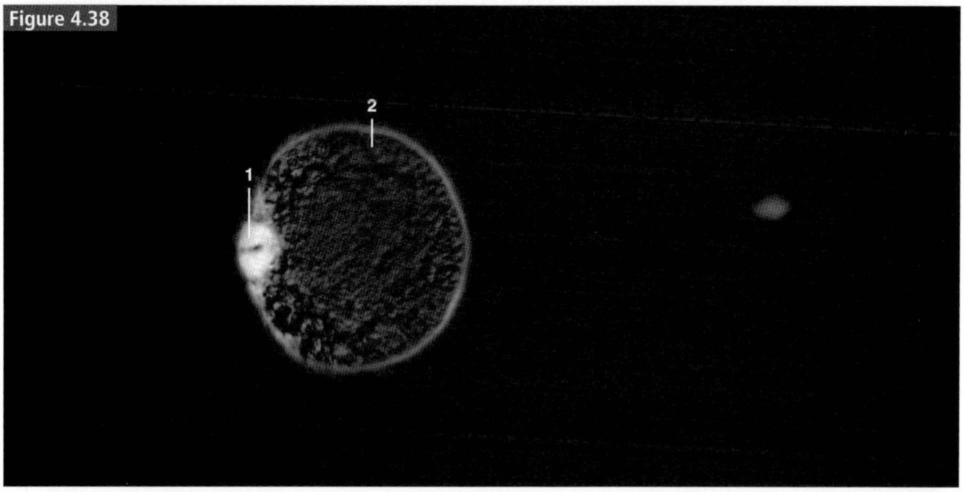

Figure 4.38

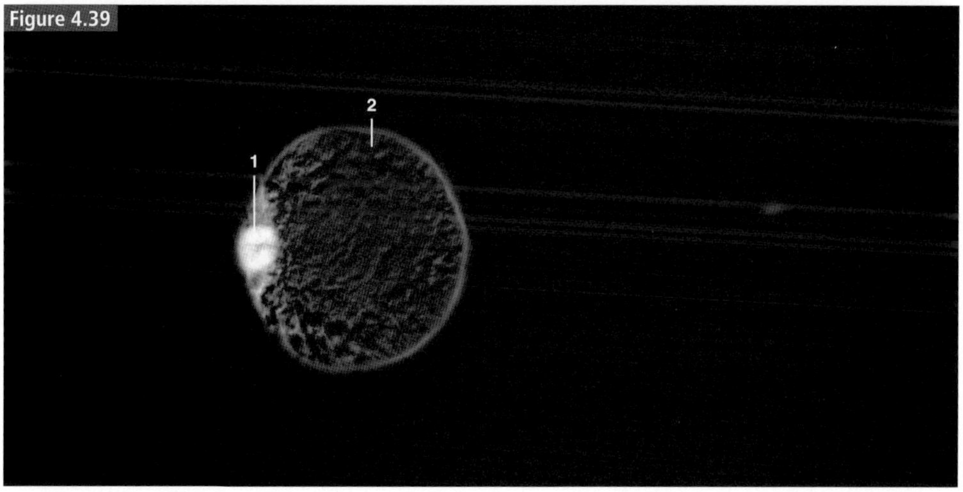

Figure 4.39

1. hindbrain	**2.** yolk

Figure 4.40

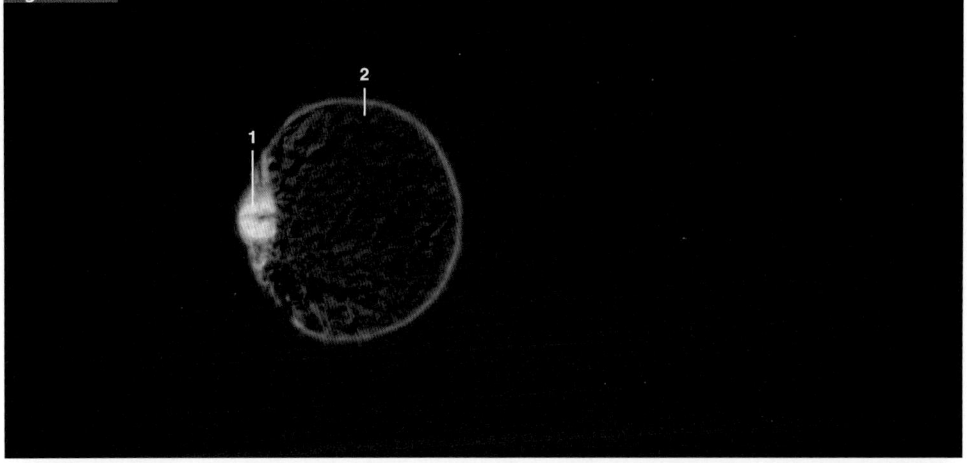

Figure 4.41

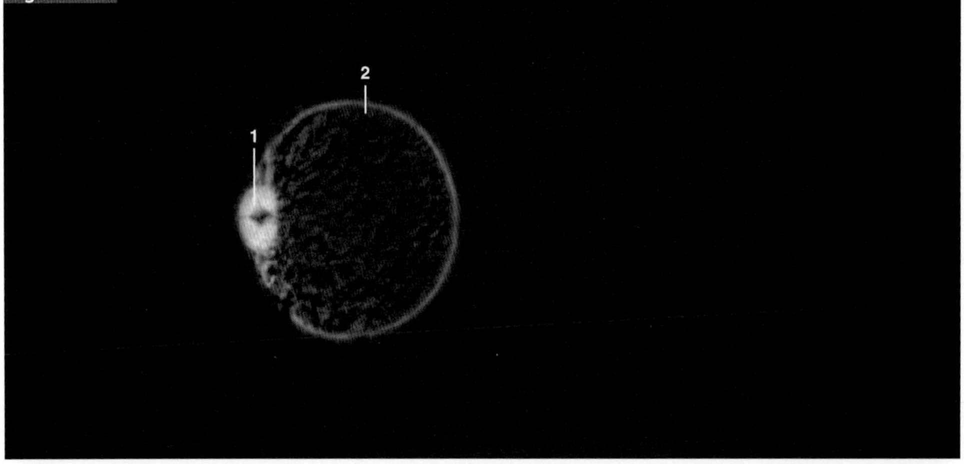

Figure 4.42

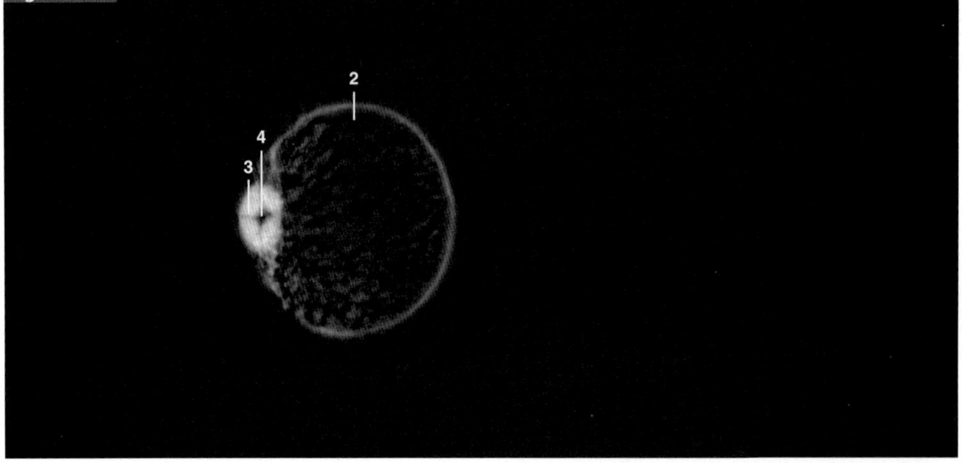

1. hindbrain	**2.** yolk	**3.** optic tectum	**4.** tectal ventricle

Figure 4.43

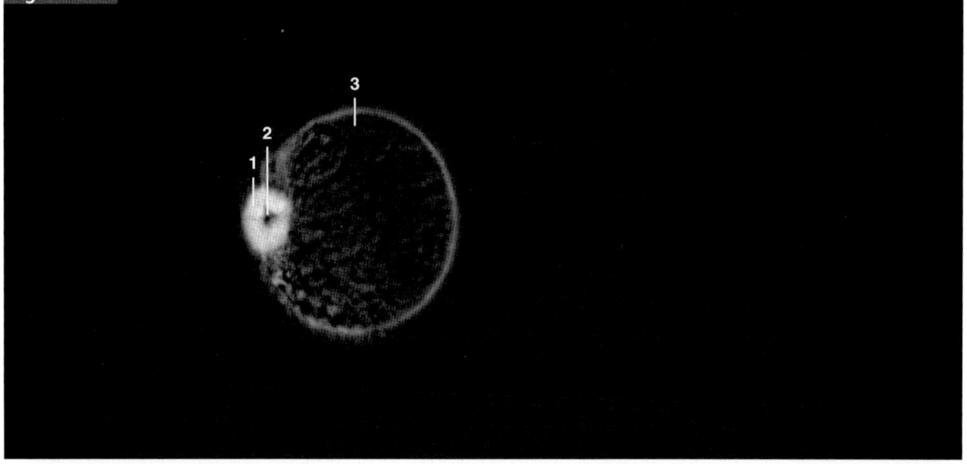

Figure 4.44

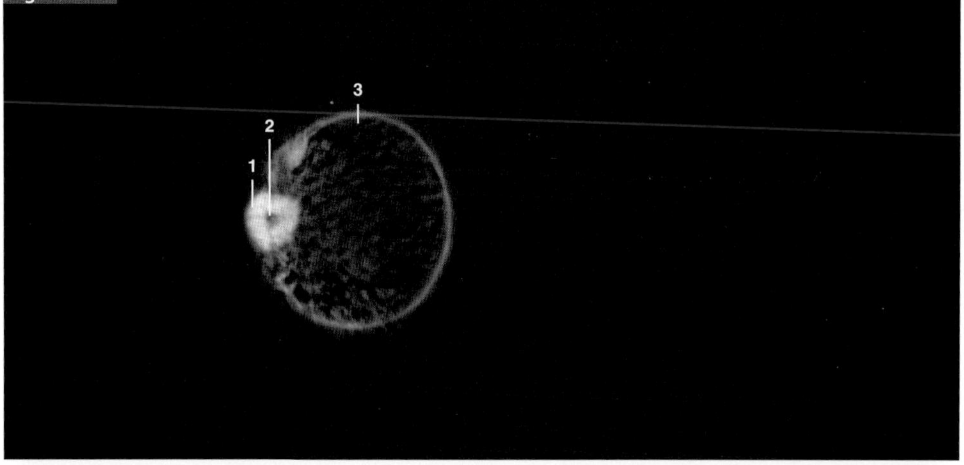

Figure 4.45

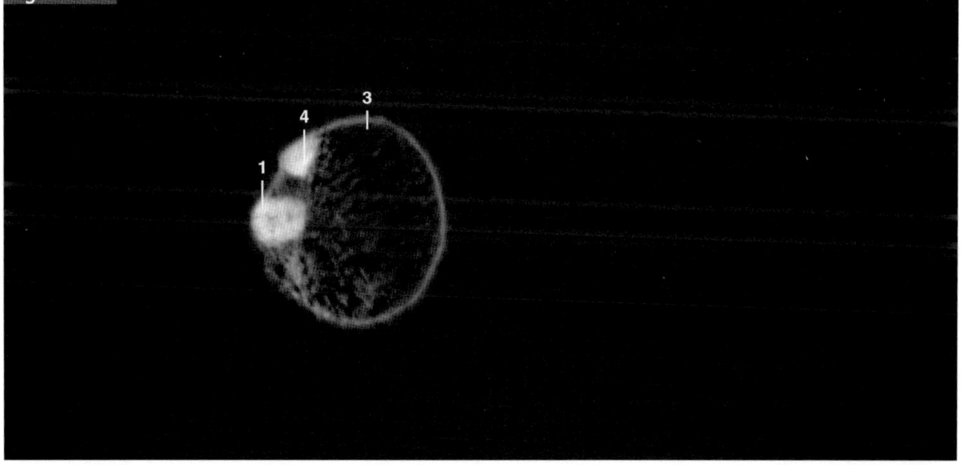

1. optic tectum	**2.** tectal ventricle	**3.** yolk	**4.** optic cup

Figure 4.46

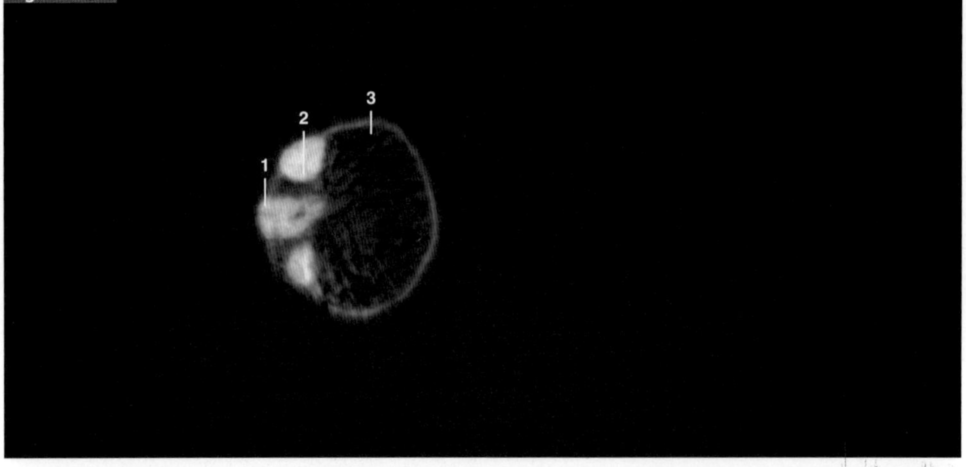

Figure 4.47

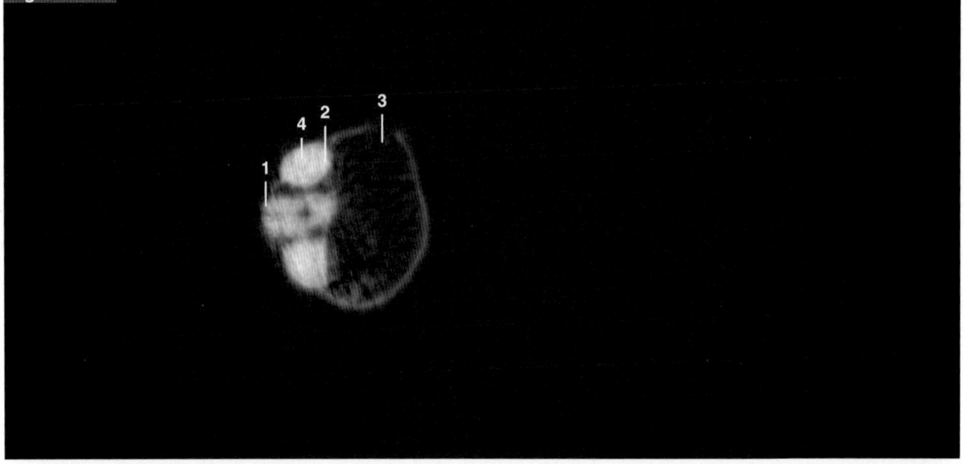

Figure 4.48

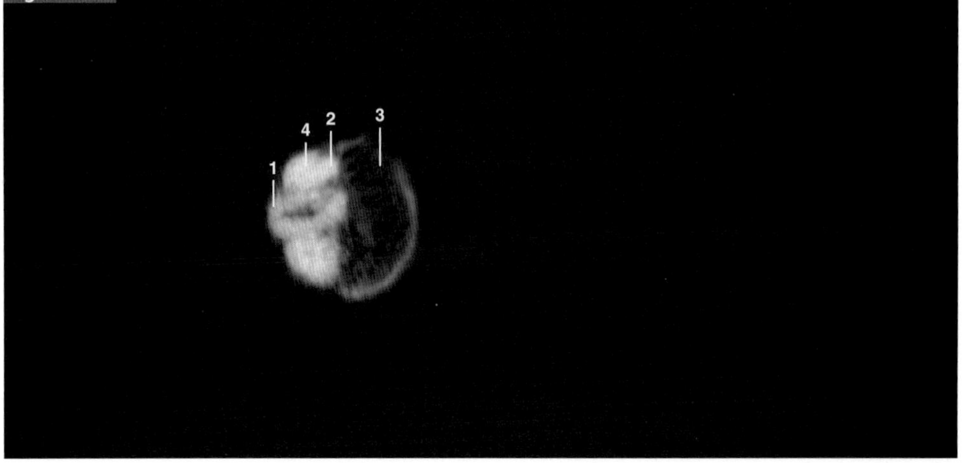

1. forebrain	**2.** optic cup	**3.** yolk	**4.** lens

Figure 4.49

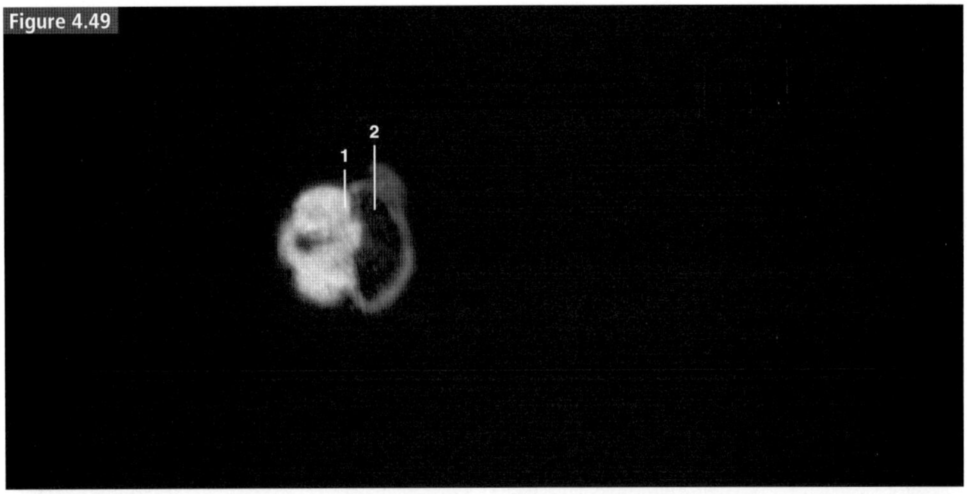

Figure 4.50

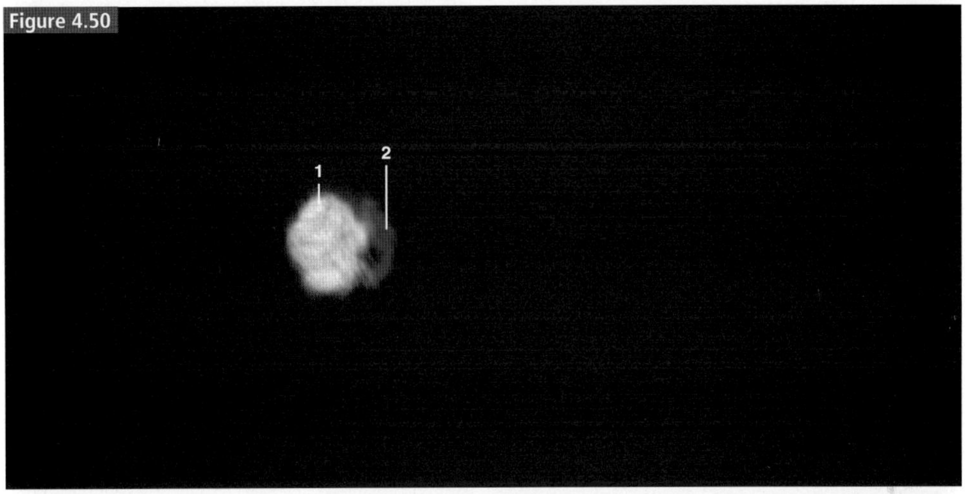

Figure 4.51

1. optic cup	**2.** yolk

Transverse Sections

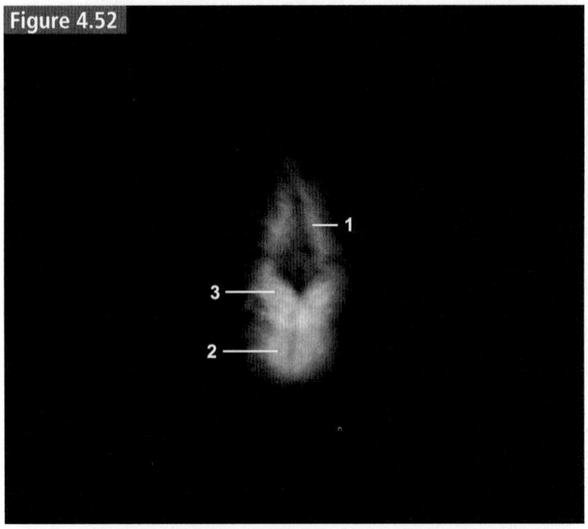

Figure 4.52

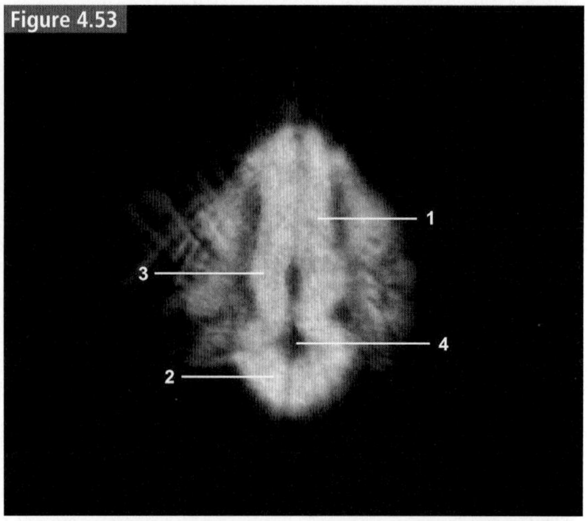

Figure 4.53

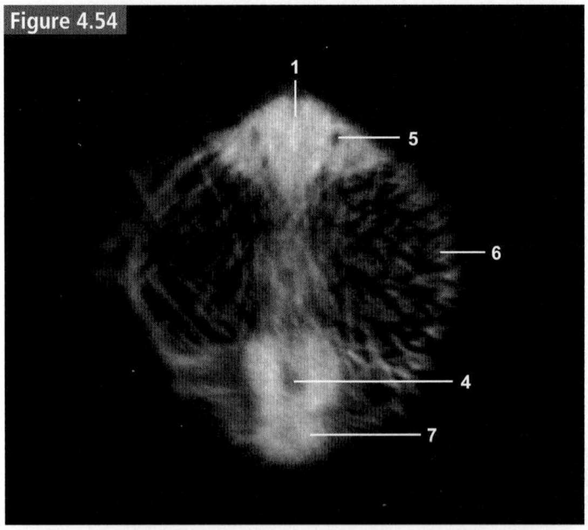

Figure 4.54

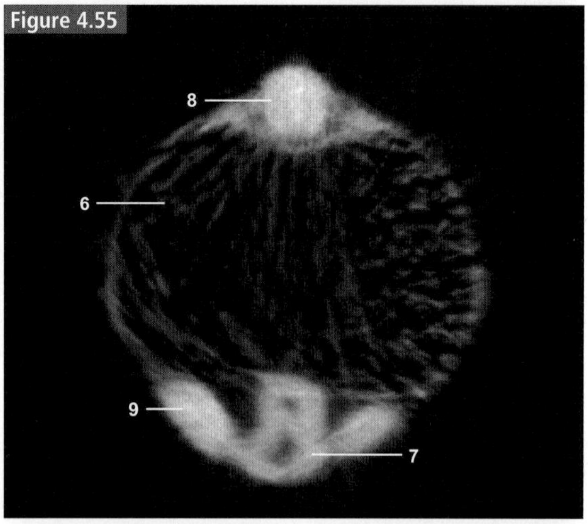

Figure 4.55

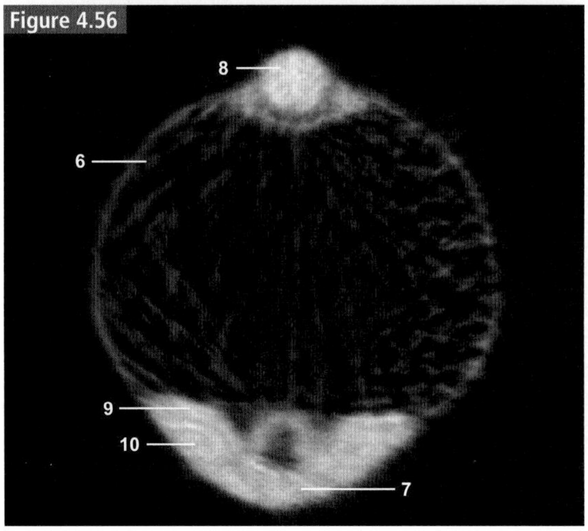

Figure 4.56

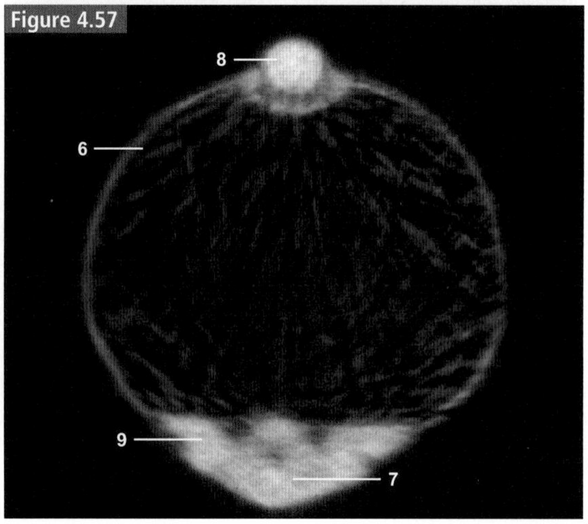

Figure 4.57

1. hindbrain	**3.** cerebellum	**5.** otic vesicle	**7.** forebrain	**9.** optic cup
2. optic tectum	**4.** tectal ventricle	**6.** yolk	**8.** neural tube	**10.** lens

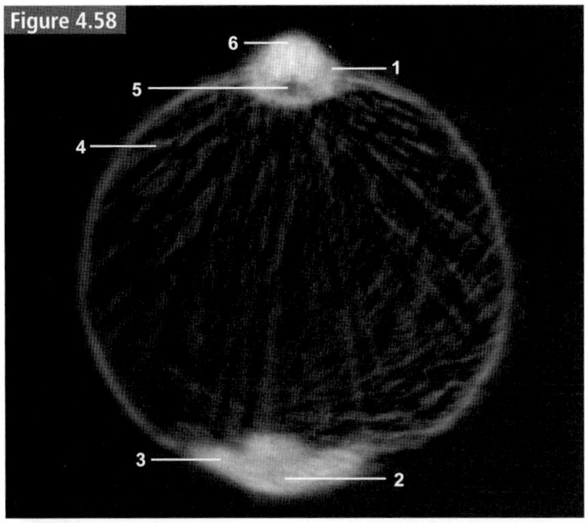

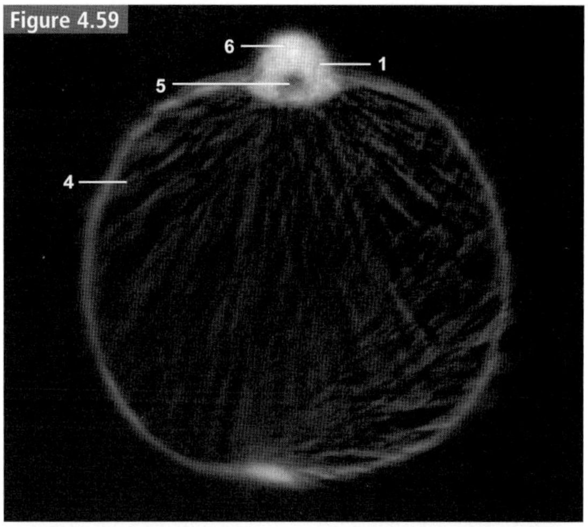

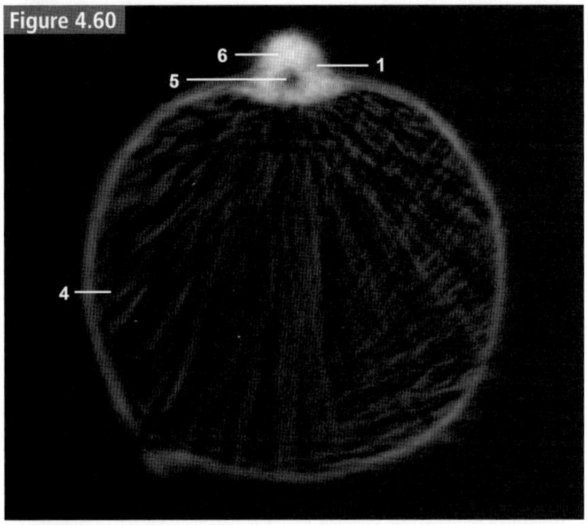

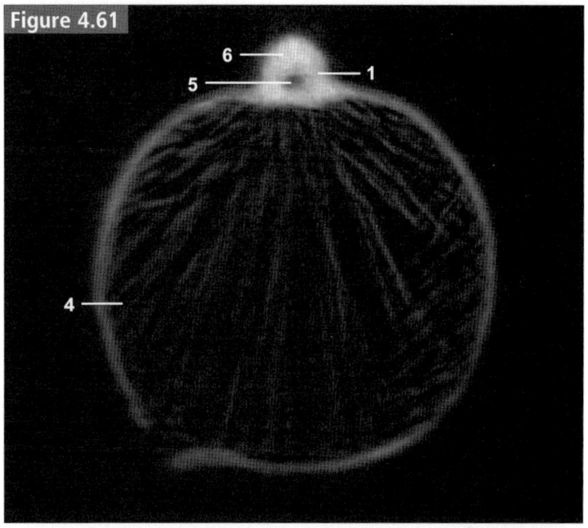

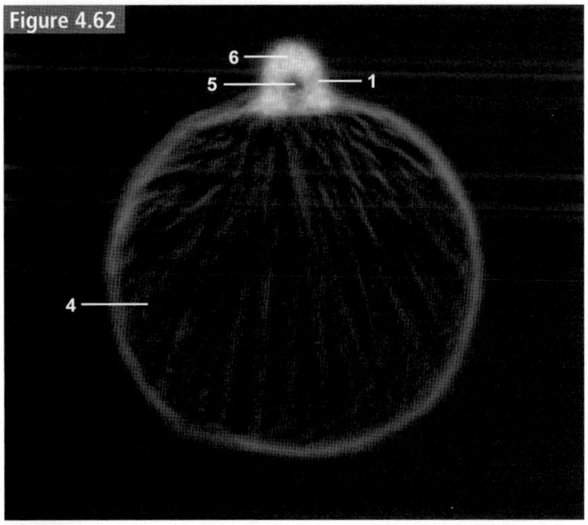

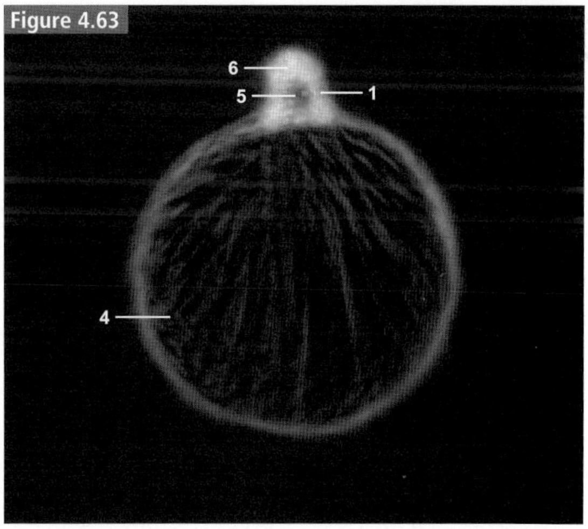

1. myotome	**2.** forebrain	**3.** optic cup	**4.** yolk	**5.** notochord	**6.** neural tube

Figure 4.64

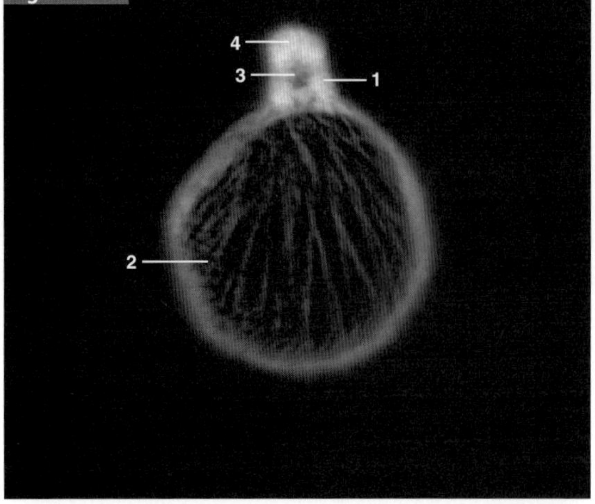

Figure 4.65

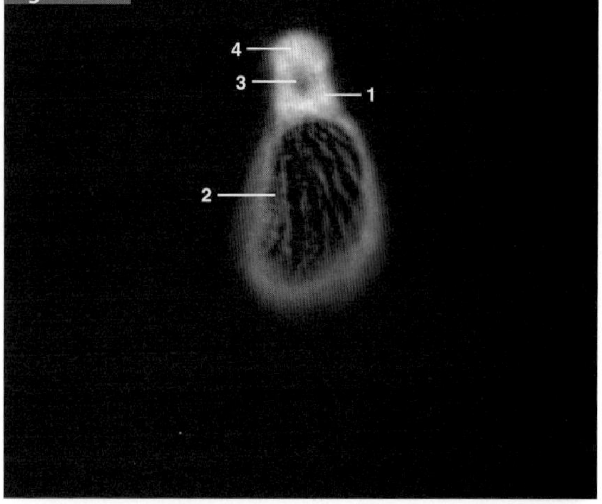

Figure 4.66

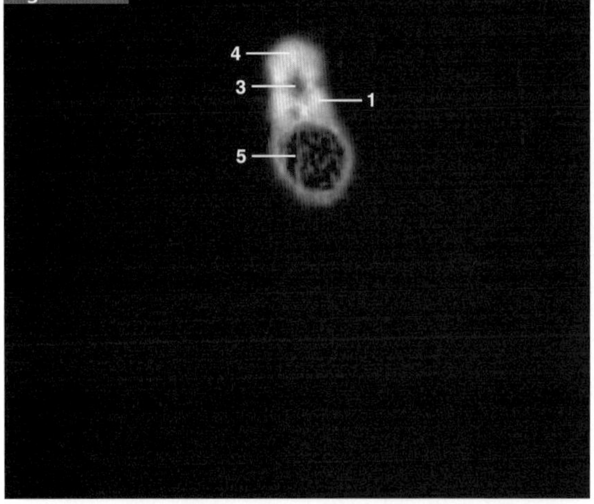

Figure 4.67

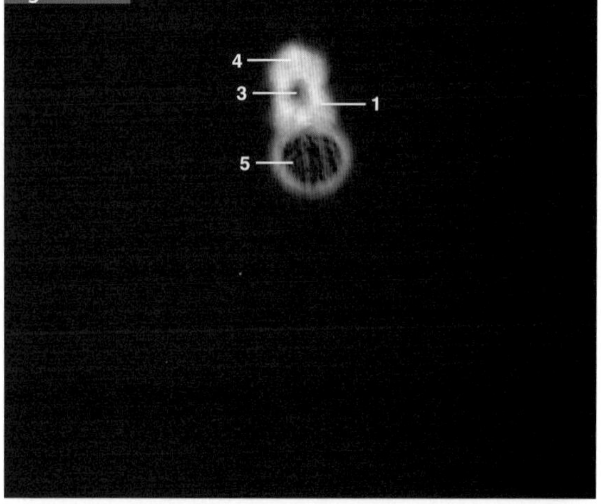

Figure 4.68

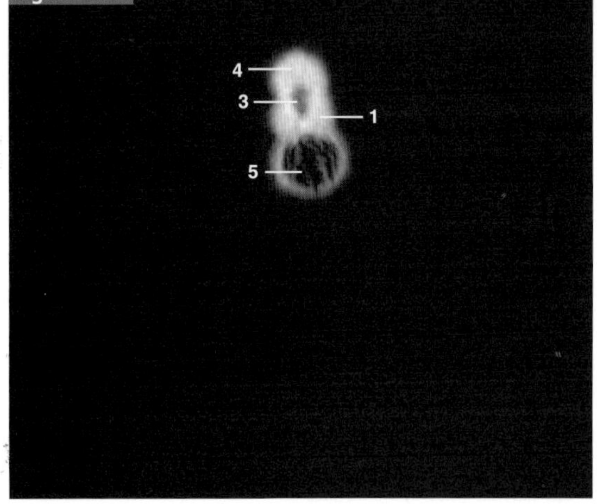

Figure 4.69

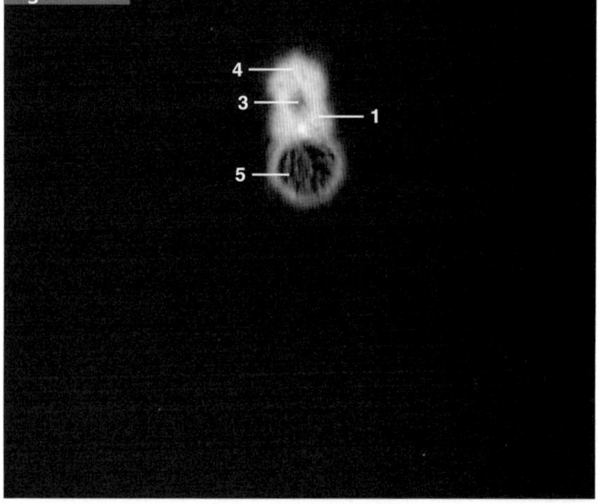

1. myotome	**2.** yolk	**3.** notochord	**4.** neural tube	**5.** yolk extension

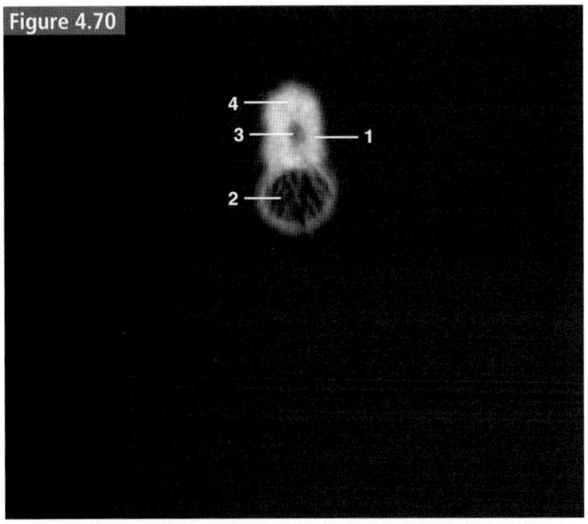

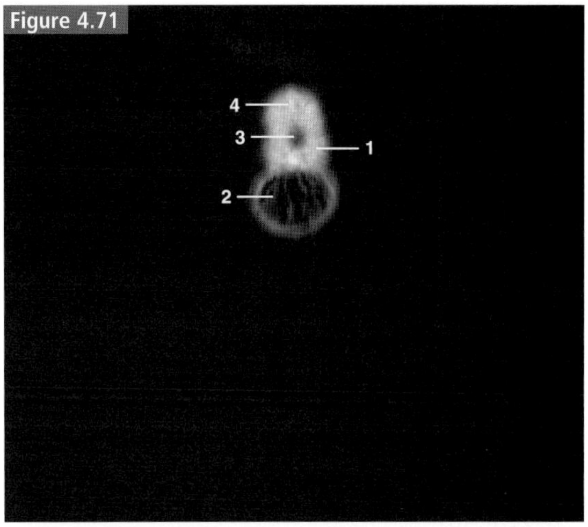

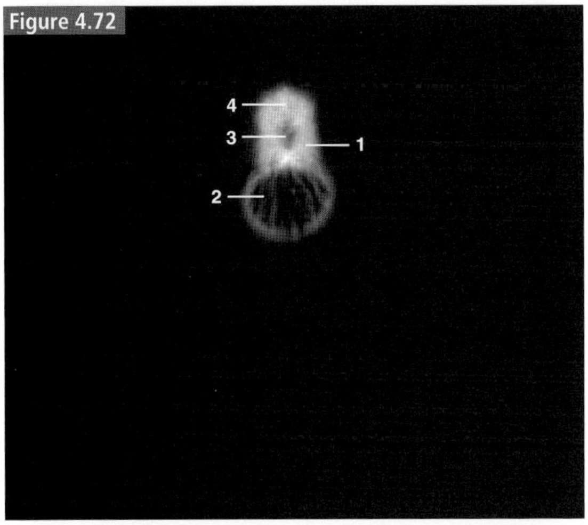

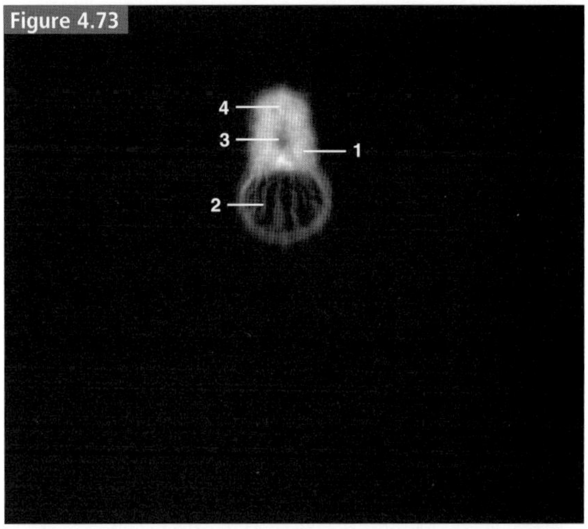

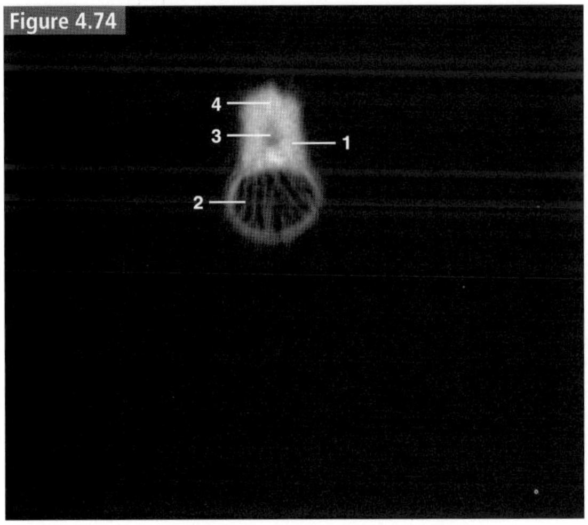

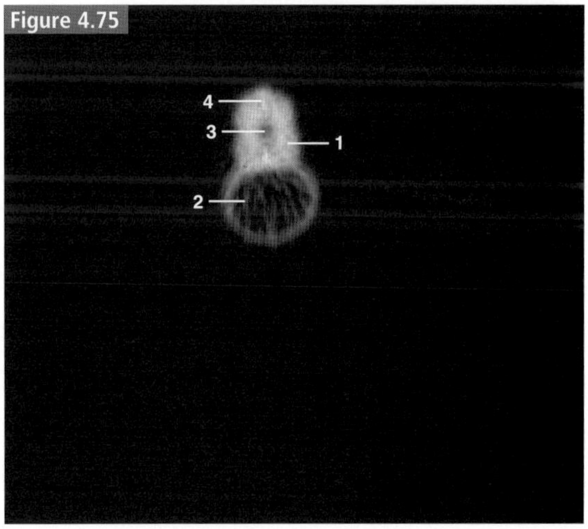

1. myotome	**2.** yolk extension	**3.** notochord	**4.** neural tube

Figure 4.76

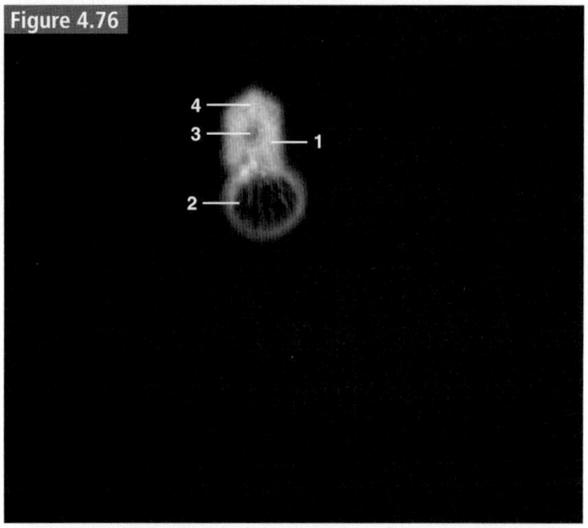

Figure 4.77

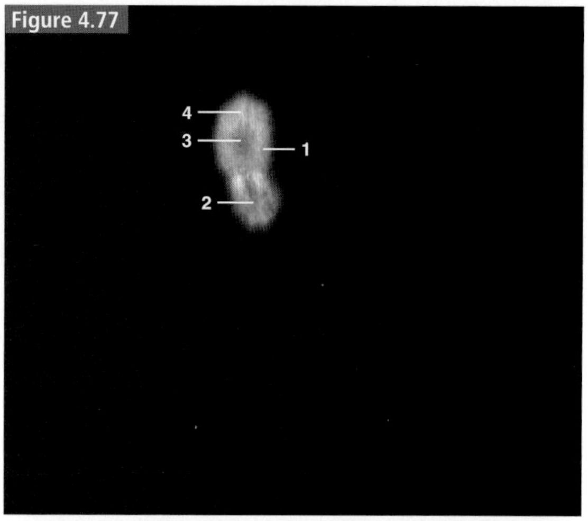

Figure 4.78

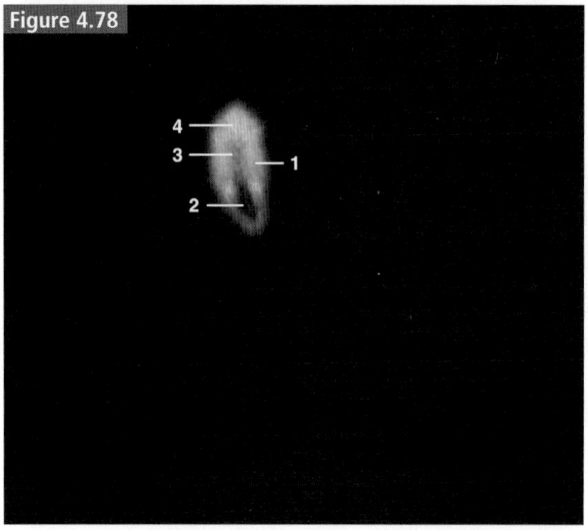

Figure 4.79

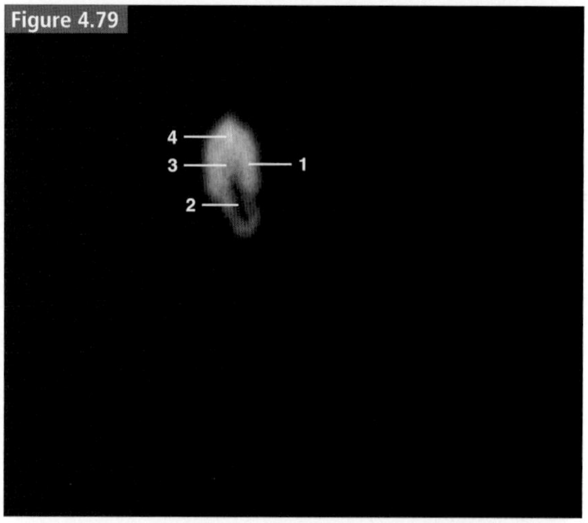

Figure 4.80

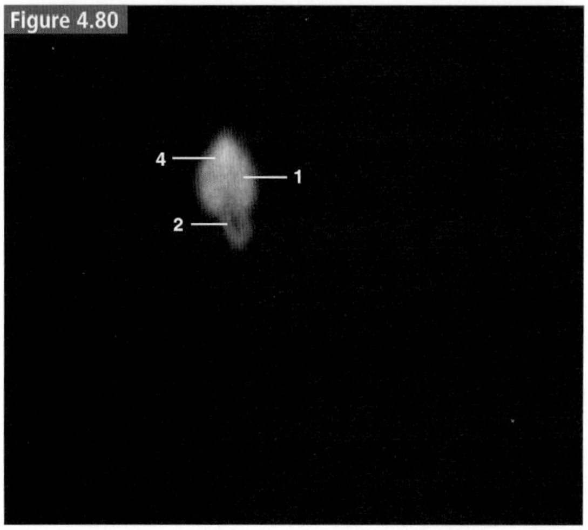

Figure 4.81

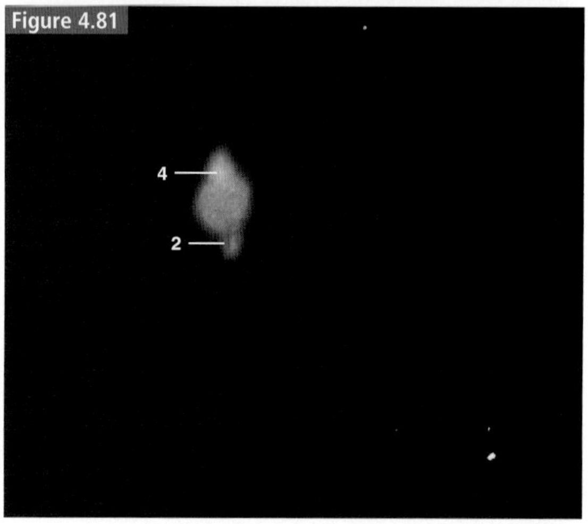

1. myotome **2.** yolk extension **3.** notochord **4.** neural tube

CHAPTER 5

48 HOURS

Sagittal Sections

Figure 5.1

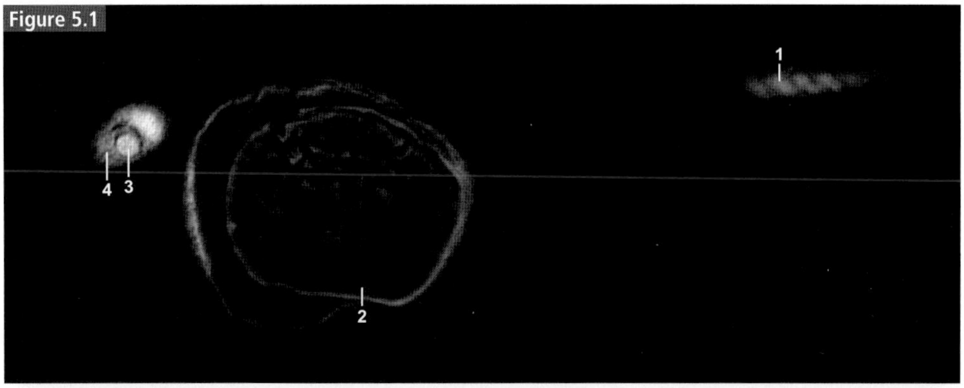

Figure 5.2

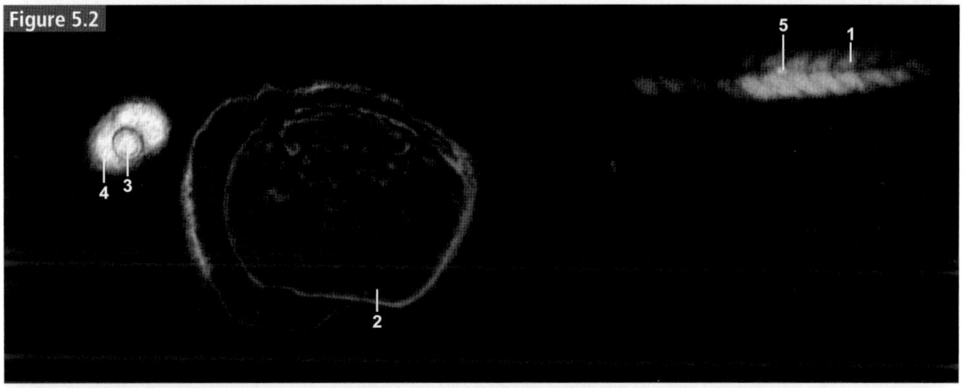

Figure 5.3

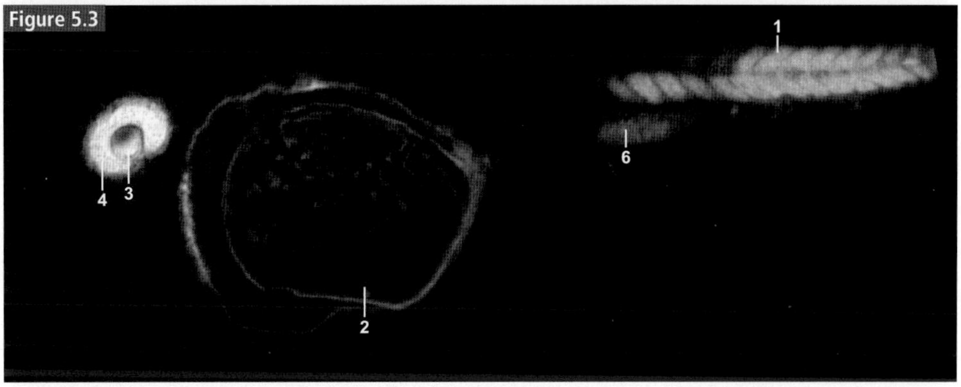

1. myotome	**2.** yolk	**3.** lens	**4.** optic cup	**5.** neuromasts	**6.** yolk extension

Atlas of Zebrafish Development. 10.1016/B978-0-12-374016-8.00005-4

Figure 5.4

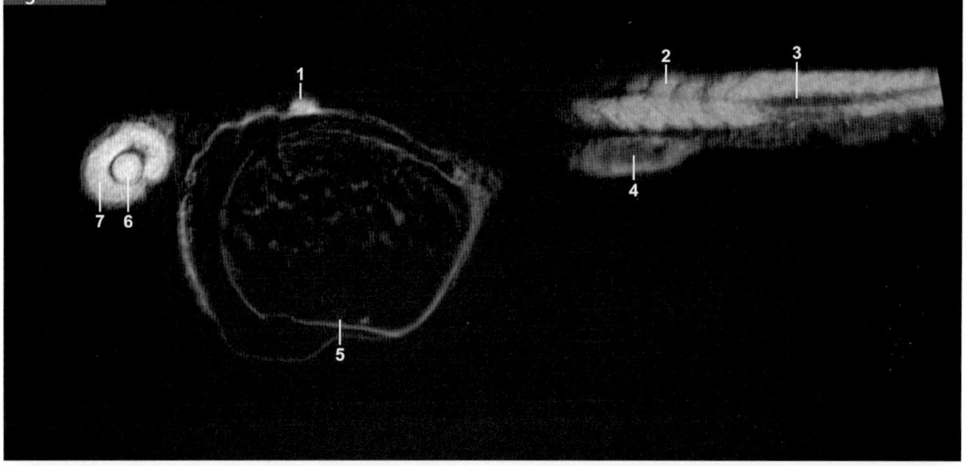

Figure 5.5

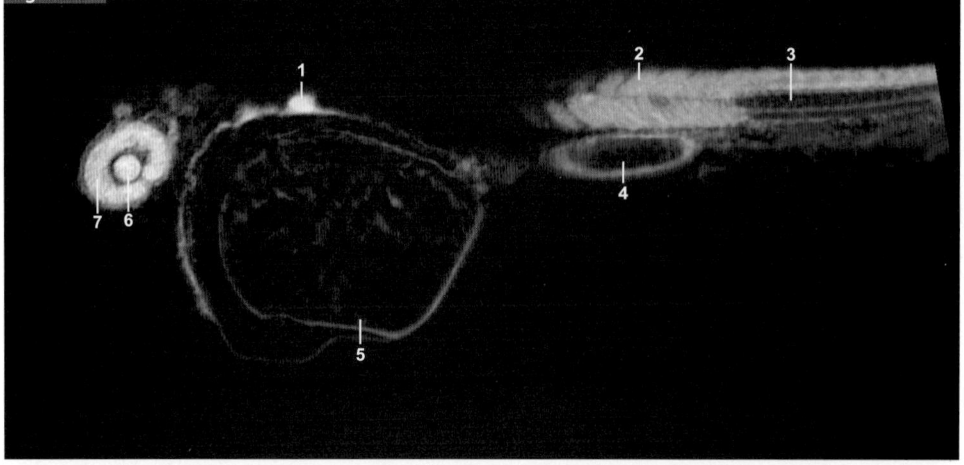

Figure 5.6

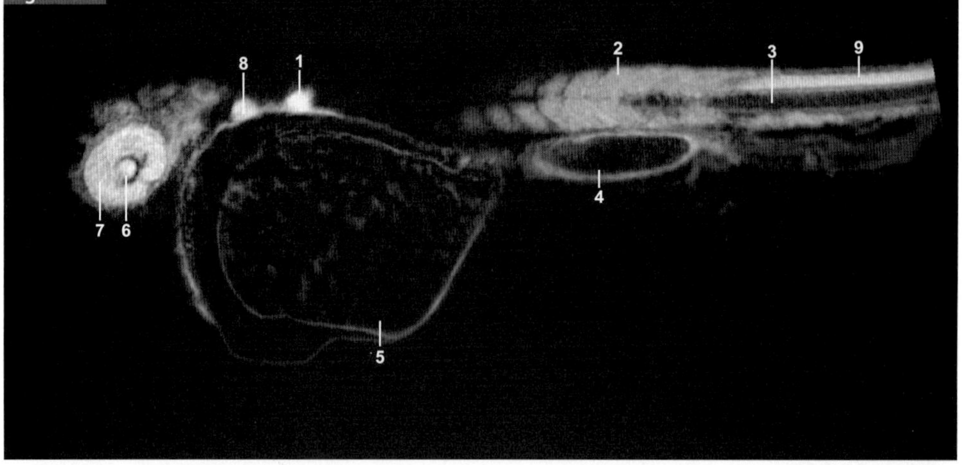

1. pectoral fin bud **2.** myotome **3.** notochord **4.** yolk extension **5.** yolk **6.** lens **7.** optic cup **8.** gill arches **9.** spinal cord

Figure 5.7

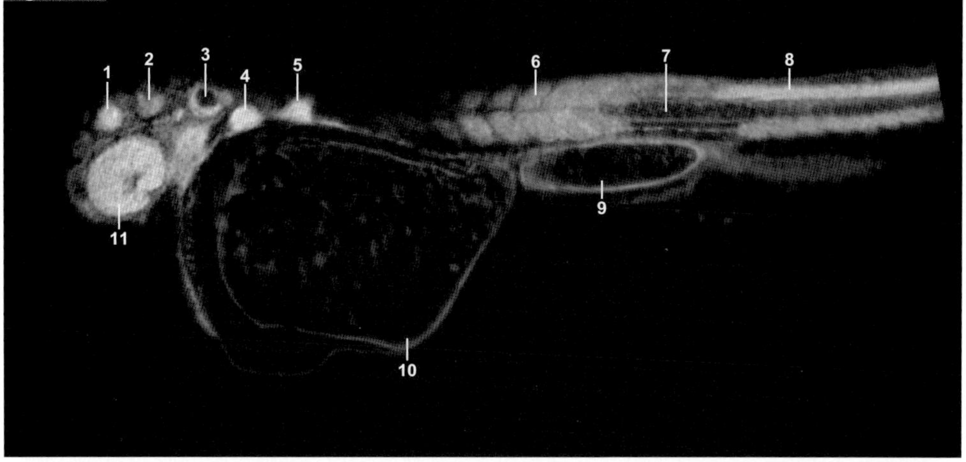

Figure 5.8

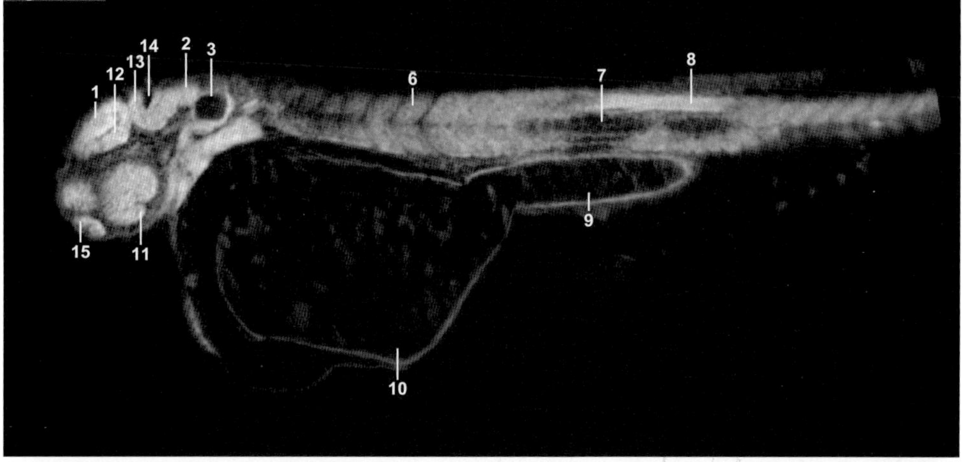

Figure 5.9

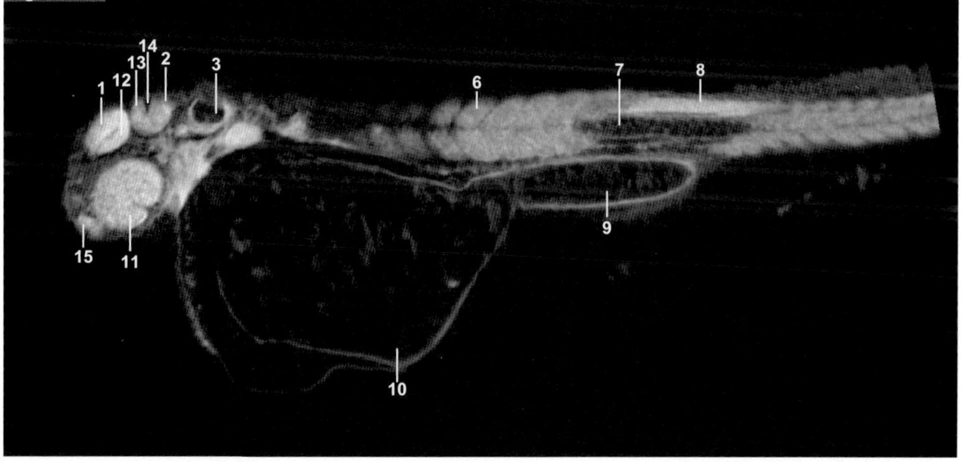

1. optic tectum	**4.** gill arches	**7.** notochord	**10.** yolk	**13.** cerebellum
2. medulla oblongata	**5.** pectoral fin bud	**8.** spinal cord	**11.** optic cup	**14.** fourth ventricle
3. otic vesicle	**6.** myotome	**9.** yolk extension	**12.** tegmentum	**15.** olfactory epithelium

Figure 5.10

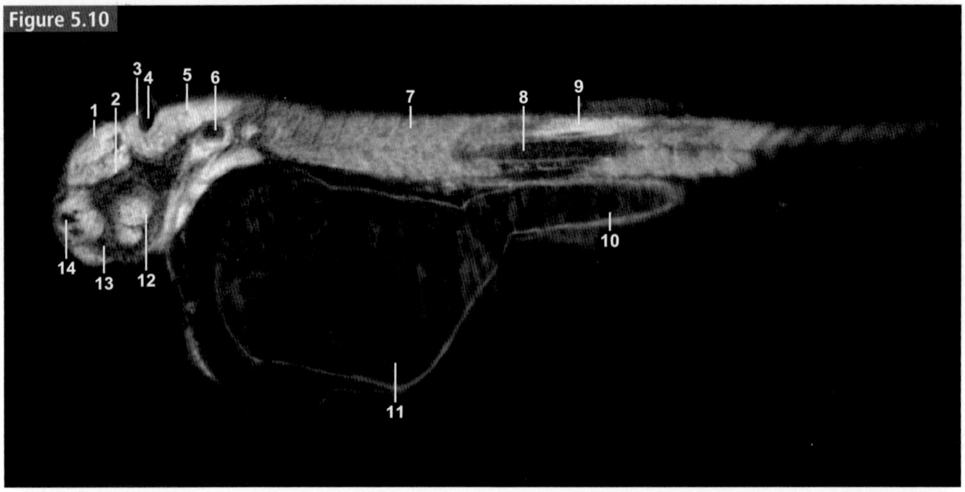

Figure 5.11

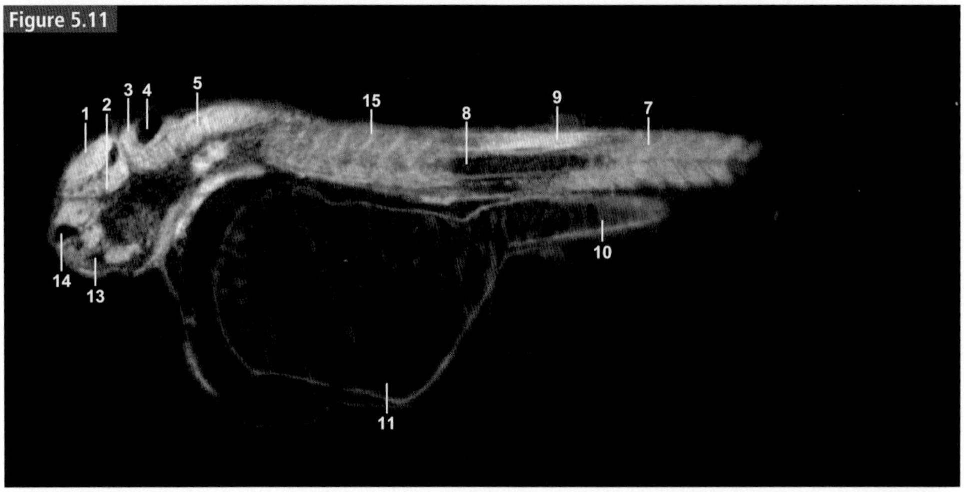

Figure 5.12

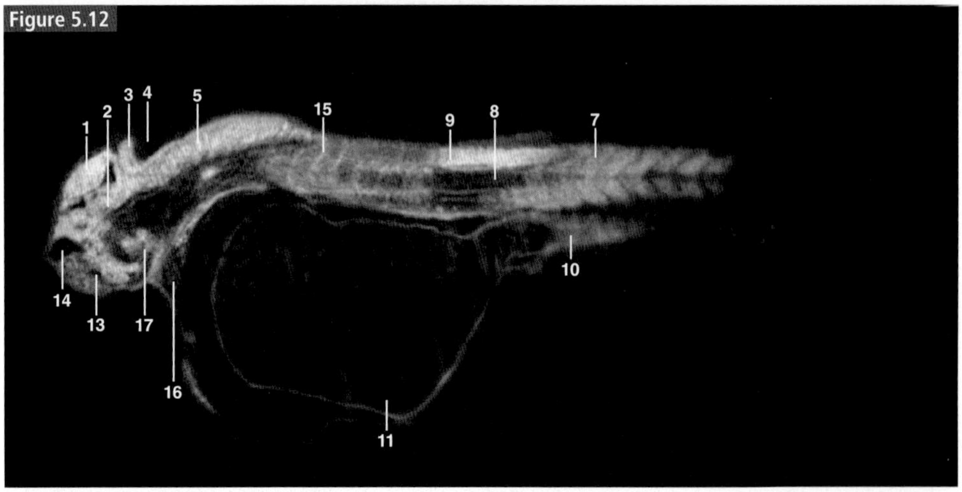

1. optic tectum	**4.** fourth ventricle	**7.** myotome	**10.** yolk extension	**13.** anterior commissure	**16.** heart
2. tegmentum	**5.** medulla oblongata	**8.** notochord	**11.** yolk	**14.** forebrain ventricle	**17.** hypothalamus
3. cerebellum	**6.** otic vesicle	**9.** spinal cord	**12.** optic cup	**15.** vertical myoseptum	

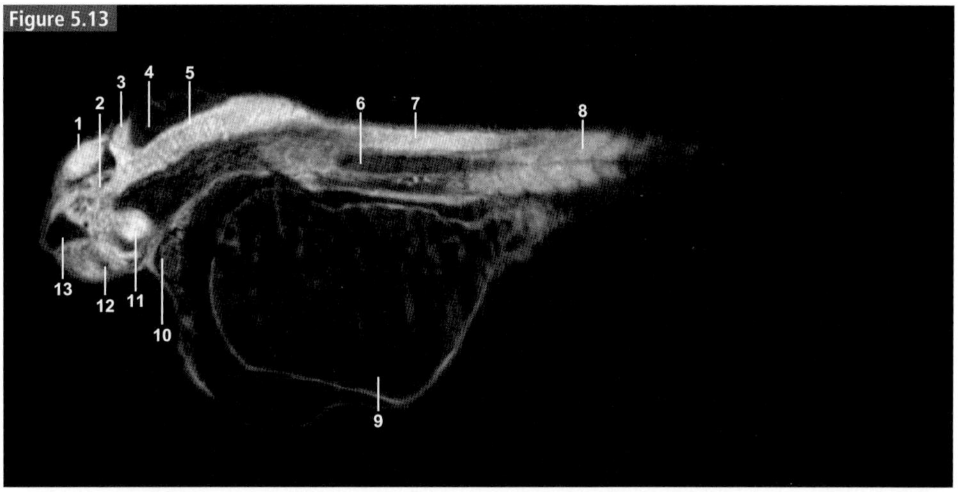

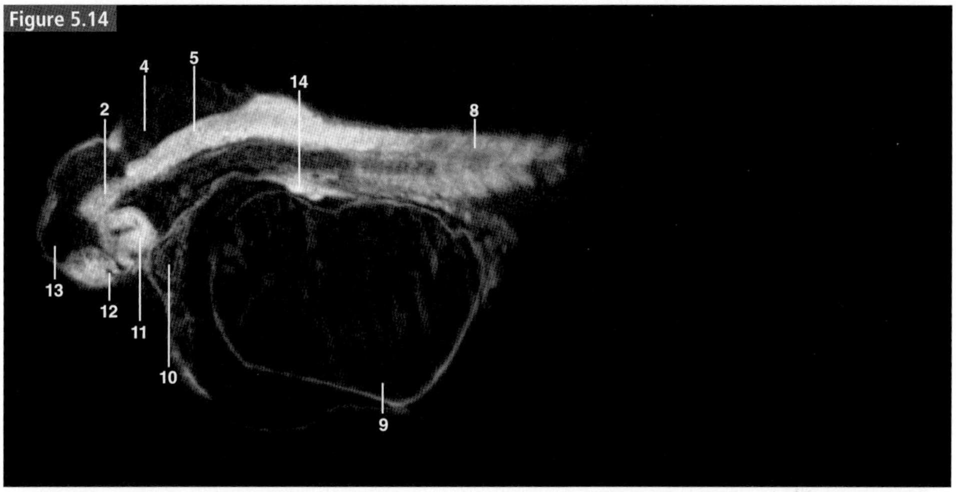

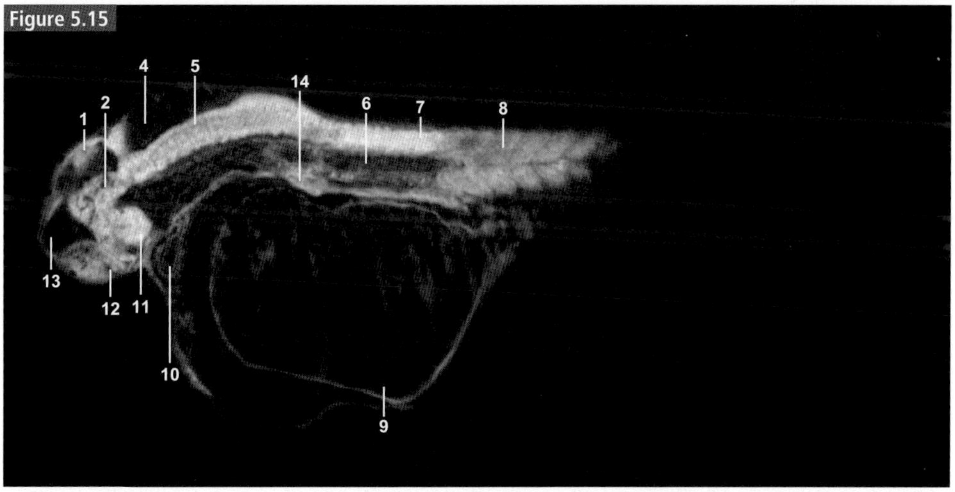

1. optic tectum	**4.** fourth ventricle	**7.** spinal cord	**10.** heart	**13.** forebrain ventricle
2. tegmentum	**5.** medulla oblongata	**8.** myotome	**11.** hypothalamus	**14.** intestine
3. cerebellum	**6.** notochord	**9.** yolk	**12.** anterior commissure	

Figure 5.16

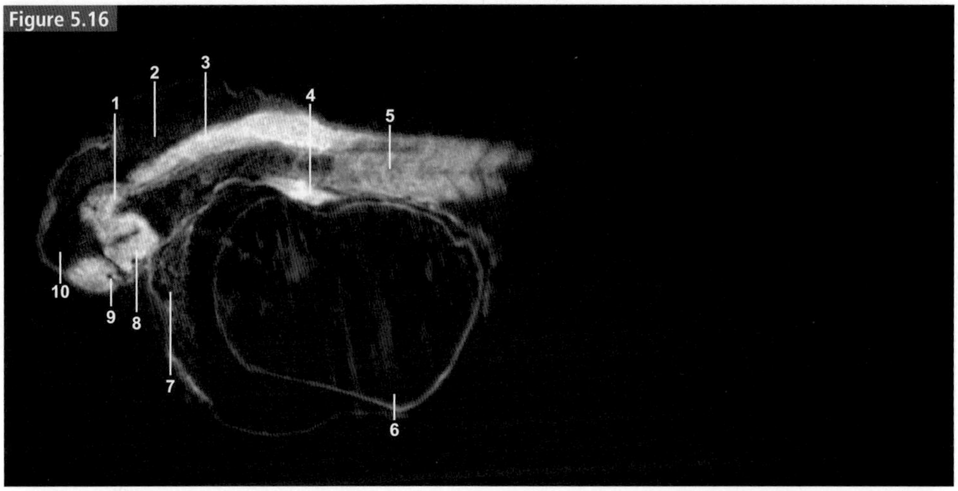

Figure 5.17

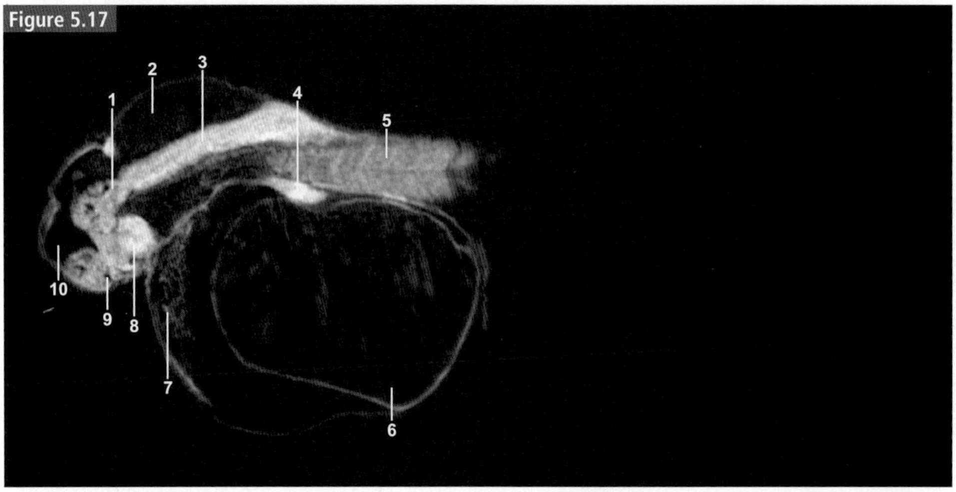

Figure 5.18

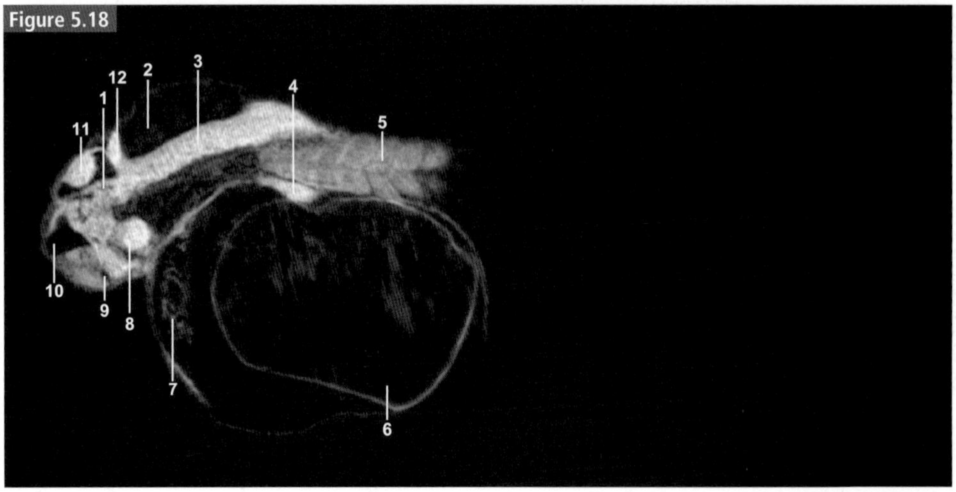

1. tegmentum	**3.** medulla oblongata	**5.** myotome	**7.** heart	**9.** anterior commissure	**11.** optic tectum
2. fourth ventricle	**4.** liver	**6.** yolk	**8.** hypothalamus	**10.** forebrain ventricle	**12.** cerebellum

Figure 5.19

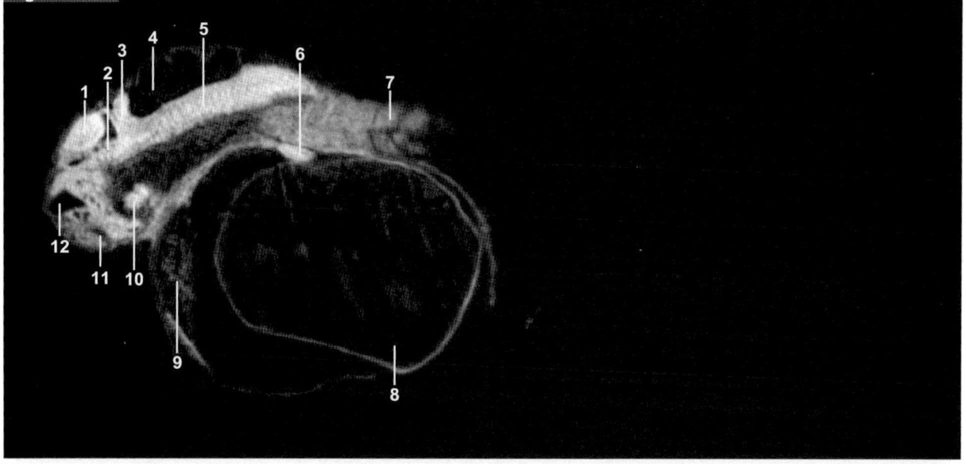

Figure 5.20

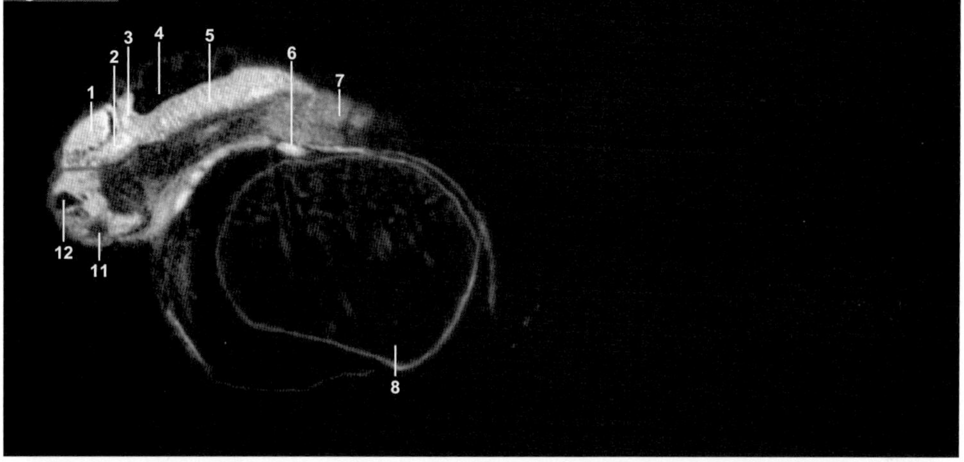

Figure 5.21

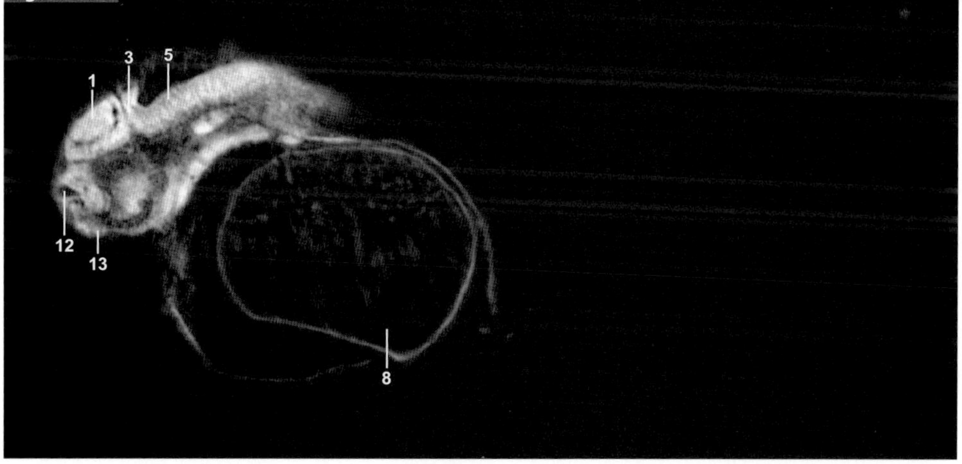

1. optic tectum	**4.** fourth ventricle	**7.** myotome	**10.** hypothalamus	**13.** olfactory epithelium
2. tegmentum	**5.** medulla oblongata	**8.** yolk	**11.** anterior commissure	
3. cerebellum	**6.** liver	**9.** heart	**12.** forebrain ventricle	

Figure 5.22

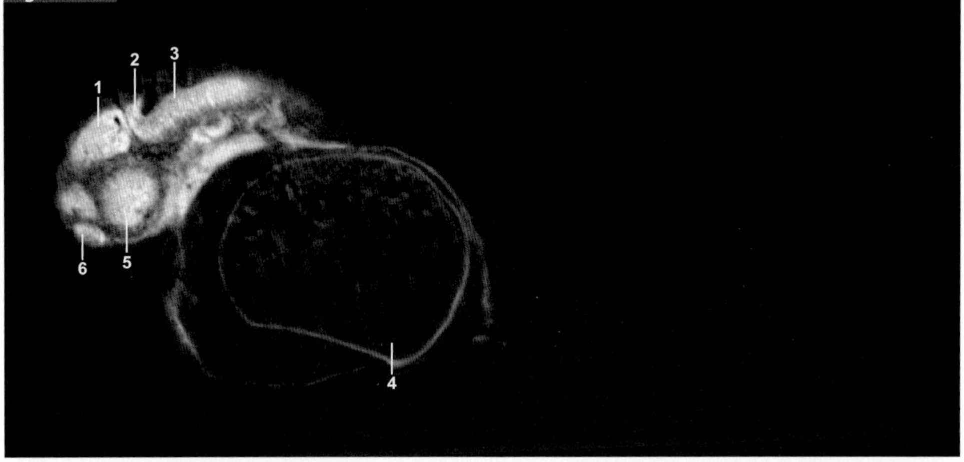

Figure 5.23

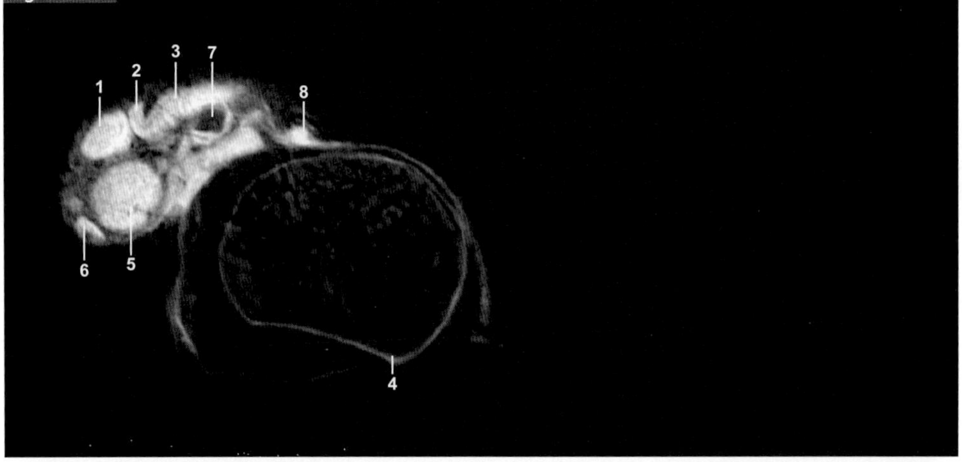

Figure 5.24

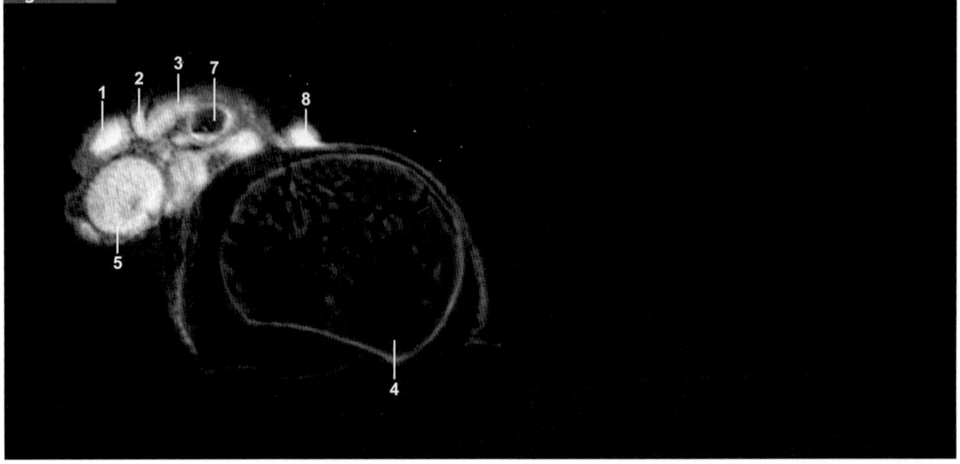

1. optic tectum	**3.** medulla oblongata	**5.** optic cup	**7.** otic vesicle
2. cerebellum	**4.** yolk	**6.** olfactory epithelium	**8.** pectoral fin bud

Figure 5.25

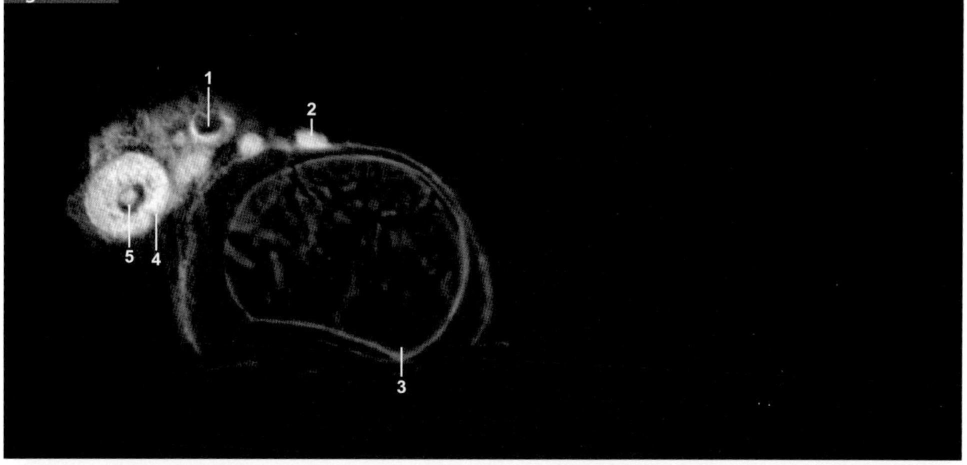

Figure 5.26

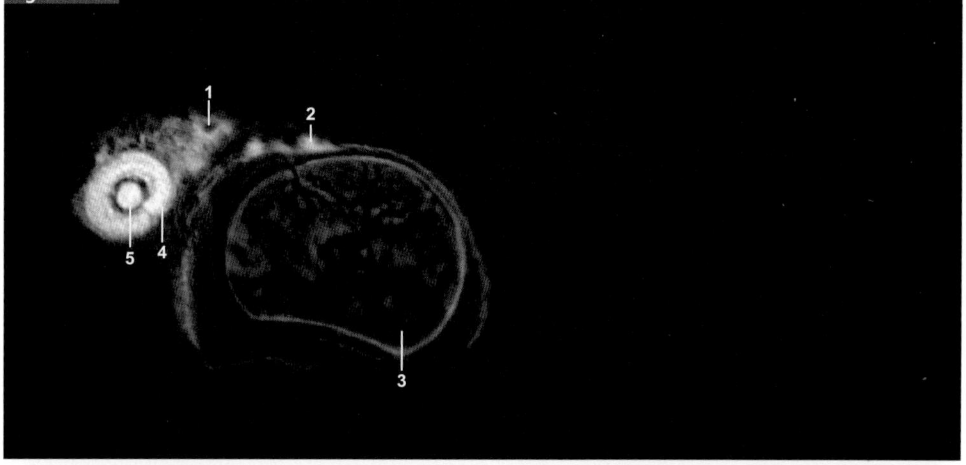

Figure 5.27

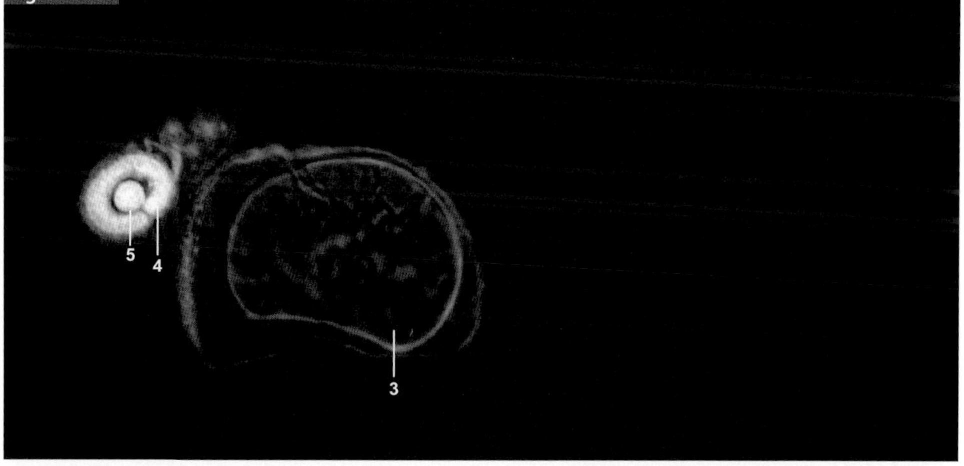

1. otic vesicle	**2.** pectoral fin bud	**3.** yolk	**4.** optic cup	**5.** lens

Coronal Sections

Figure 5.28

Figure 5.29

Figure 5.30

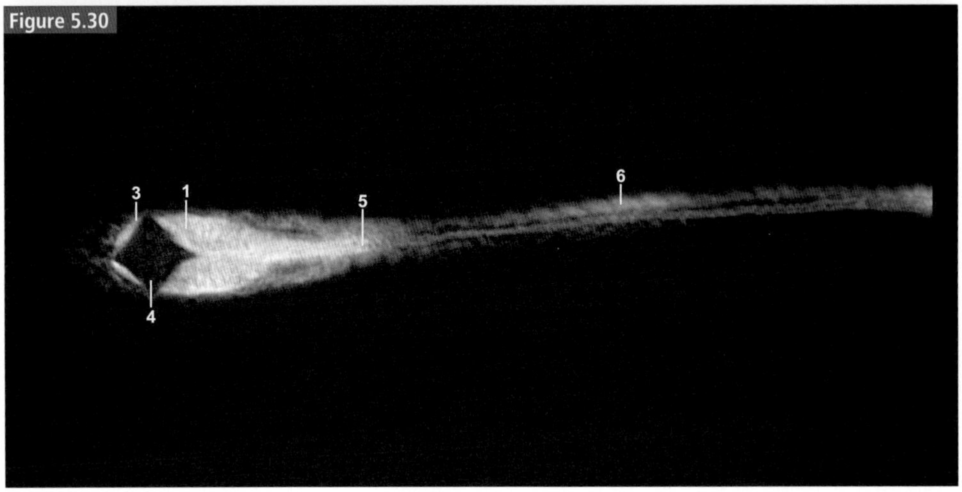

1. medulla oblongata	**2.** median fin fold	**3.** cerebellum	**4.** fourth ventricle	**5.** spinal cord	**6.** myotome

Figure 5.31

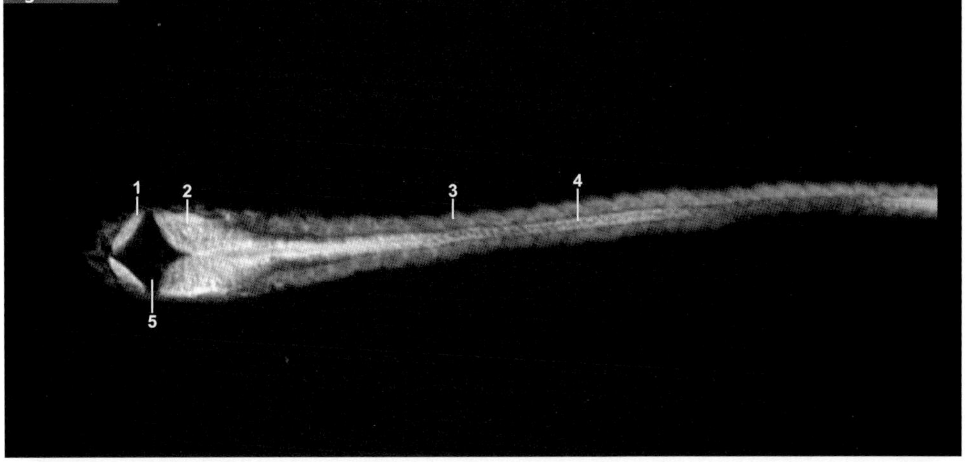

Figure 5.32

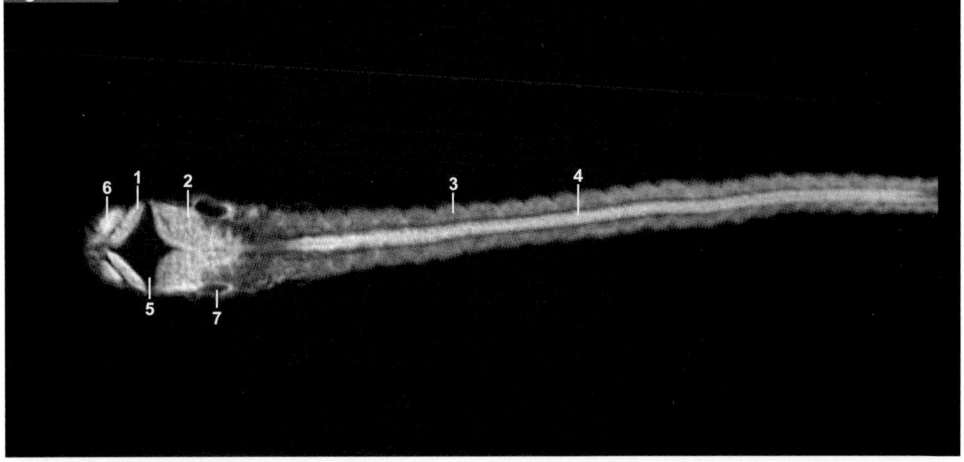

Figure 5.33

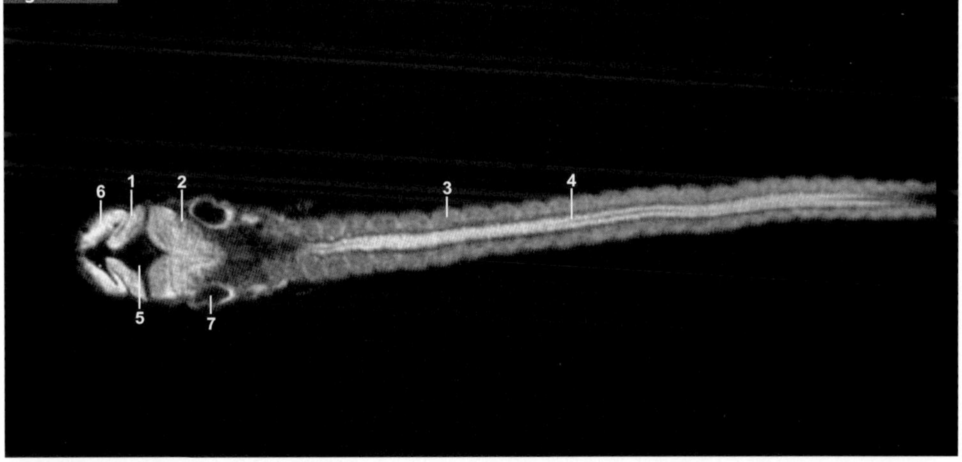

1. cerebellum	**2.** medulla oblongata	**3.** myotome	**4.** spinal cord	**5.** fourth ventricle	**6.** optic tectum	**7.** otic vesicle

Figure 5.34

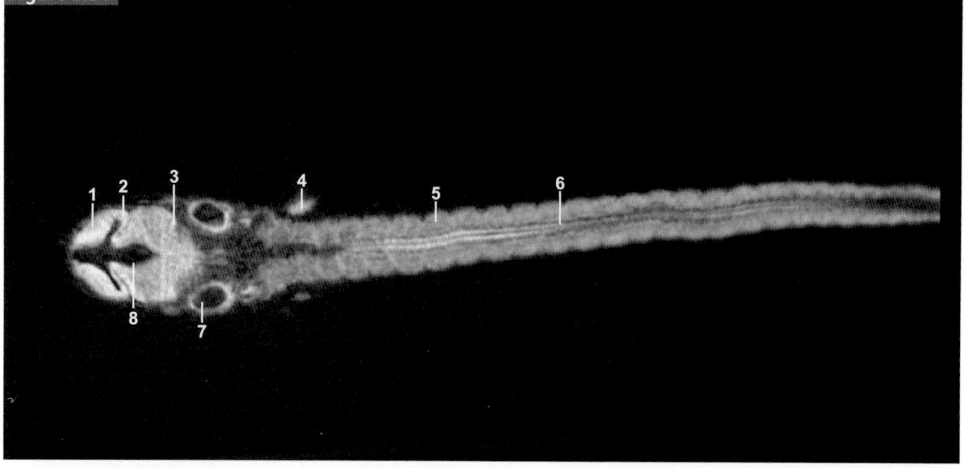

Figure 5.35

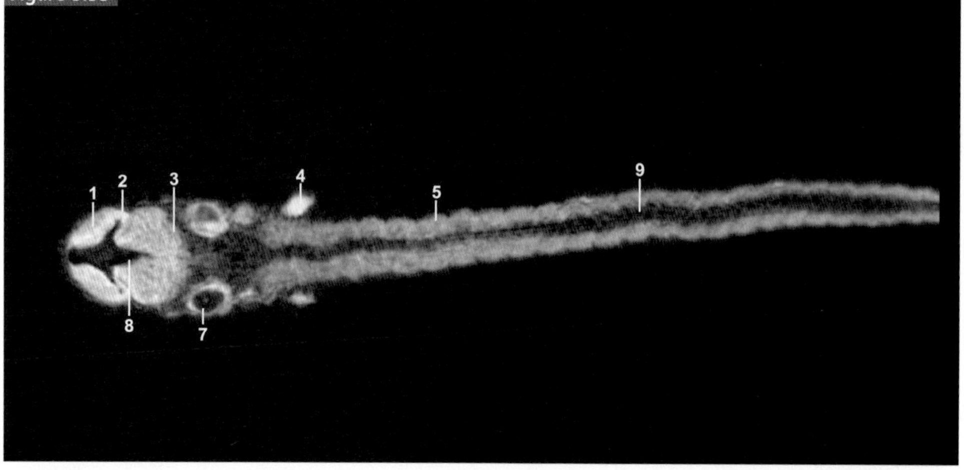

Figure 5.36

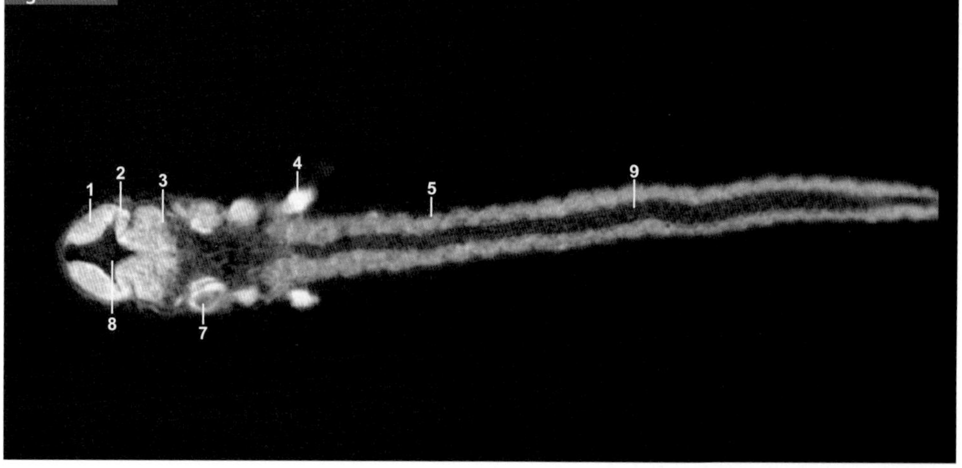

1. optic tectum	**4.** pectoral fin bud	**7.** otic vesicle
2. cerebellum	**5.** myotome	**8.** fourth ventricle
3. medulla oblongata	**6.** spinal cord	**9.** notochord

Figure 5.37

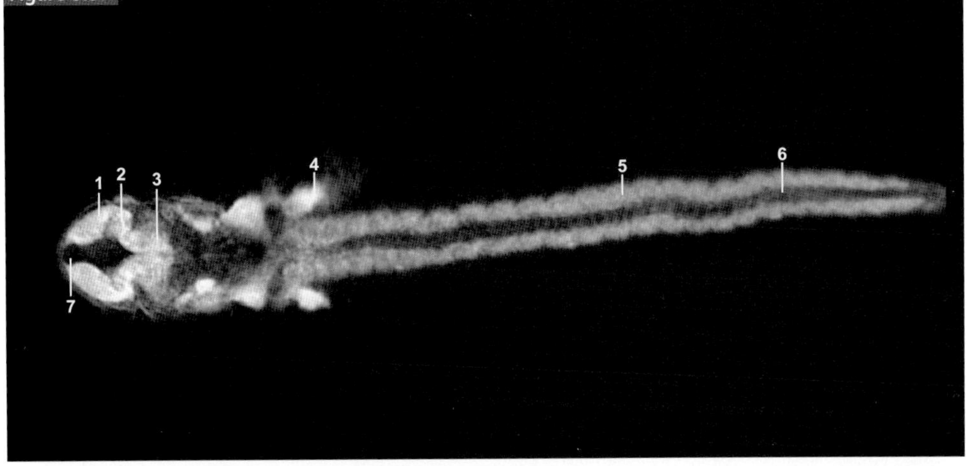

Figure 5.38

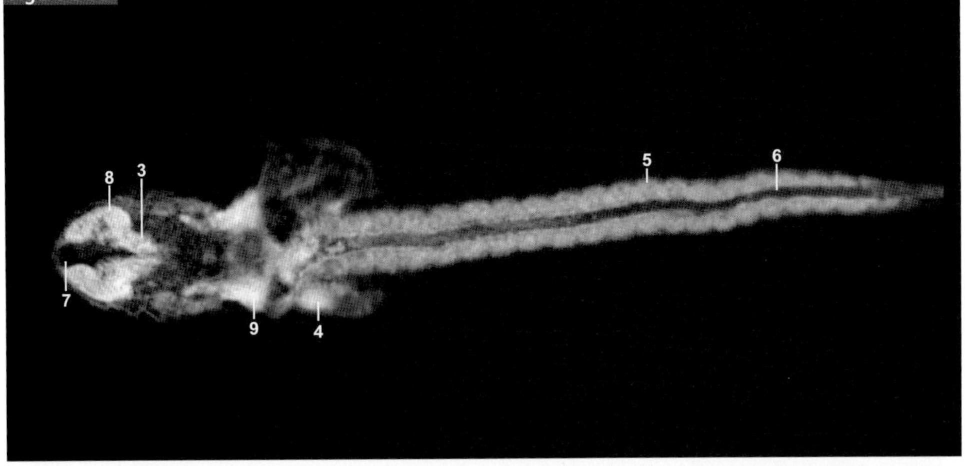

Figure 5.39

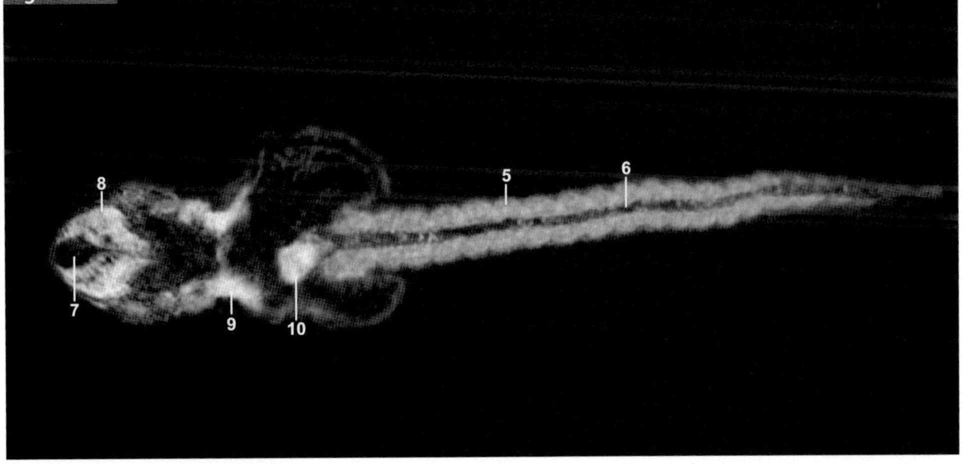

1. optic tectum	**3.** medulla oblongata	**5.** myotome	**7.** forebrain ventricle	**9.** gill arches
2. cerebellum	**4.** pectoral fin bud	**6.** notochord	**8.** tegmentum	**10.** liver

Figure 5.40

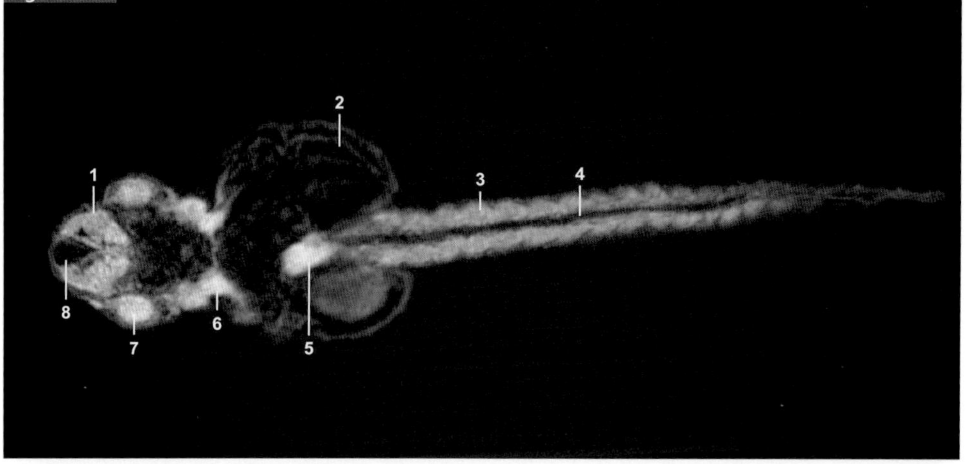

Figure 5.41

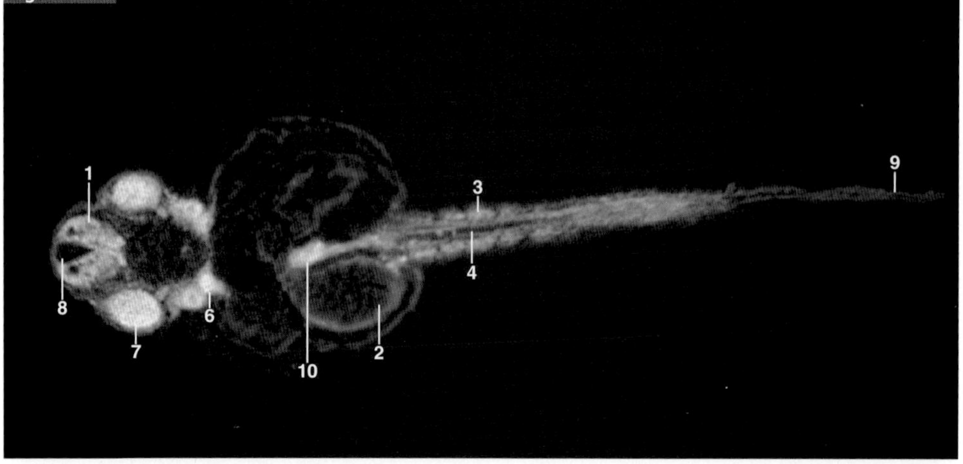

Figure 5.42

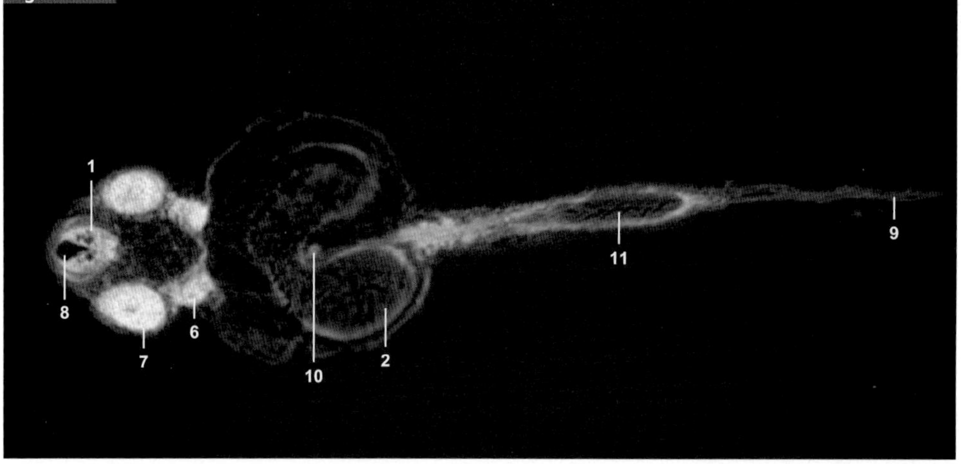

1. tegmentum
2. yolk
3. myotome
4. notochord
5. liver
6. gill arches
7. optic cup
8. forebrain ventricle
9. median fin fold
10. intestine
11. yolk extension

Figure 5.43

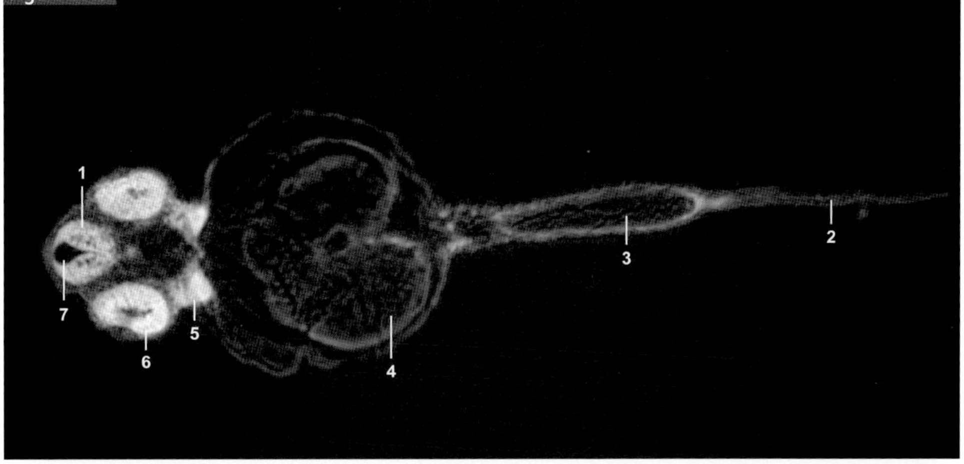

Figure 5.44

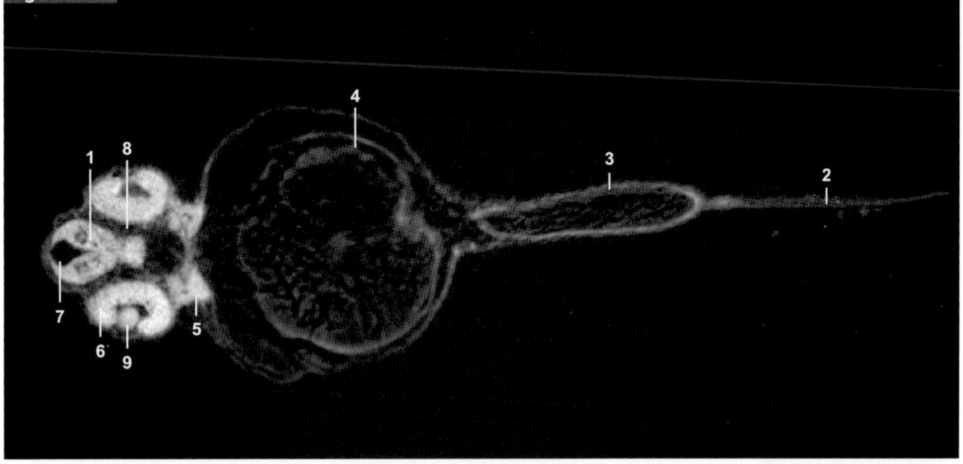

Figure 5.45

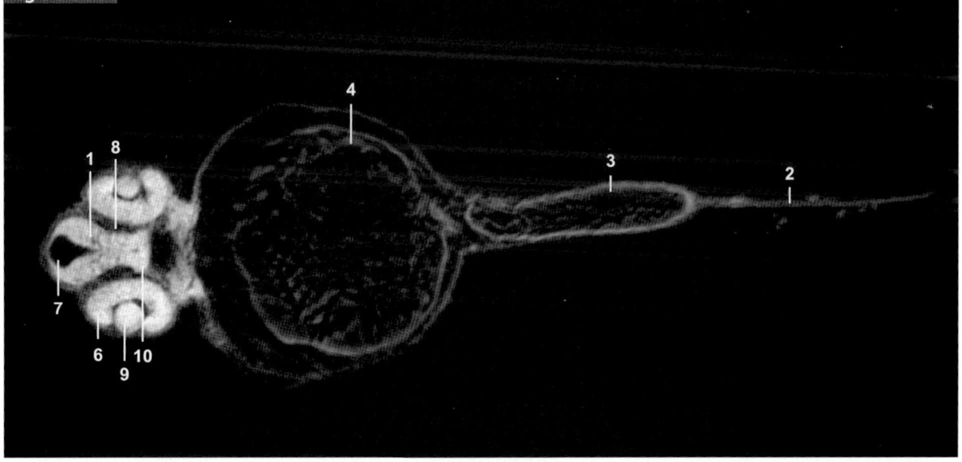

1. tegmentum	**3.** yolk extension	**5.** gill arches	**7.** forebrain ventricle	**9.** lens
2. median fin fold	**4.** yolk	**6.** optic cup	**8.** optic chiasm	**10.** hypothalamus

Figure 5.46

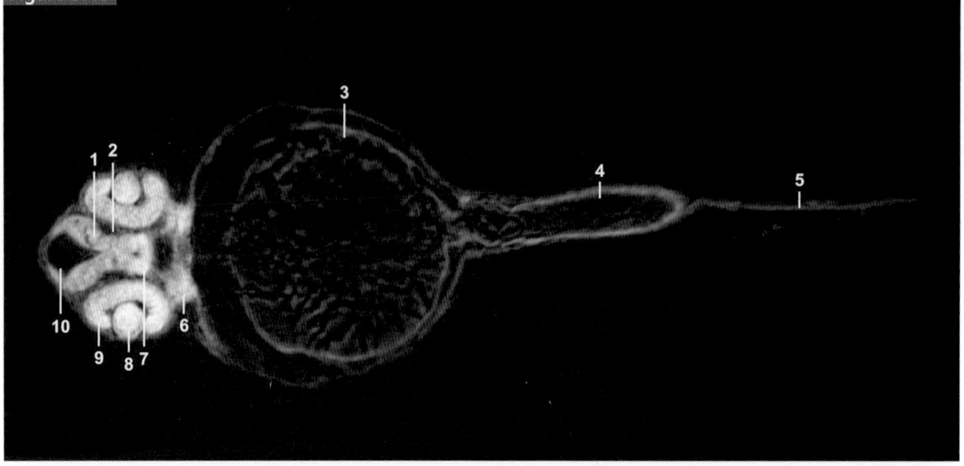

Figure 5.47

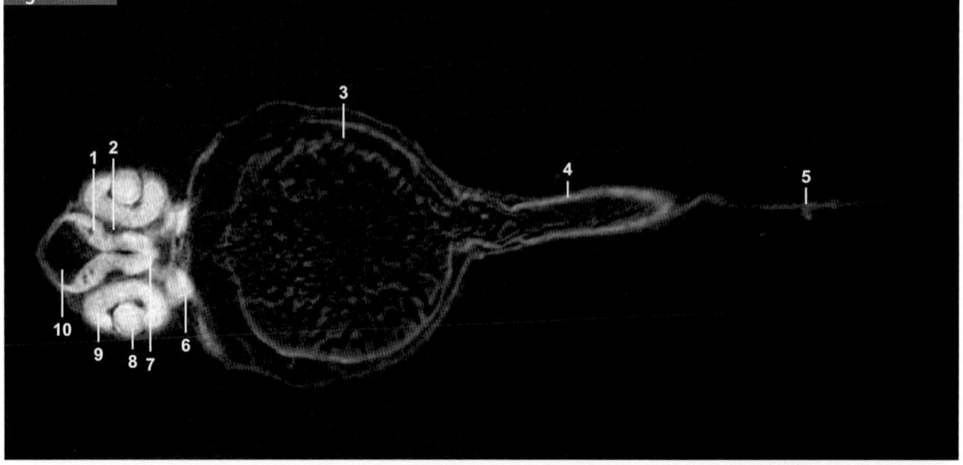

Figure 5.48

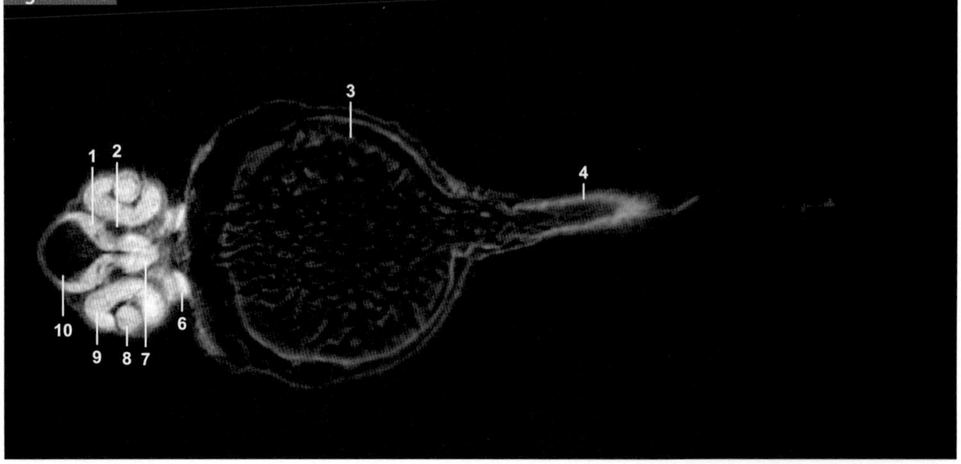

1. tegmentum	**3.** yolk	**5.** median fin fold	**7.** hypothalamus	**9.** optic cup
2. optic chiasm	**4.** yolk extension	**6.** gill arches	**8.** lens	**10.** forebrain ventricle

Figure 5.49

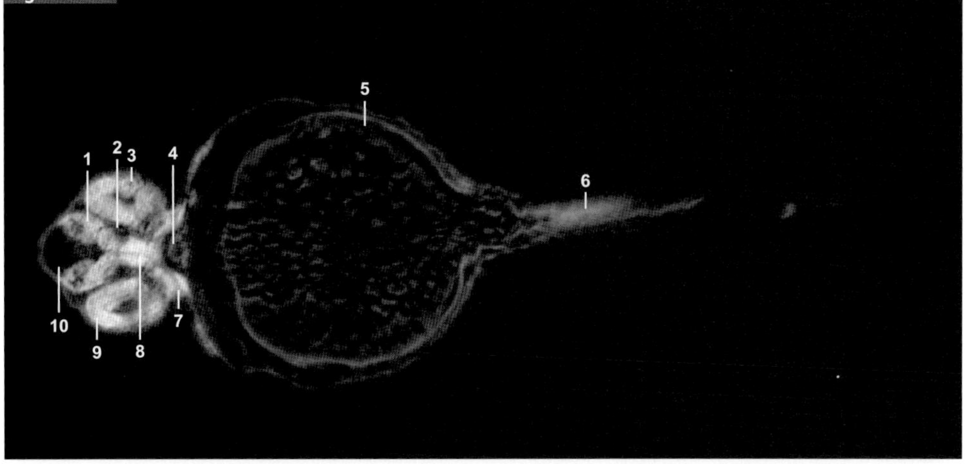

Figure 5.50

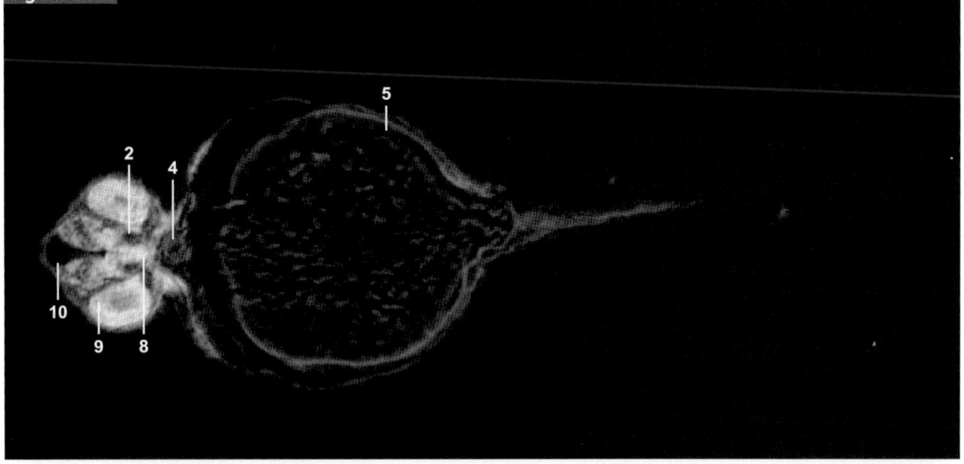

Figure 5.51

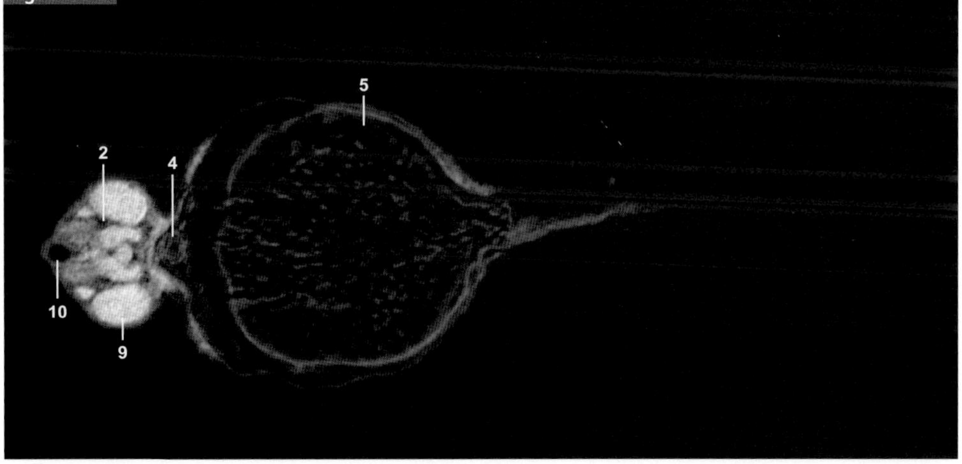

1. tegmentum	**3.** lens	**5.** yolk	**7.** gill arches	**9.** optic cup
2. optic chiasm	**4.** heart	**6.** yolk extension	**8.** hypothalamus	**10.** forebrain ventricle

Figure 5.52

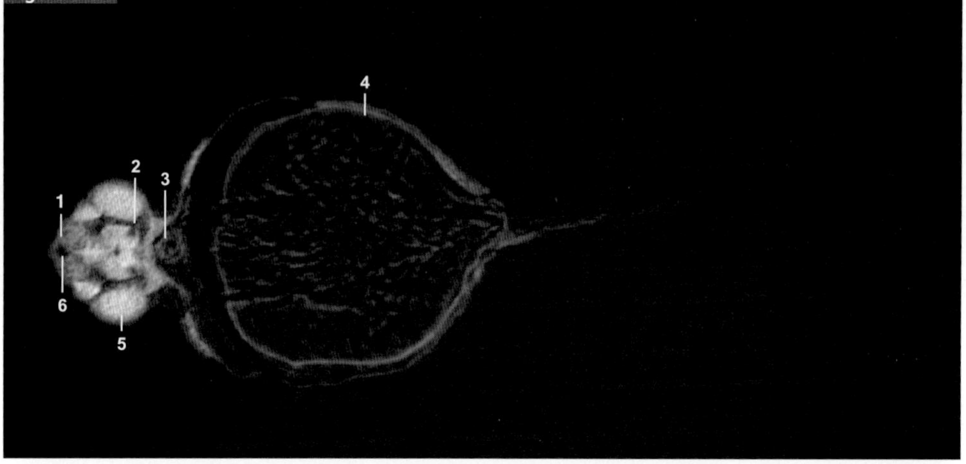

Figure 5.53

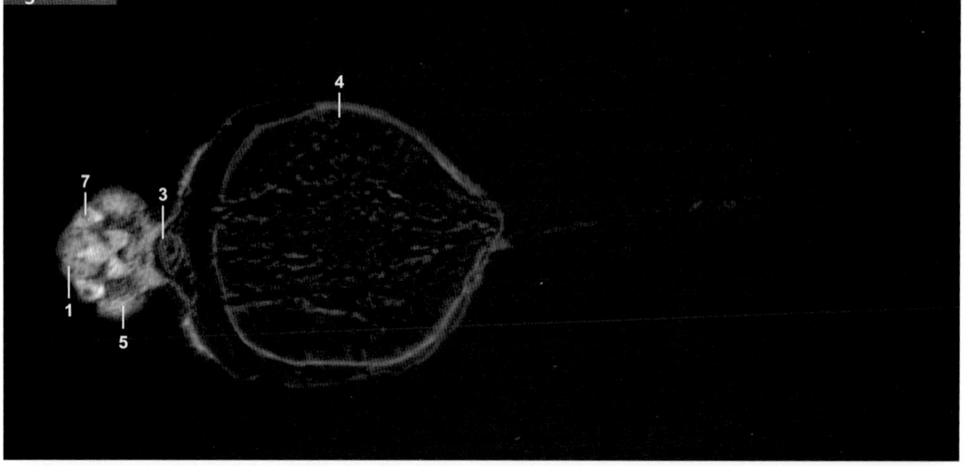

Figure 5.54

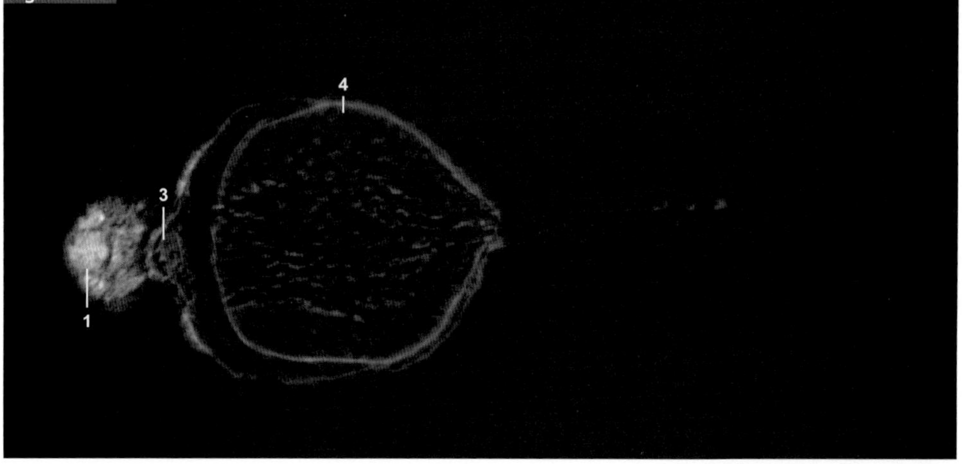

1. olfactory bulb	**3.** heart	**5.** optic cup	**6.** forebrain ventricle	**7.** olfactory pit
2. optic chiasm	**4.** yolk			

Transverse Sections

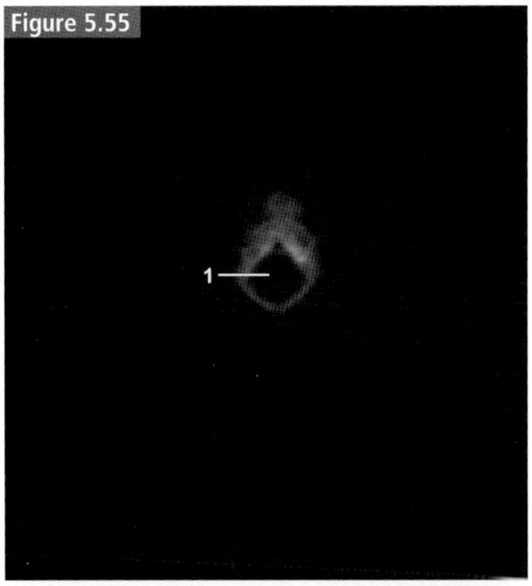

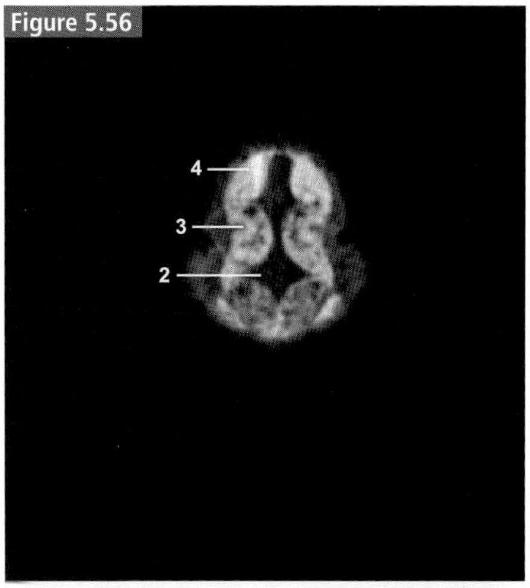

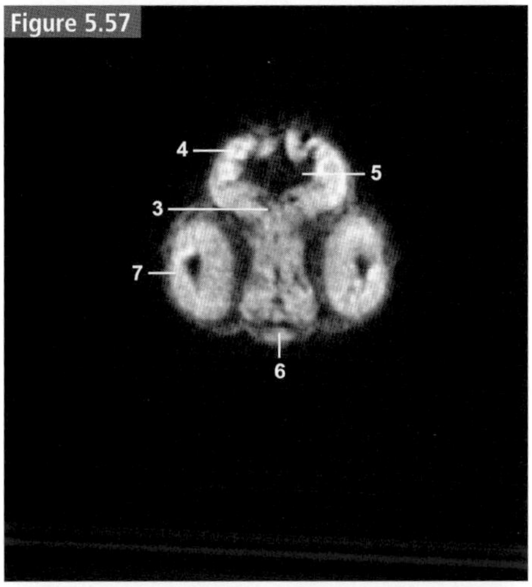

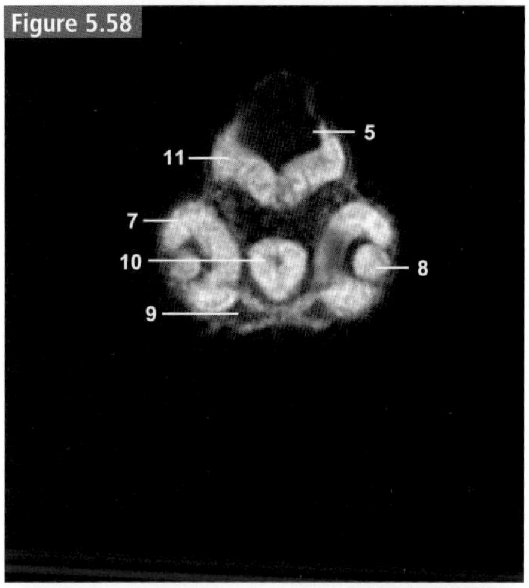

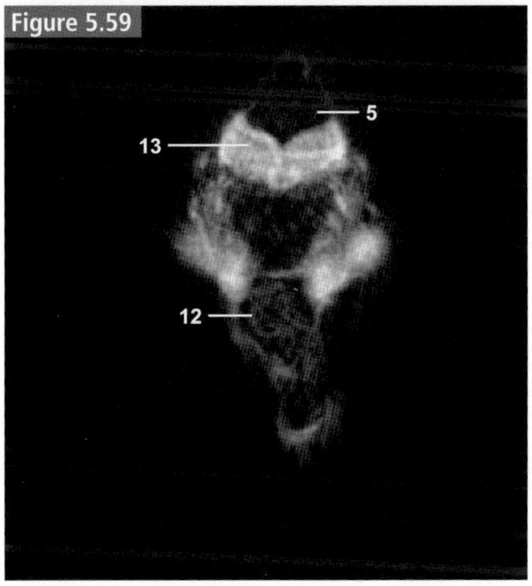

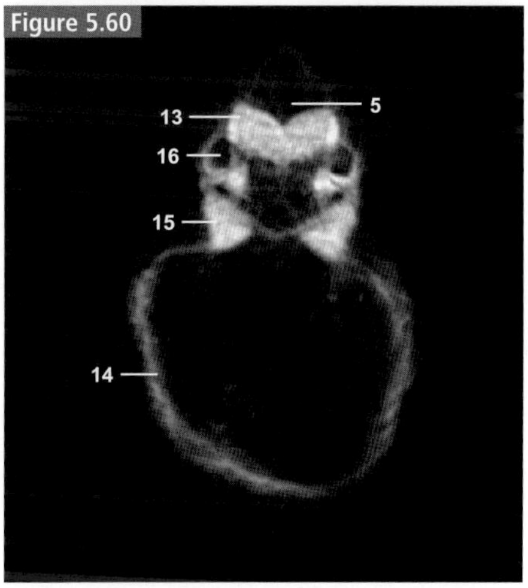

1. forebrain ventricle	**5.** fourth ventricle	**9.** anterior commissure	**13.** medulla oblongata
2. tectal ventricle	**6.** olfactory bulb	**10.** hypothalamus	**14.** yolk
3. tegmentum	**7.** optic cup	**11.** cerebellum	**15.** gill arches
4. optic tectum	**8.** lens	**12.** heart	**16.** otic vesicle

Figure 5.61

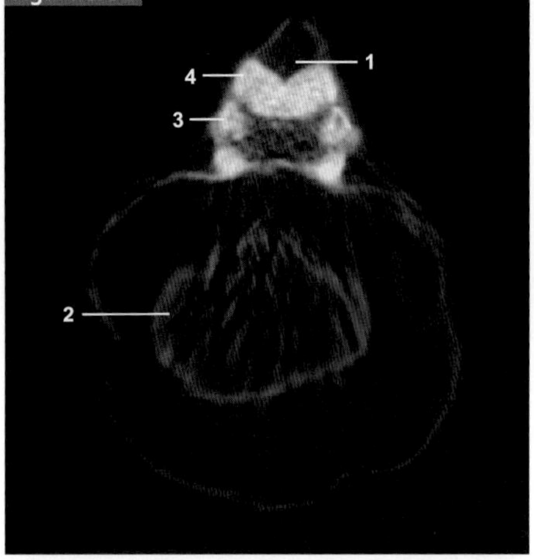

Figure 5.62

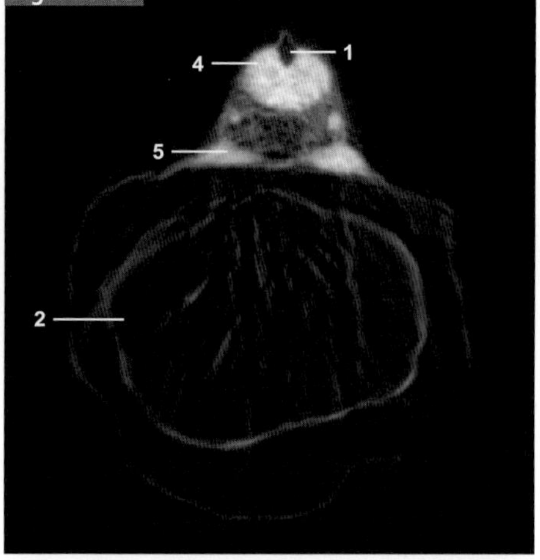

Figure 5.63

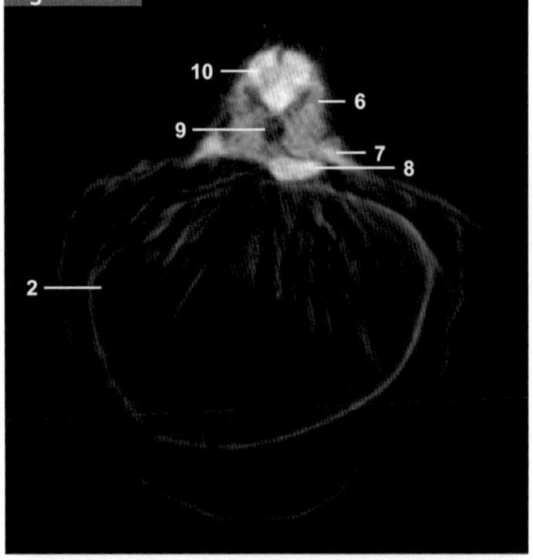

Figure 5.64

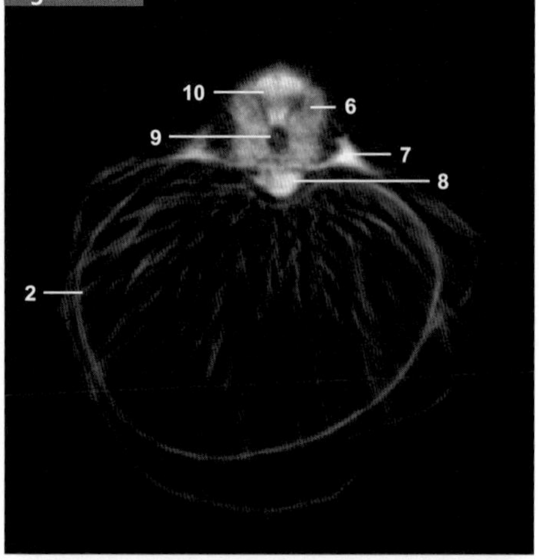

Figure 5.65

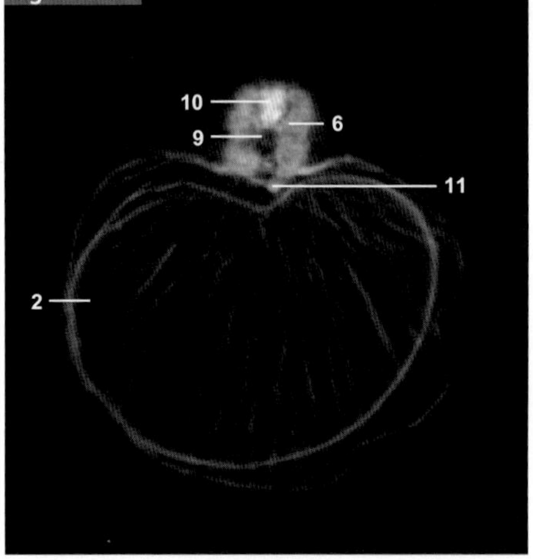

Figure 5.66

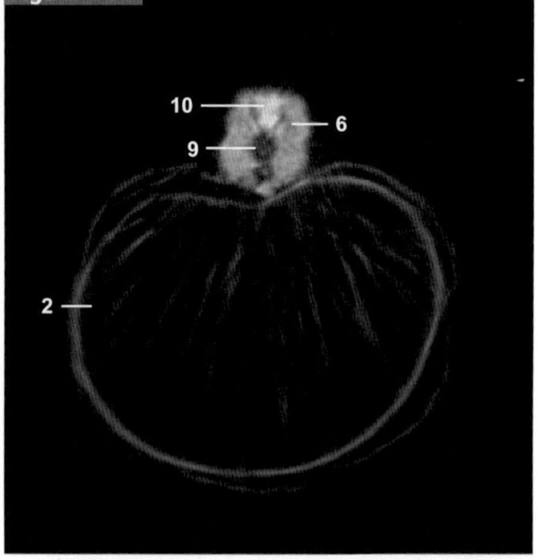

1. fourth ventricle	**4.** medulla oblongata	**7.** pectoral fin bud	**10.** spinal cord
2. yolk	**5.** gill arches	**8.** liver	**11.** intestine
3. otic vesicle	**6.** myotome	**9.** notochord	

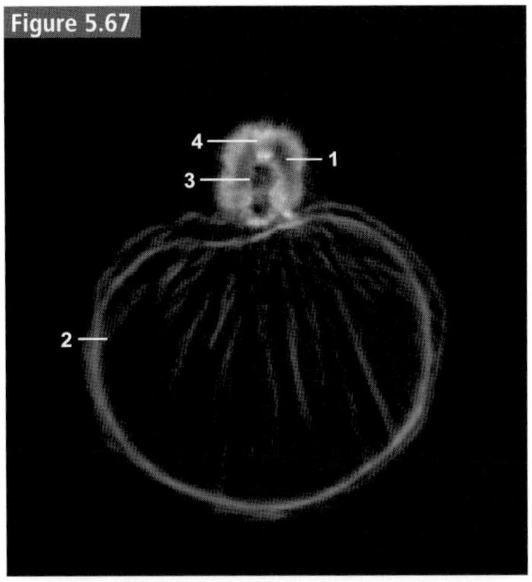

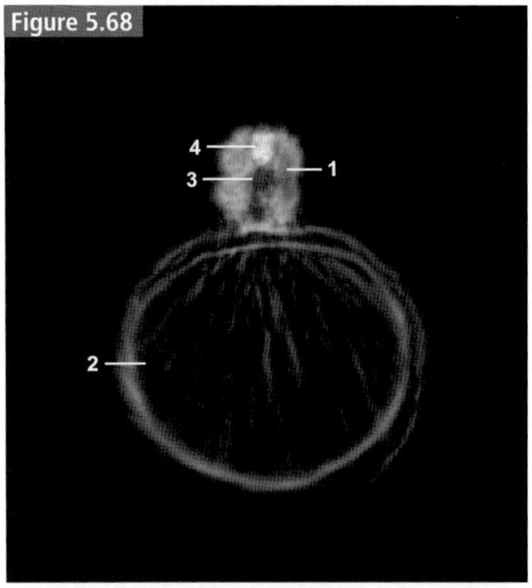

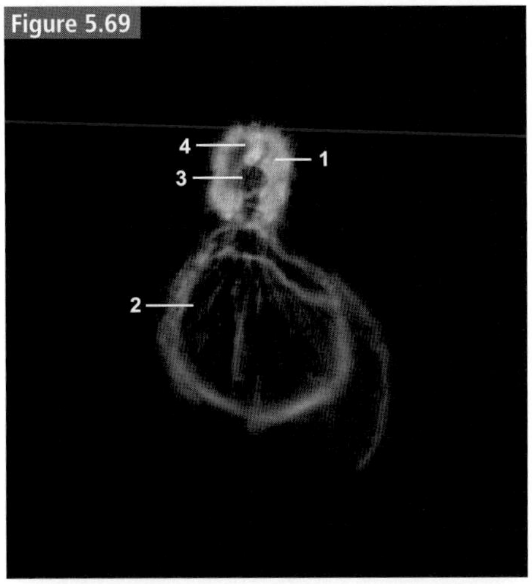

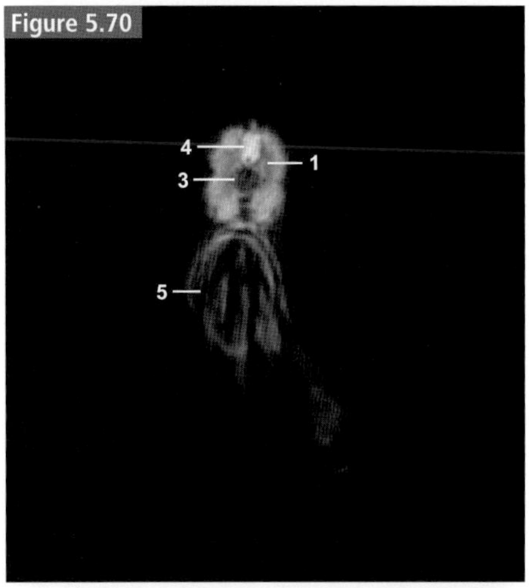

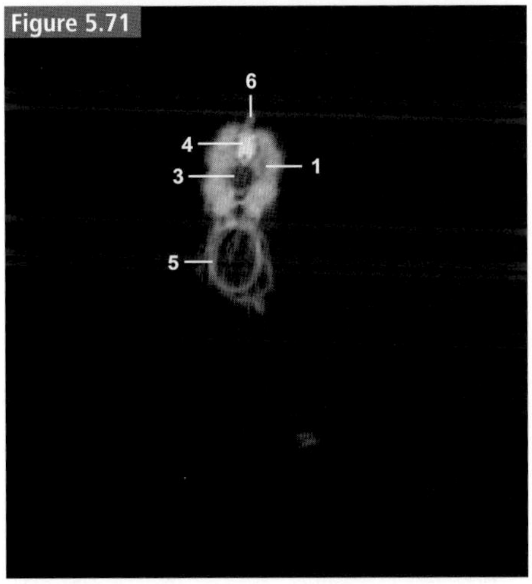

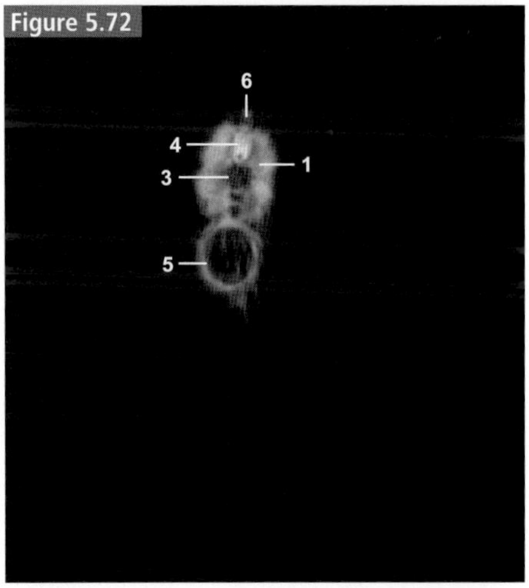

1. myotome	**2.** yolk	**3.** notochord	**4.** spinal cord	**5.** yolk extension	**6.** median fin fold

Figure 5.73

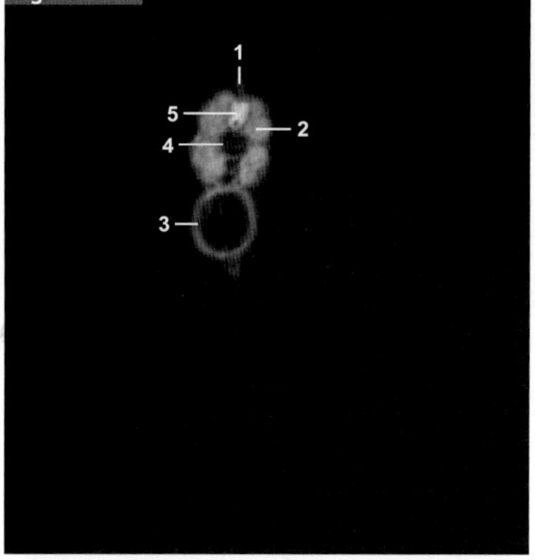

Figure 5.74

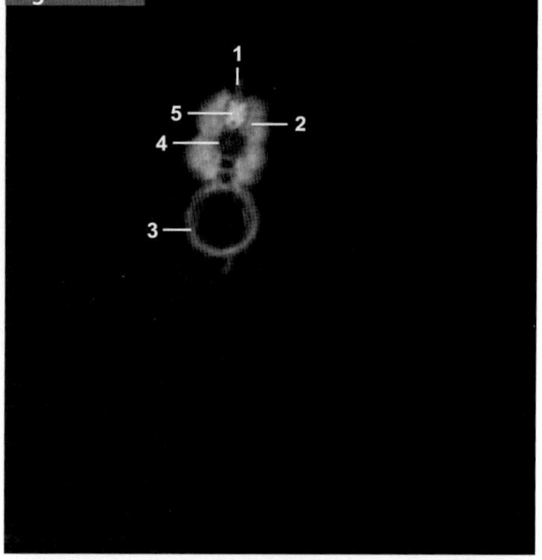

Figure 5.75

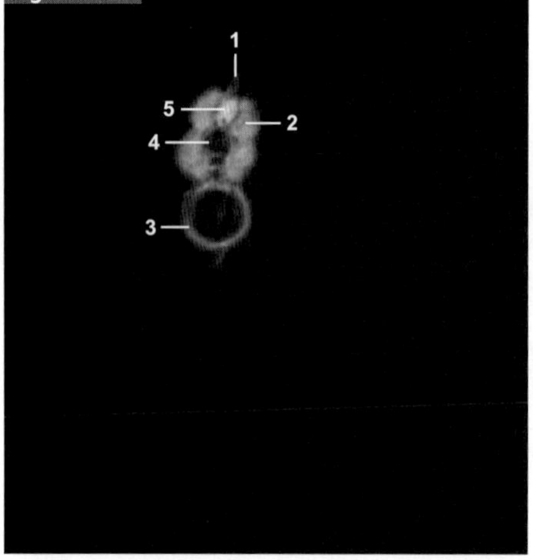

Figure 5.76

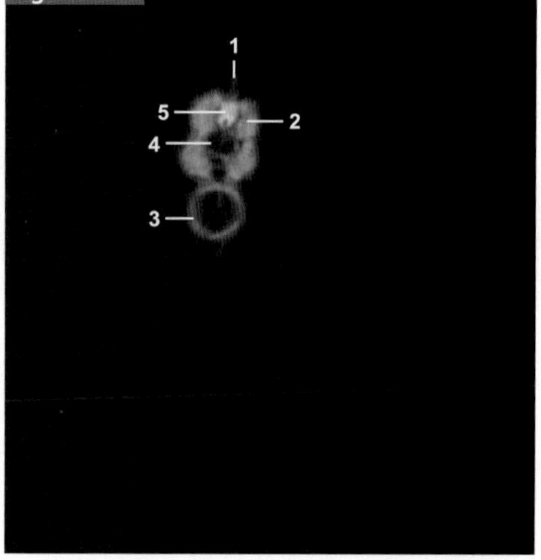

Figure 5.77

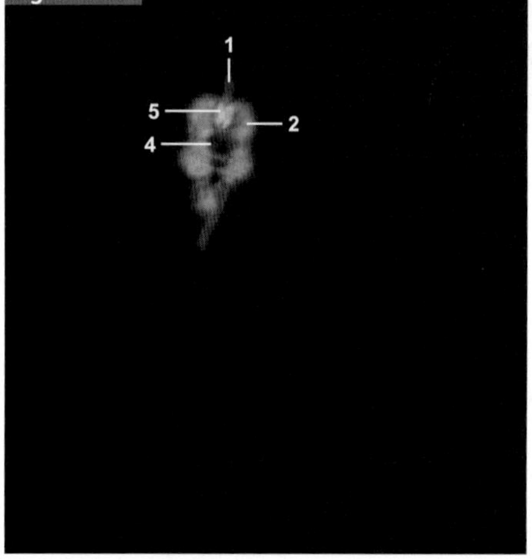

Figure 5.78

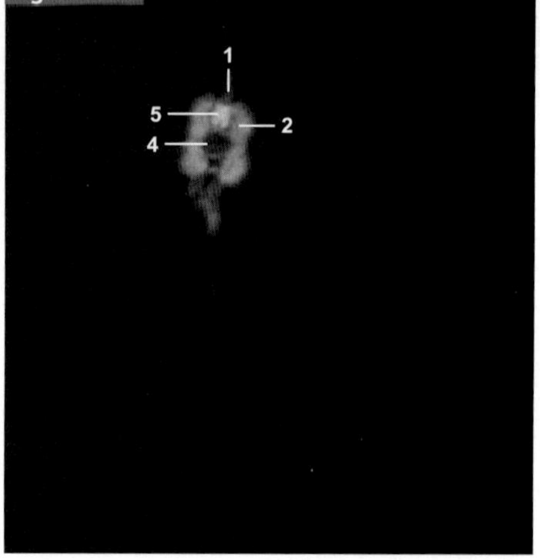

1. median fin fold	**2.** myotome	**3.** yolk extension	**4.** notochord	**5.** spinal cord

CHAPTER 6

72 HOURS

Sagittal Sections

Figure 6.1

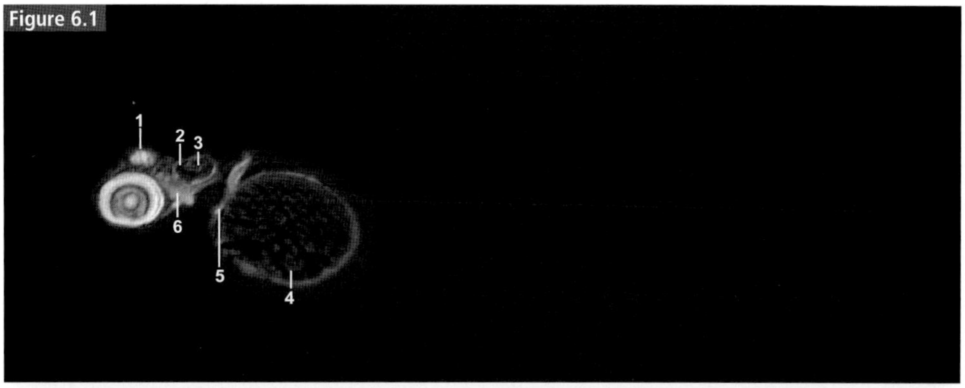

Figure 6.2

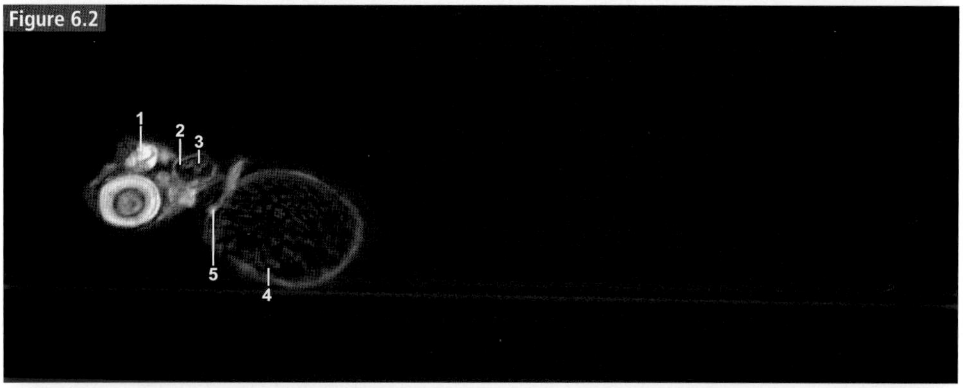

Figure 6.3

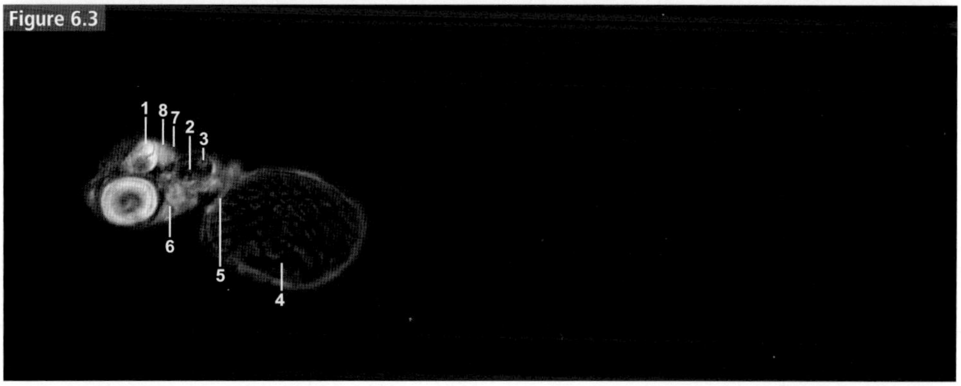

1. optic tectum	**3.** immature otolith	**5.** pectoral fin	**7.** cerebellum
2. otic vesicle	**4.** yolk	**6.** hyosymplectic cartilage	**8.** medulla oblongata

Atlas of Zebrafish Development. 10.1016/B978-0-12-374016-8.00006-6

Figure 6.4

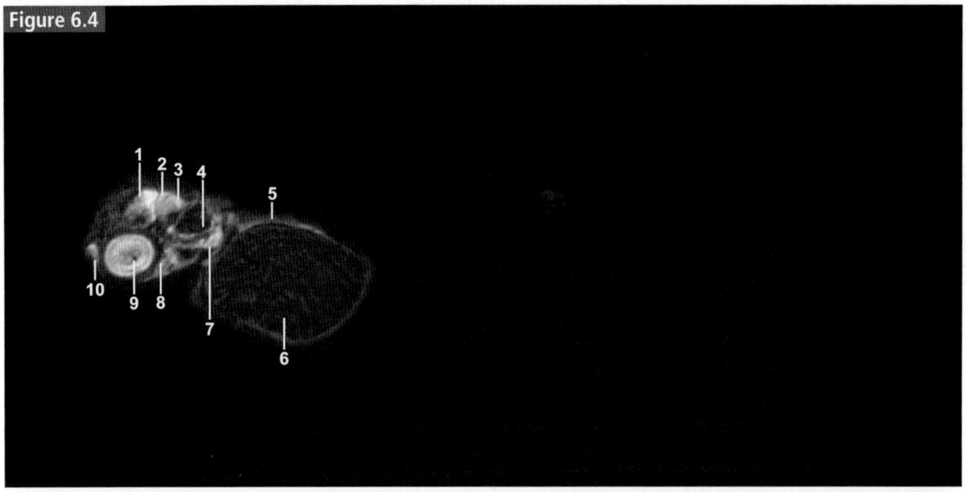

Figure 6.5

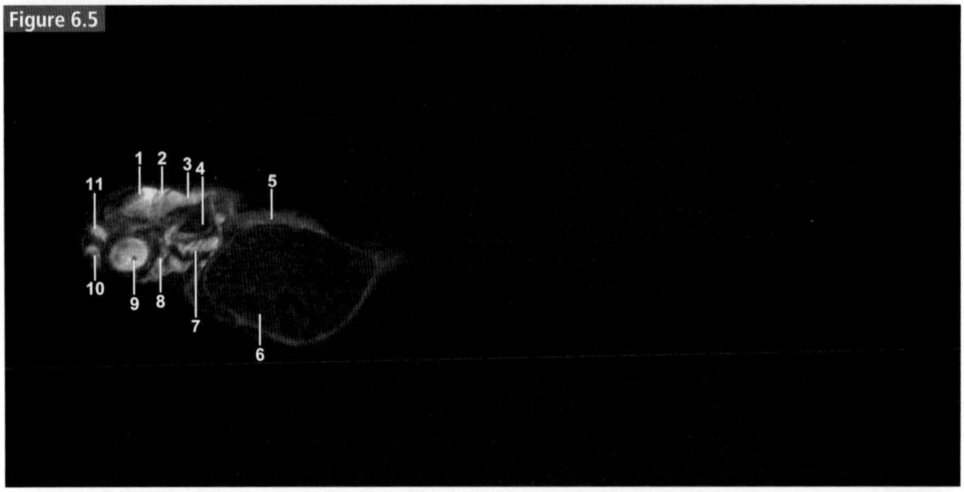

Figure 6.6

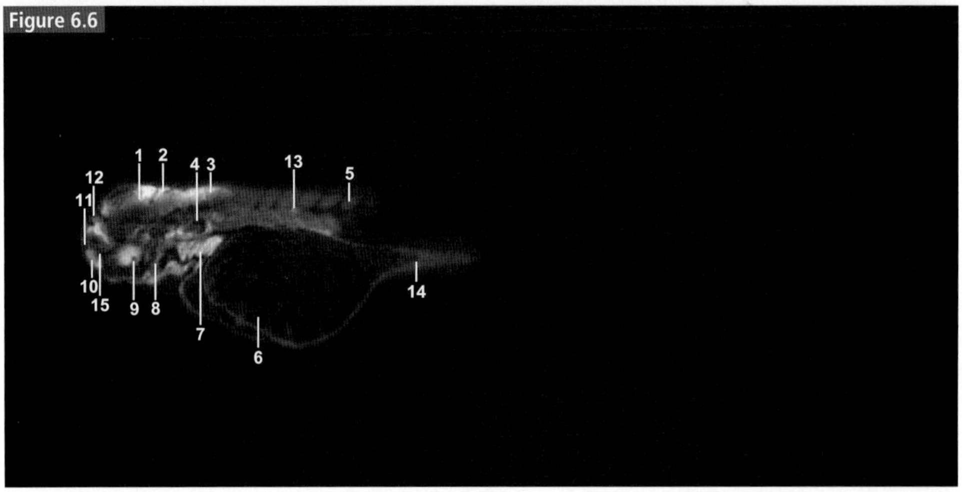

1. optic tectum	**4.** otic vesicle	**7.** gill arches	**10.** olfactory pit	**13.** neuromasts
2. cerebellum	**5.** myotome	**8.** hyosymplectic cartilage	**11.** olfactory epithelium	**14.** extension
3. medulla oblongata	**6.** yolk extension	**9.** cranial nerve II	**12.** pallium	**15.** anterior commissure

Figure 6.7

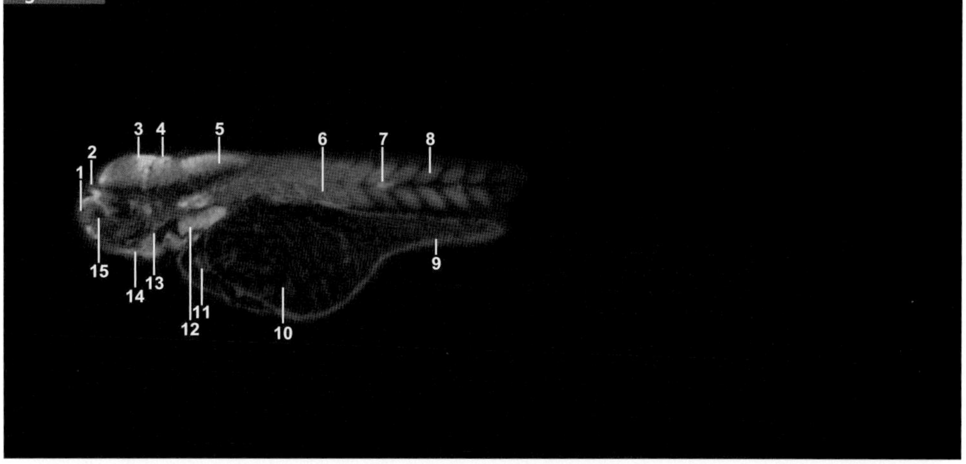

Figure 6.8

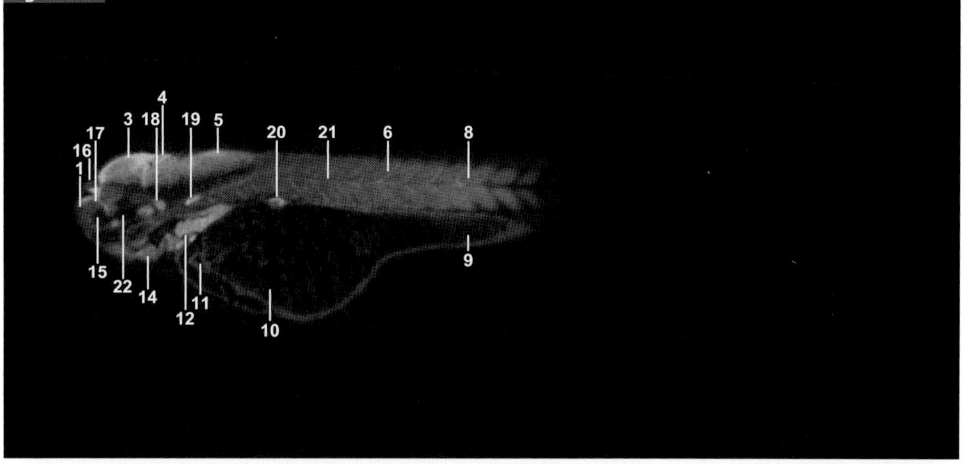

Figure 6.9

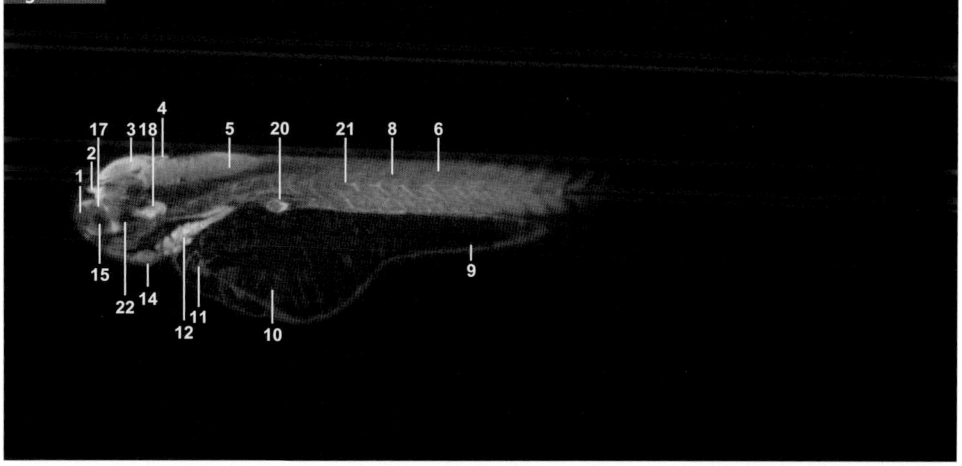

1. olfactory bulb	**6.** vertical myoseptum	**11.** heart	**16.** epiphysis	**21.** horizontal myoseptum
2. pallium	**7.** neuromasts	**12.** gill arches	**17.** telencephalon	**22.** optic chiasm
3. optic tectum	**8.** myotome	**13.** hyosymplectic cartilage	**18.** hypothalamus	
4. cerebellum	**9.** extension	**14.** palatoquadrate cartilage	**19.** optic ganglion	
5. medulla oblongata	**10.** yolk extension	**15.** anterior commissure	**20.** pancreas	

Figure 6.10

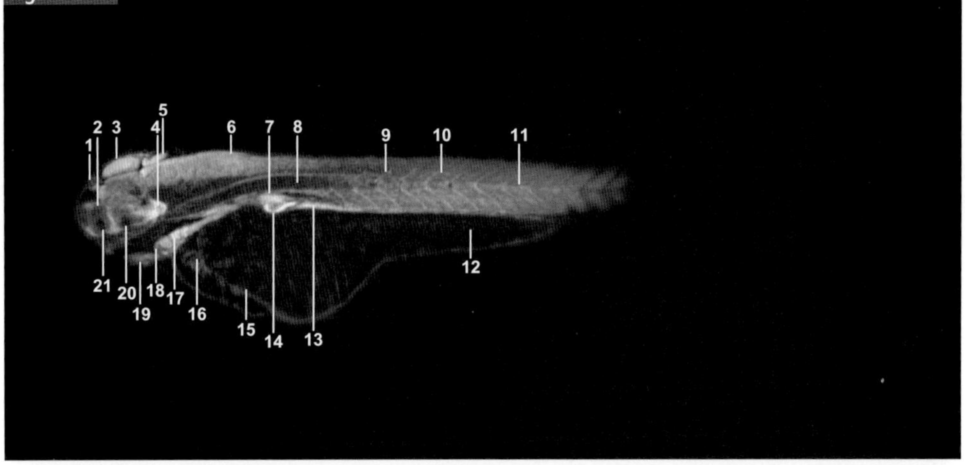

Figure 6.11

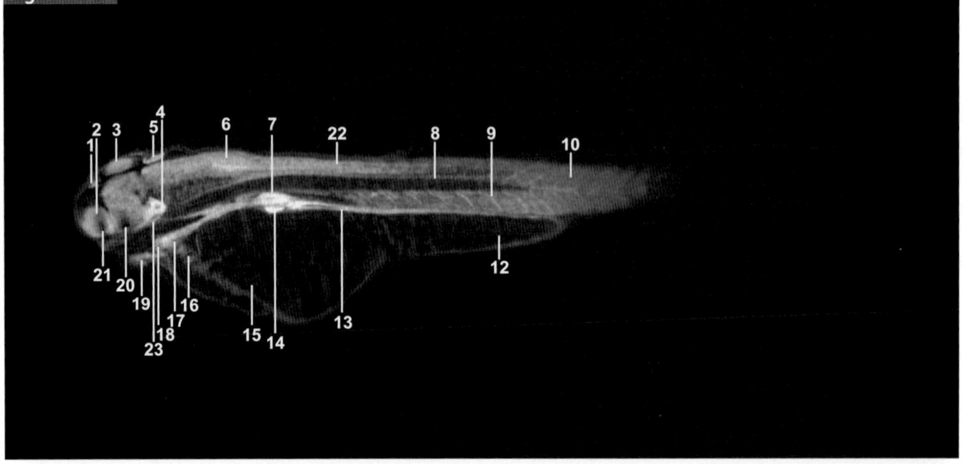

Figure 6.12

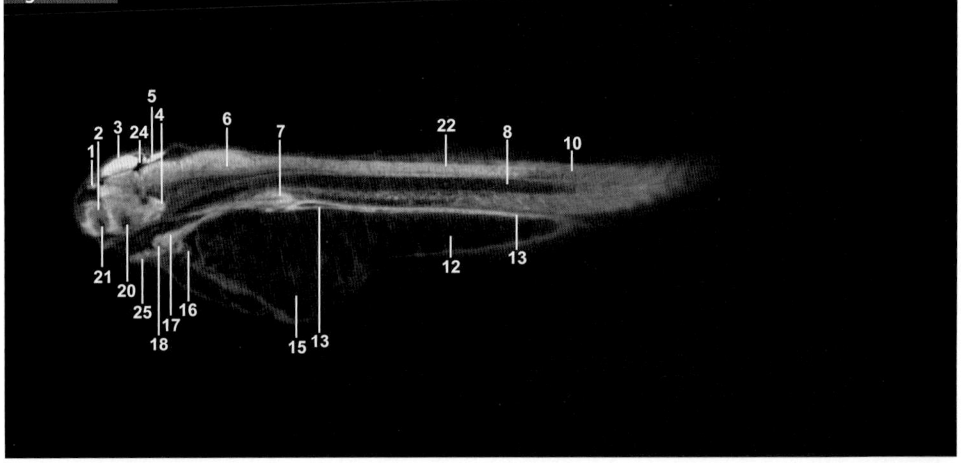

1. epiphysis	**6.** medulla oblongata	**11.** horizontal myoseptum	**16.** heart	**21.** anterior commissure
2. telencephalon	**7.** swim bladder	**12.** yolk extension	**17.** gill arches	**22.** spinal cord
3. optic tectum	**8.** notochord	**13.** gut	**18.** ceratohyal	**23.** hypophysis
4. hypothalamus	**9.** vertical myoseptum	**14.** pancreas	**19.** basihyal	**24.** tectal ventricle
5. cerebellum	**10.** myotome	**15.** yolk	**20.** optic chiasm	**25.** palatoquadrate cartilage

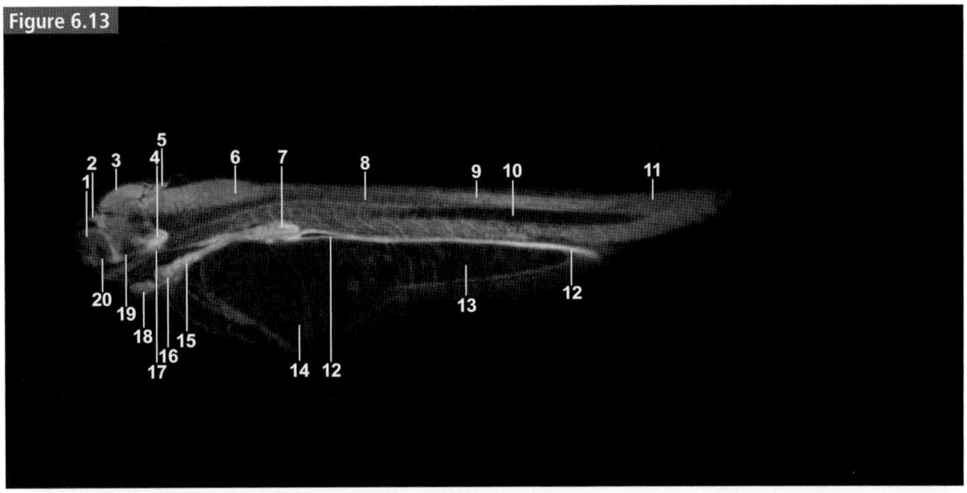

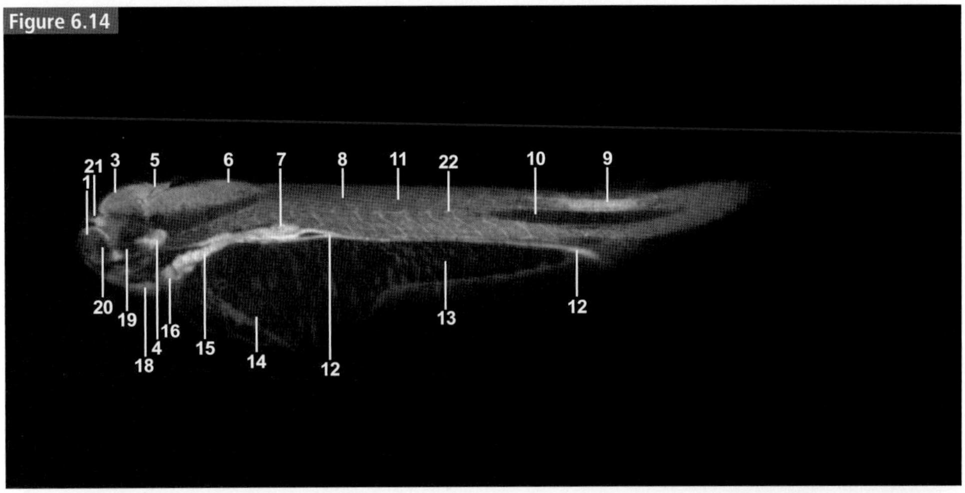

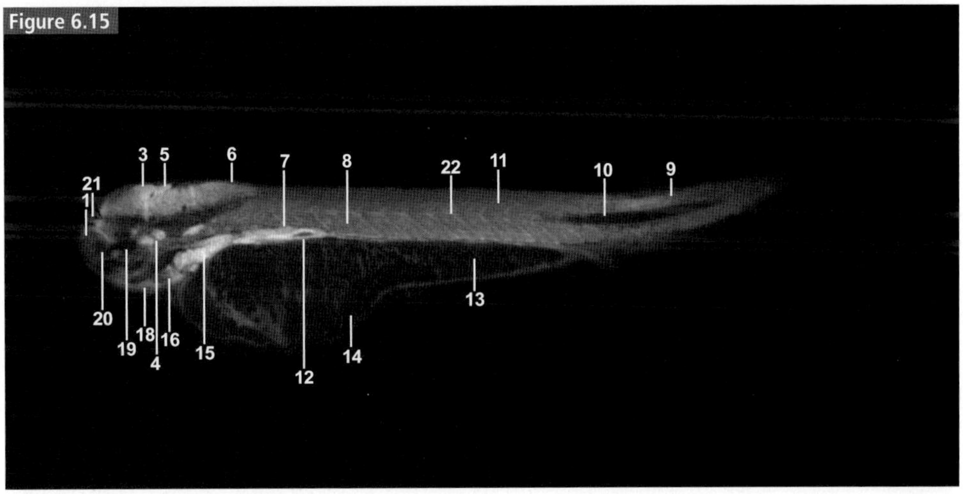

1. telencephalon	**6.** medulla oblongata	**11.** myotome	**16.** ceratohyal	**21.** pallium
2. epiphysis	**7.** swim bladder	**12.** gut	**17.** hypophysis	**22.** horizontal myoseptum
3. optic tectum	**8.** vertical myoseptum	**13.** yolk extension	**18.** palatoquadrate cartilage	
4. hypothalamus	**9.** spinal cord	**14.** yolk	**19.** optic chiasm	
5. cerebellum	**10.** notochord	**15.** gill arches	**20.** anterior commissure	

Figure 6.16

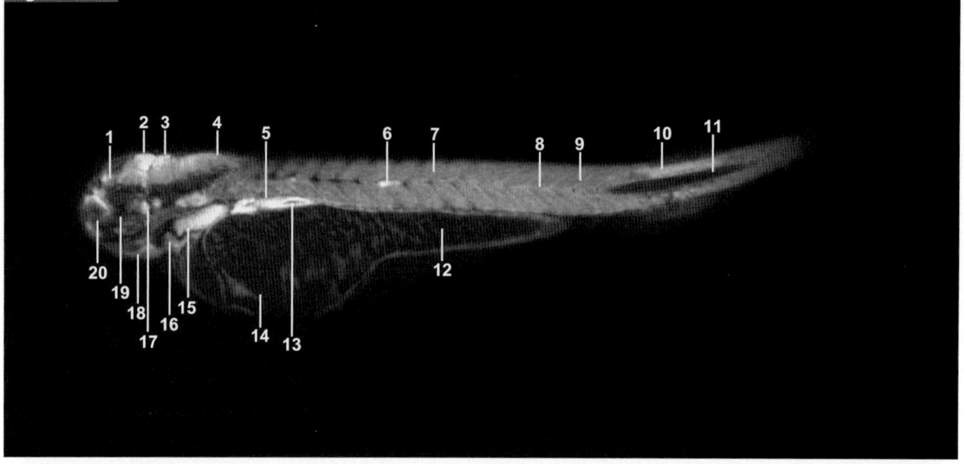

Figure 6.17

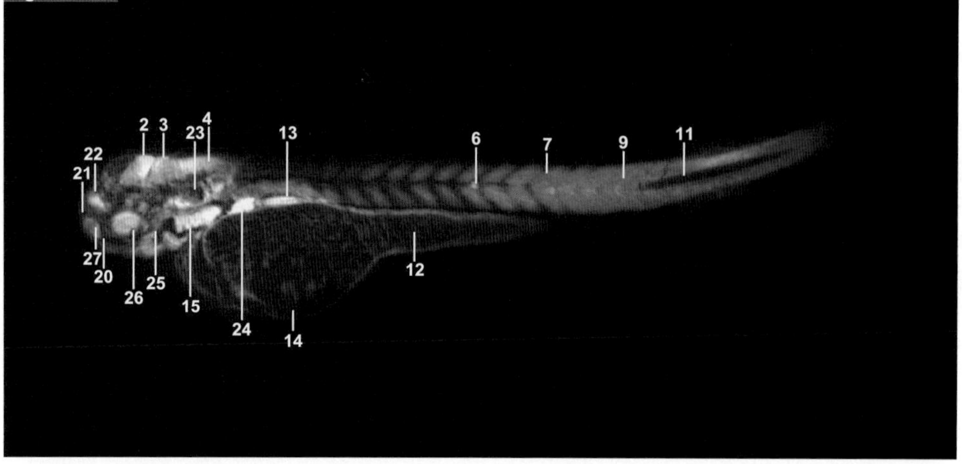

Figure 6.18

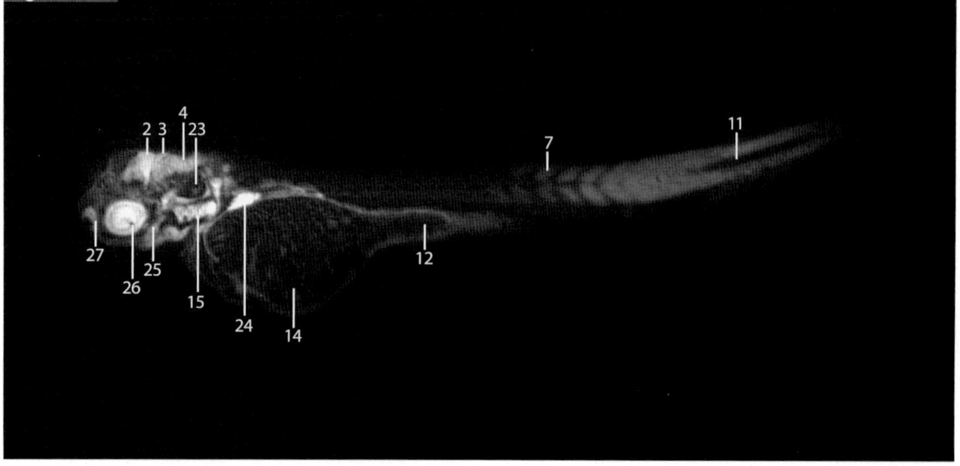

1. epiphysis
2. optic tectum
3. cerebellum
4. medulla oblongata
5. swim bladder
6. neuromasts
7. myotome
8. horizontal myoseptum
9. vertical myoseptum
10. spinal cord
11. notochord
12. yolk extension
13. gut
14. yolk
15. gill arches
16. ceratohyal
17. hypothalamus
18. palatoquadrate cartilage
19. optic chiasm
20. anterior commissure
21. olfactory bulb
22. pallium
23. otic vesicle
24. liver
25. hyosymplectic cartilage
26. cranial nerve II
27. olfactory epithelium

Figure 6.19

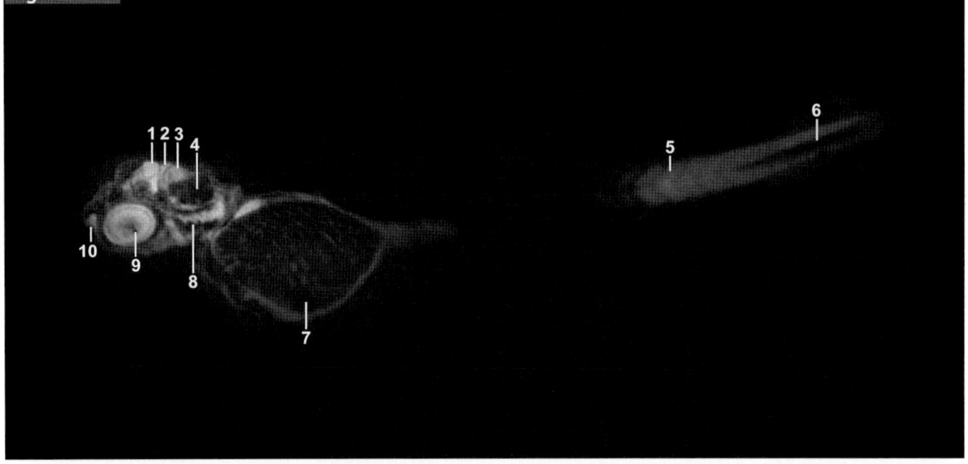

Figure 6.20

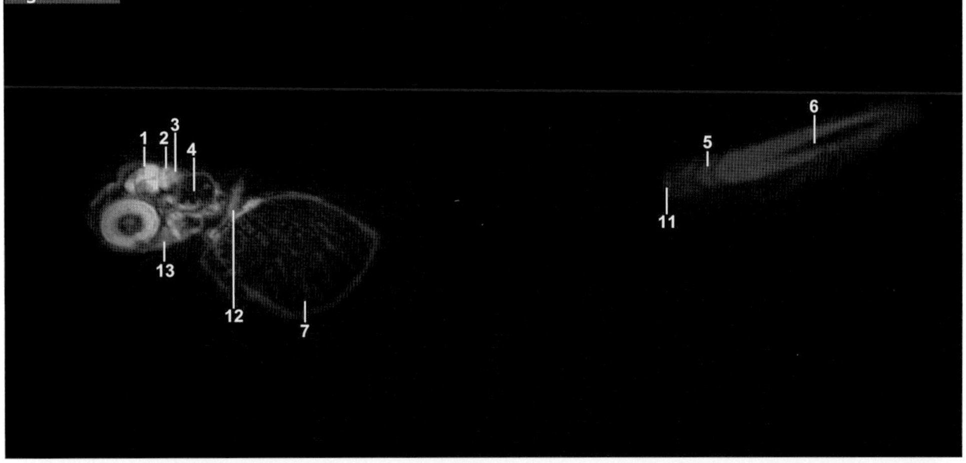

Figure 6.21

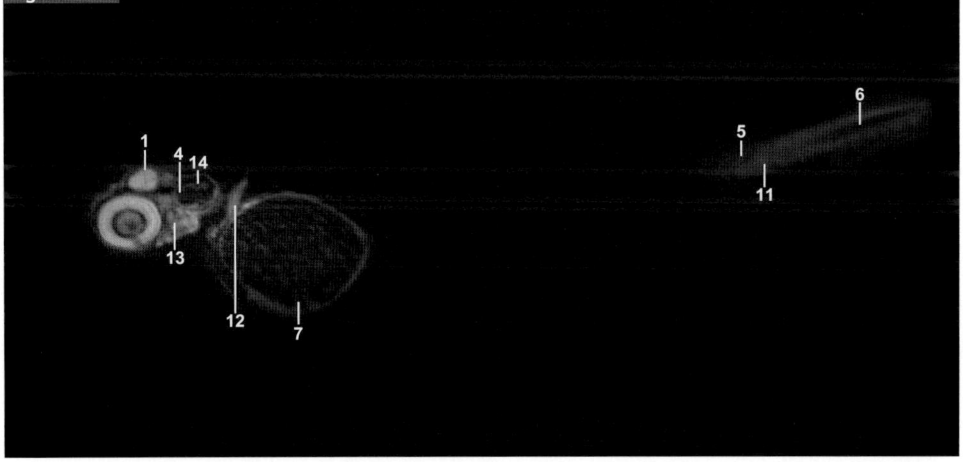

1. optic tectum	**4.** otic vesicle	**7.** yolk	**10.** olfactory epithelium	**13.** hyosymplectic cartilage
2. cerebellum	**5.** myotome	**8.** gill arches	**11.** neuromasts	**14.** immature otolith
3. medulla oblongata	**6.** notochord	**9.** cranial nerve II	**12.** pectoral fin	

Figure 6.22

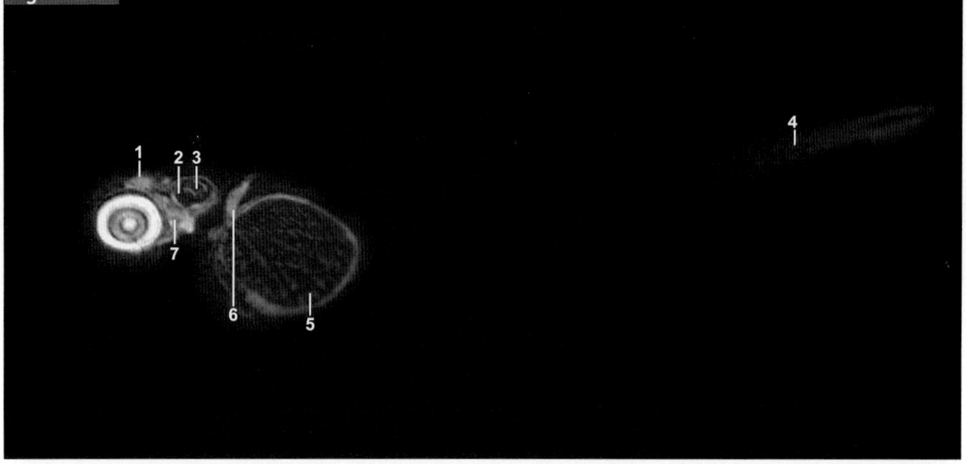

Figure 6.23

Figure 6.24

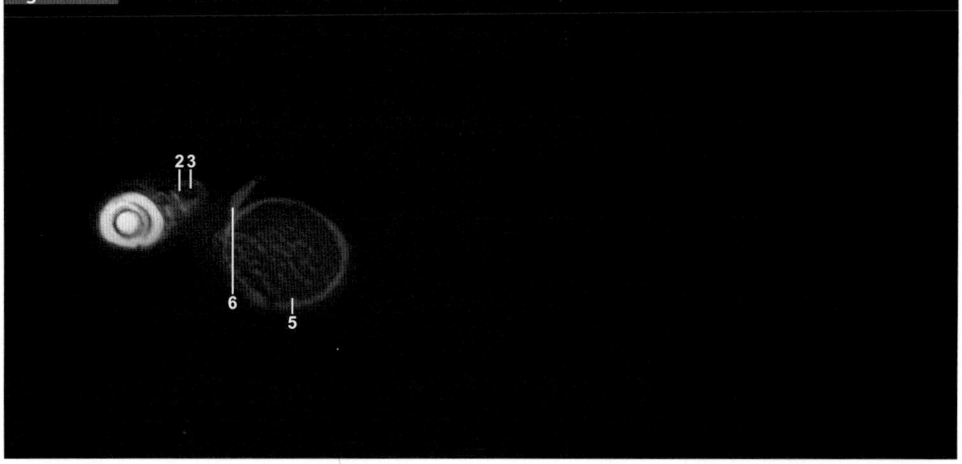

1. optic tectum	**2.** otic vesicle	**3.** immature otoliths	**4.** myotome	**5.** yolk	**6.** pectoral fin	**7.** hyosymplectic cartilage

Coronal Sections

Figure 6.25

Figure 6.26

Figure 6.27

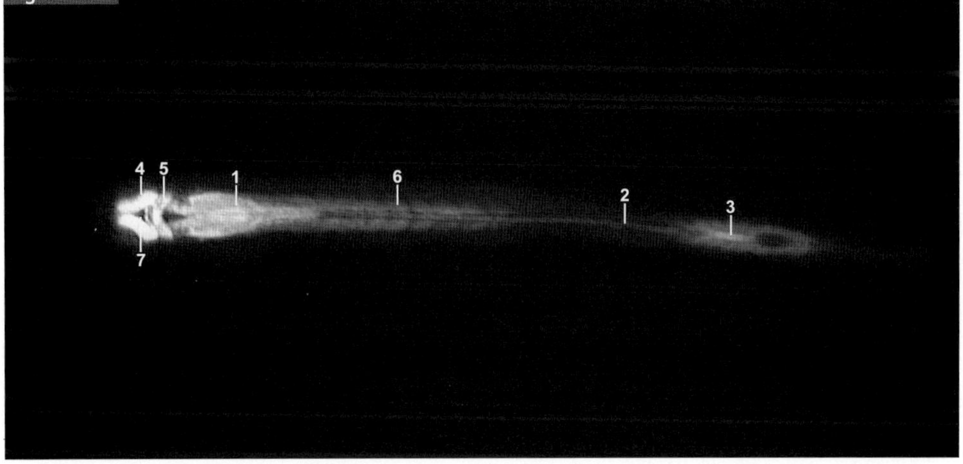

1. medulla oblongata	**2.** median fin fold	**3.** spinal cord	**4.** optic tectum	**5.** cerebellum	**6.** myotome	**7.** tectal ventricle

Figure 6.28

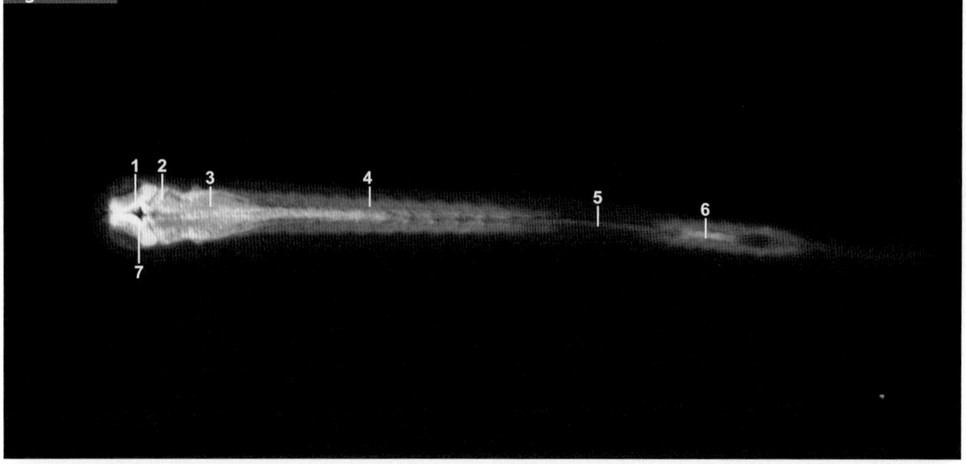

Figure 6.29

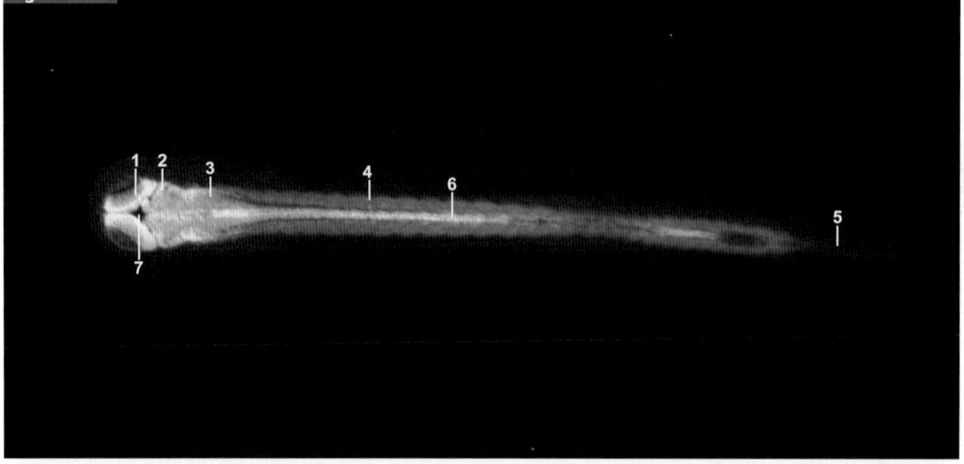

Figure 6.30

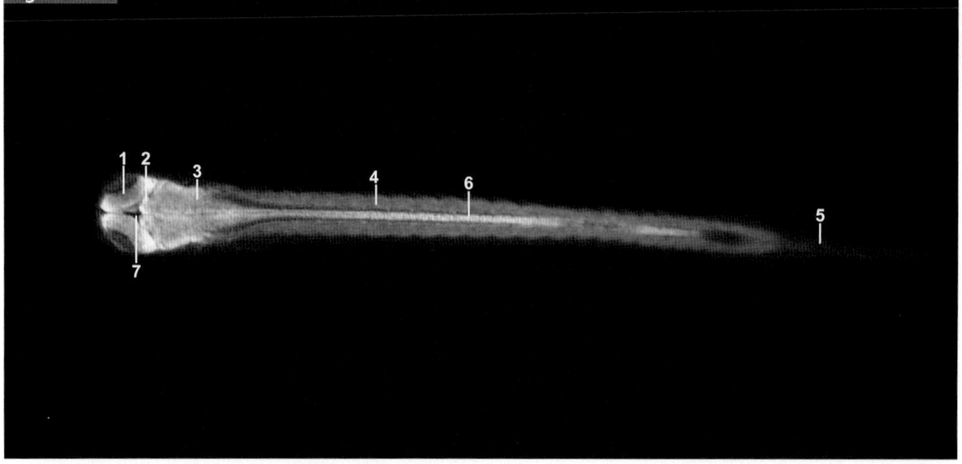

1. optic tectum	**2.** cerebellum	**3.** medulla oblongata	**4.** myotome	**5.** median fin fold	**6.** spinal cord	**7.** tectal ventricle

Figure 6.31

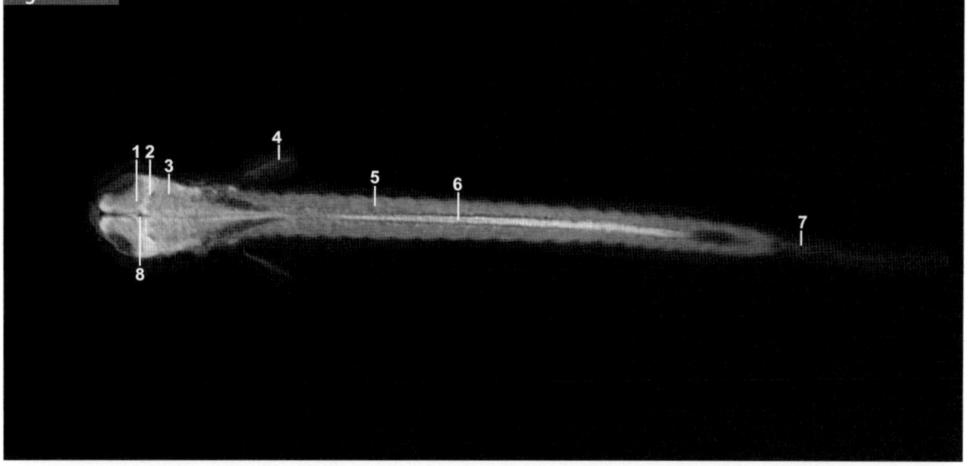

Figure 6.32

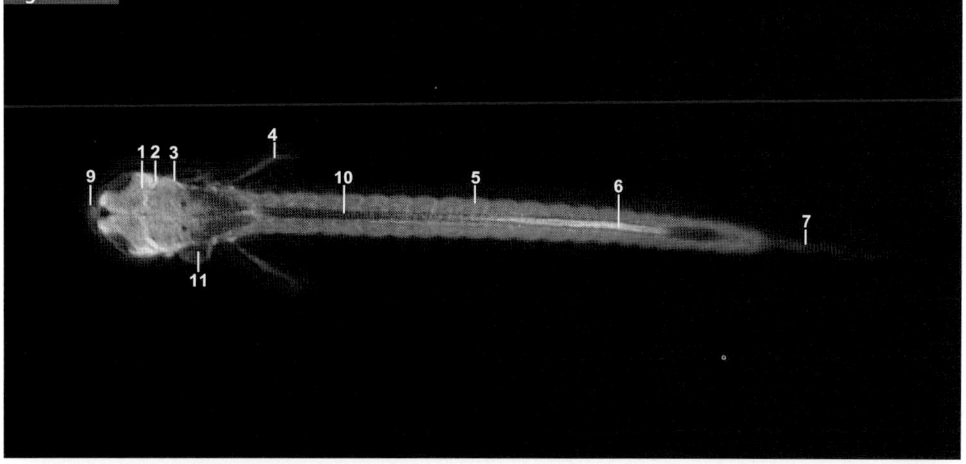

Figure 6.33

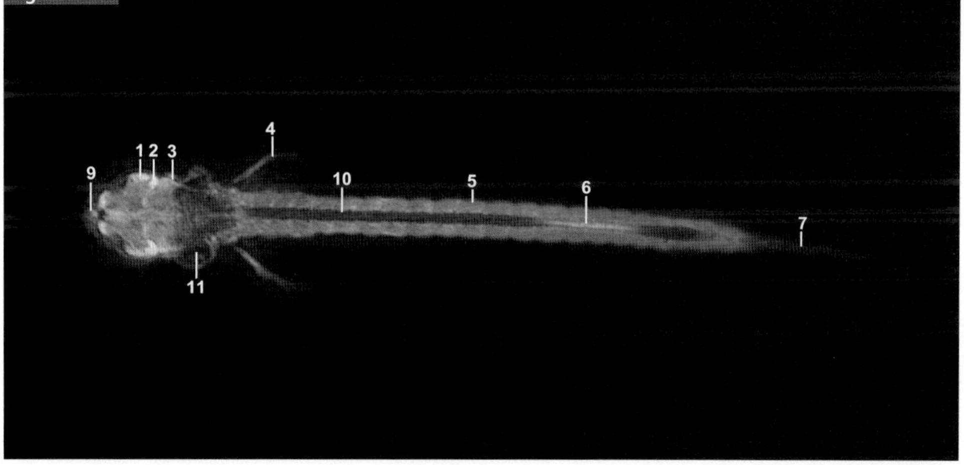

1. optic tectum	**3.** medulla oblongata	**5.** myotome	**7.** median fin fold	**9.** epiphysis	**11.** otic vesicle
2. cerebellum	**4.** pectoral fin	**6.** spinal cord	**8.** tectal ventricle	**10.** notochord	

Figure 6.34

Figure 6.35

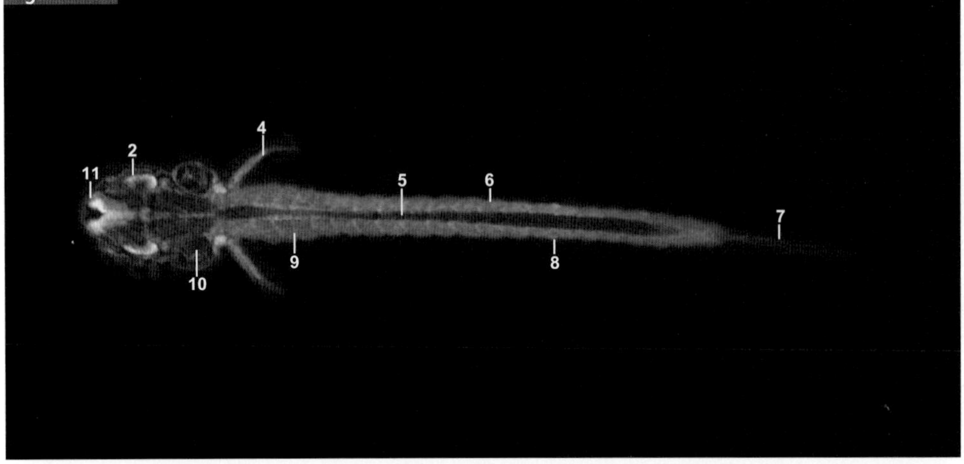

Figure 6.36

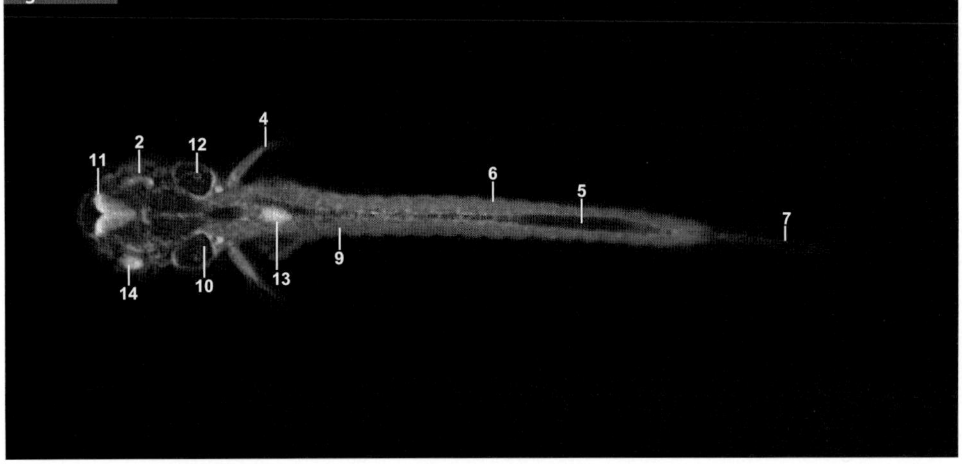

1. epiphysis	**4.** pectoral fin	**7.** median fin fold	**10.** otic vesicle	**13.** swim bladder
2. optic tectum	**5.** notochord	**8.** neuromasts	**11.** pallium	**14.** optic cup
3. medulla oblongata	**6.** myotome	**9.** vertical myoseptum	**12.** immature otolith	

Figure 6.37

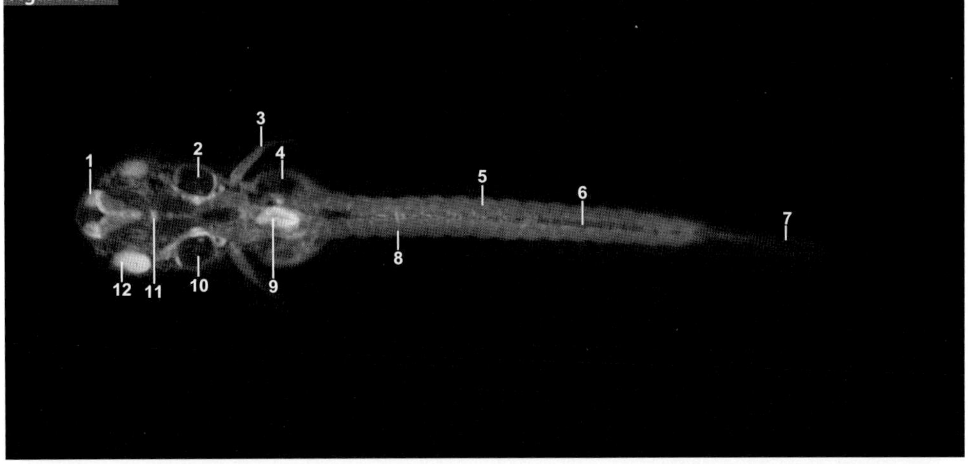

Figure 6.38

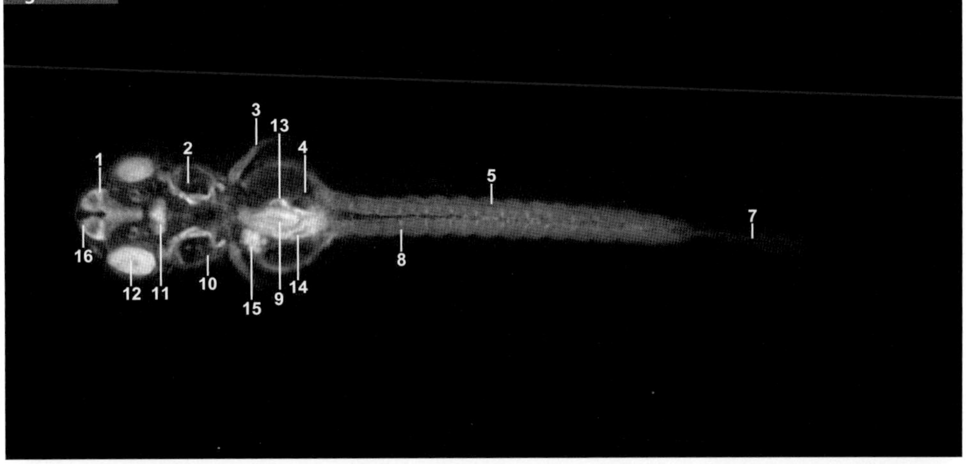

Figure 6.39

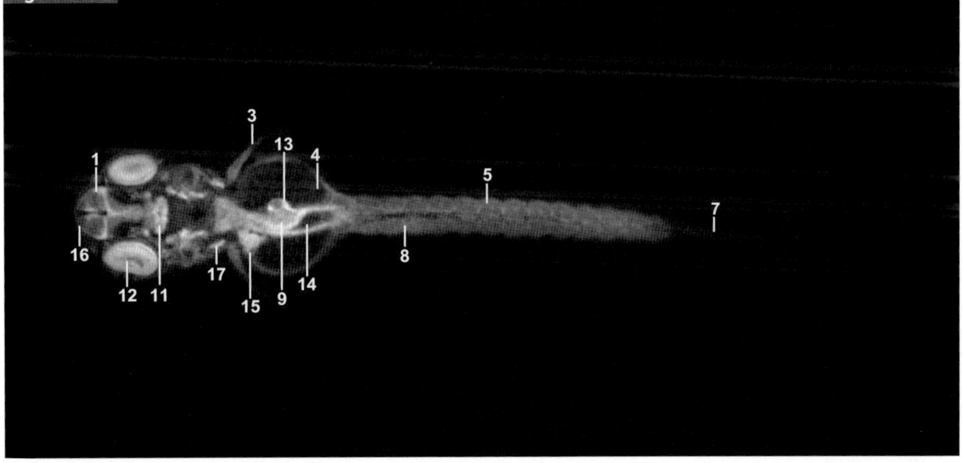

1. telencephalon
2. immature otolith
3. pectoral fin
4. yolk
5. myotome
6. notochord
7. median fin fold
8. vertical myoseptum
9. swim bladder
10. otic vesicle
11. hypothalamus
12. retina
13. pancreas
14. gut
15. liver
16. olfactory bulb
17. gill arches

Figure 6.40

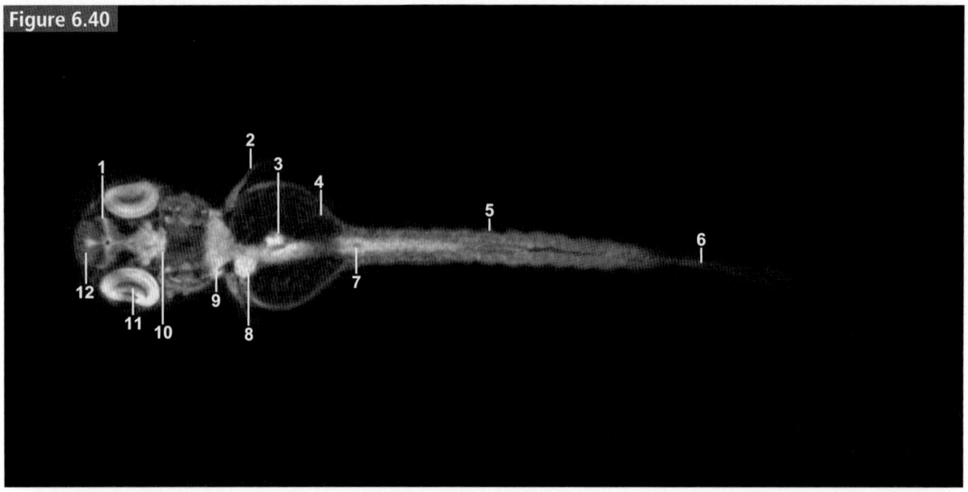

Figure 6.41

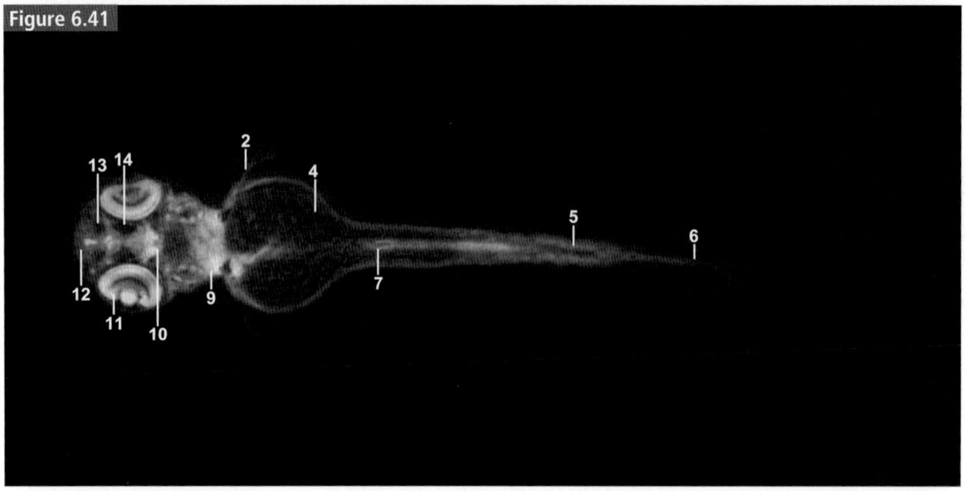

Figure 6.42

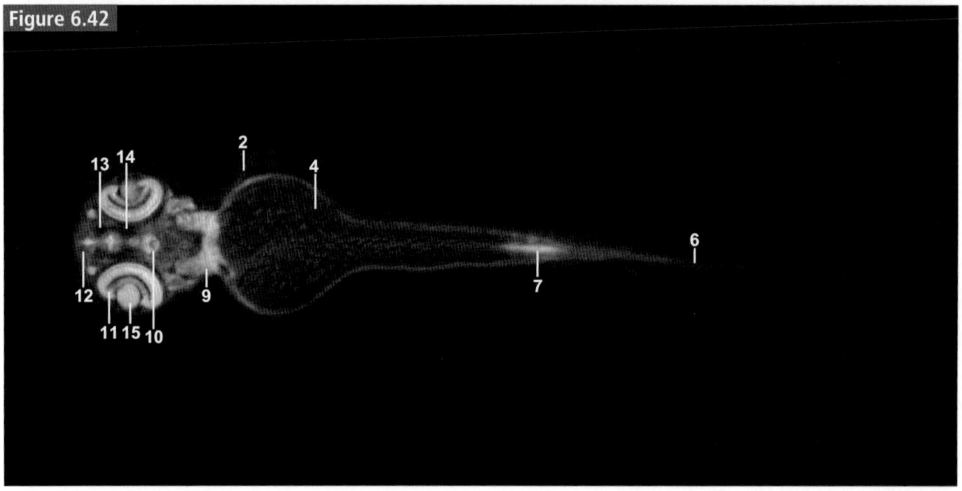

1. telencephalon
2. pectoral fin
3. pancreas
4. yolk
5. myotome
6. median fin fold
7. gut
8. liver
9. gill arches
10. hypothalamus
11. retina
12. olfactory bulb
13. anterior commissure
14. optic chiasm
15. lens

Figure 6.43

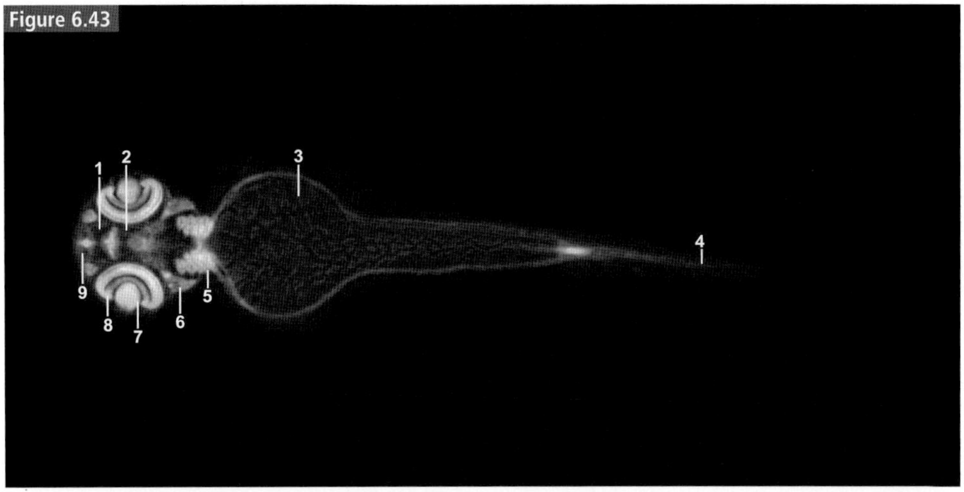

Figure 6.44

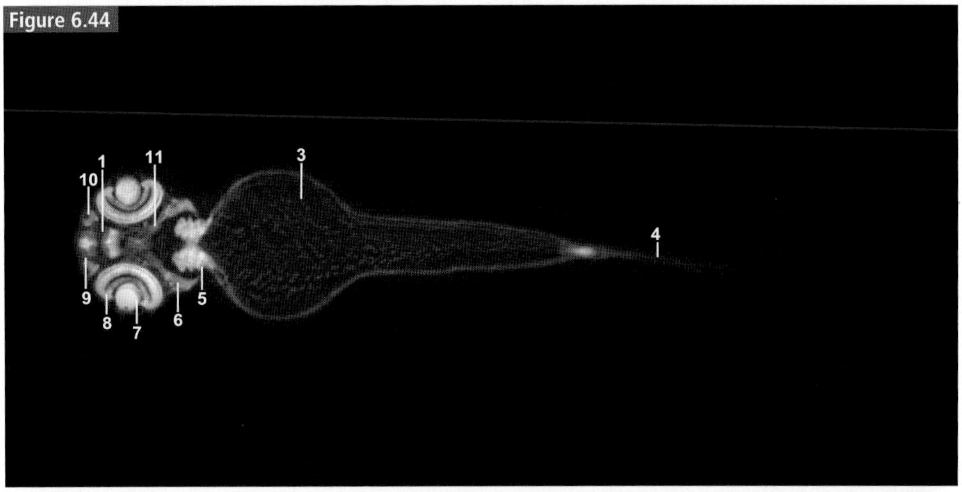

Figure 6.45

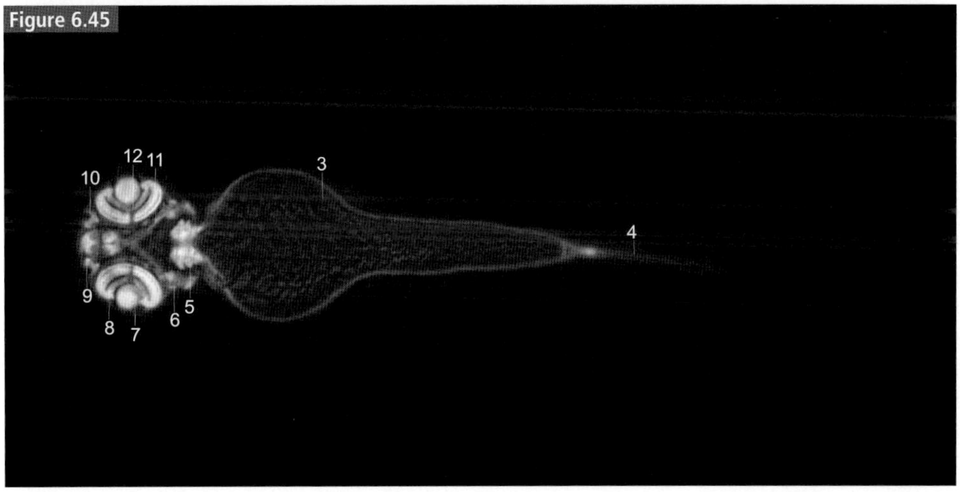

1. anterior commissure	**3.** yolk	**5.** gill arches	**7.** lens	**9.** olfactory bulb	**11.** palatoquadrate cartilage
2. optic chiasm	**4.** median fin fold	**6.** hyosymplectic cartilage	**8.** retina	**10.** olfactory pit	**12.** cranial nerve II

Figure 6.46

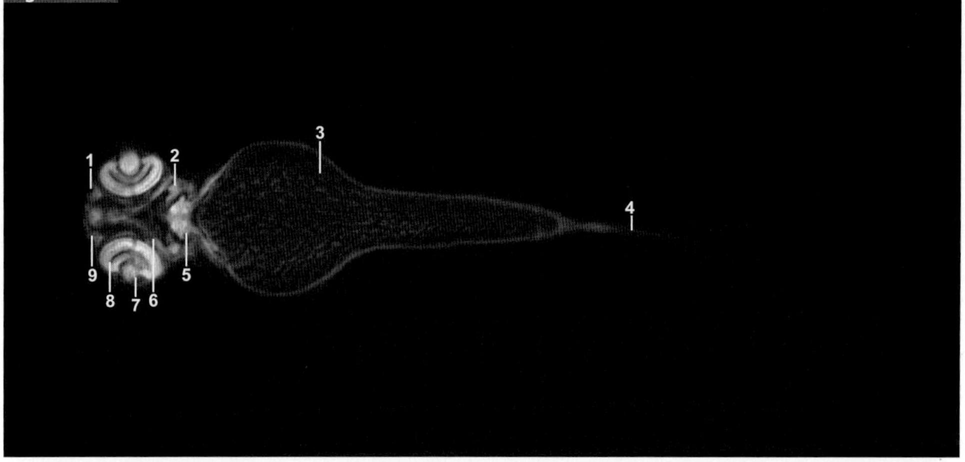

Figure 6.47

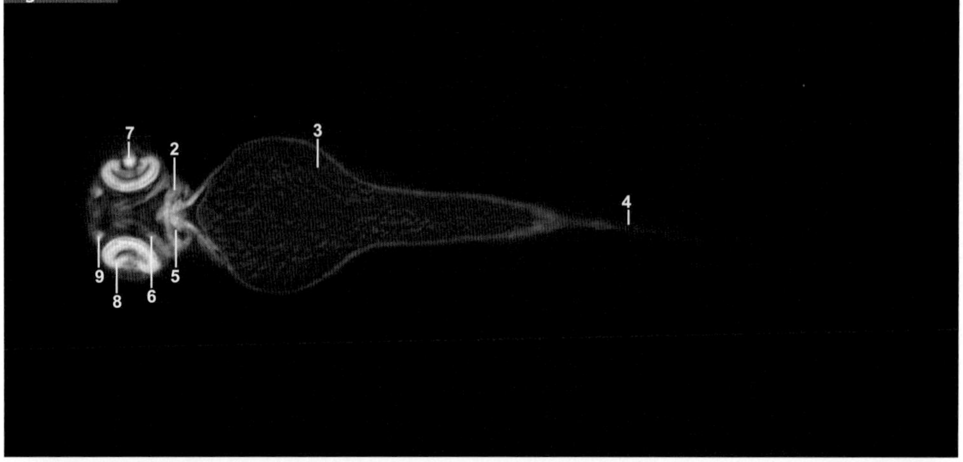

Figure 6.48

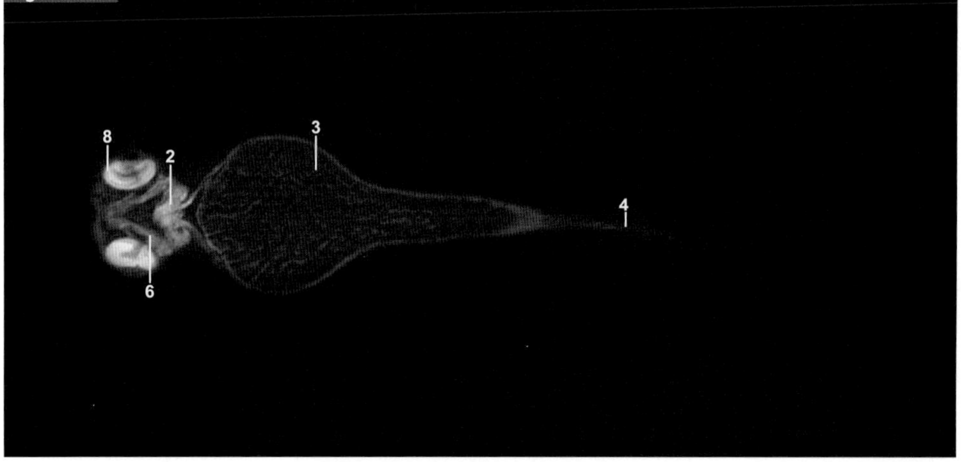

1. olfactory pit	**4.** median fin fold	**7.** lens
2. ceratohyal	**5.** gill arches	**8.** retina
3. yolk	**6.** palatoquadrate cartilage	**9.** olfactory epithelium

Figure 6.49

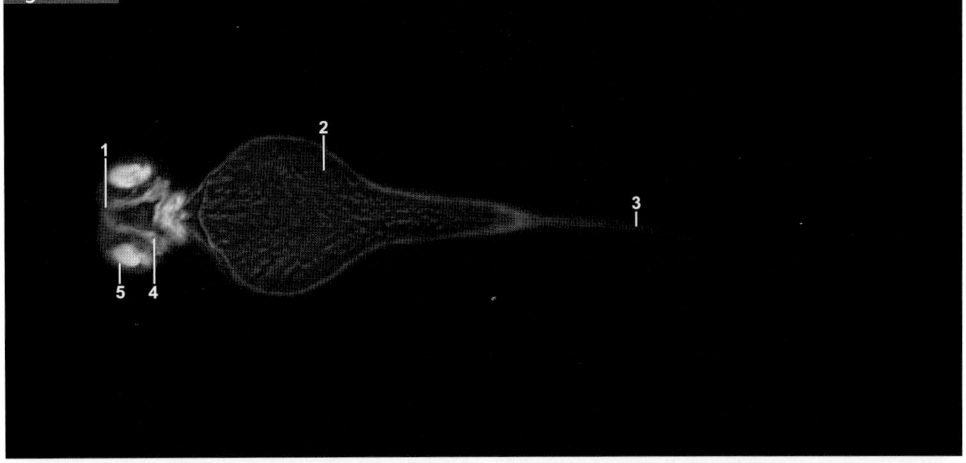

Figure 6.50

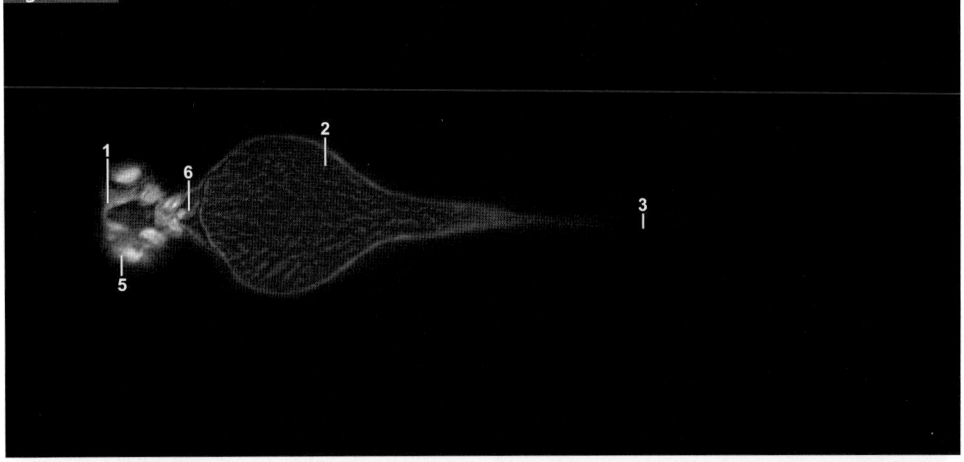

Figure 6.51

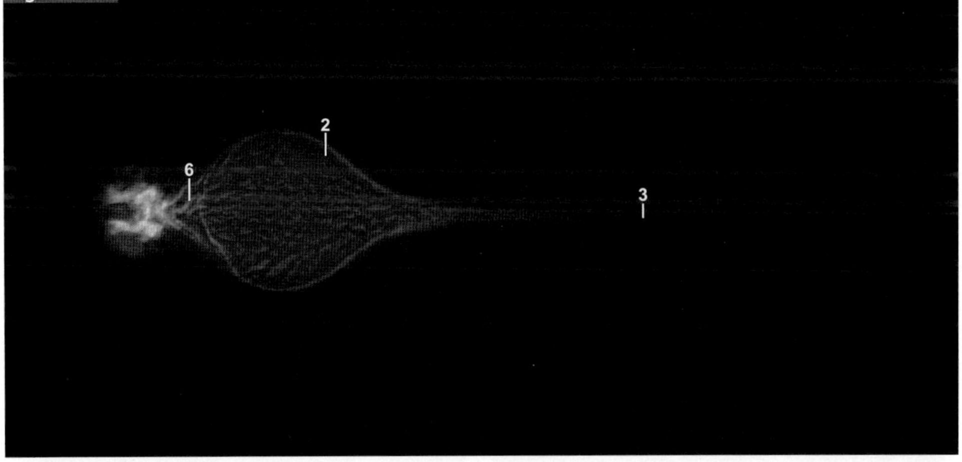

1. Meckel's cartilage	**2.** yolk	**3.** median fin fold	**4.** palatoquadrate cartilage	**5.** optic cup	**6.** heart

Transverse Sections

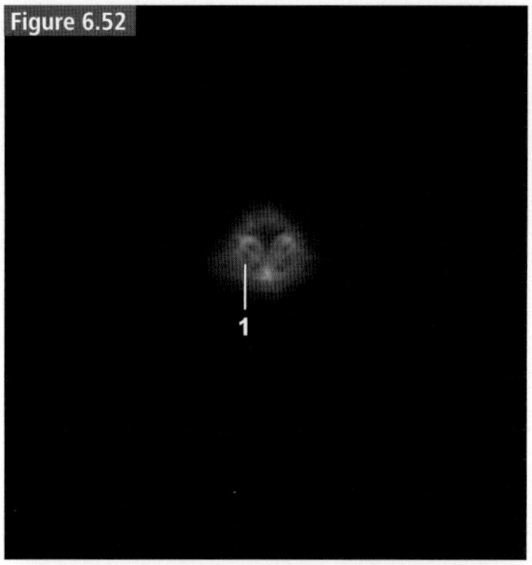

Figure 6.52

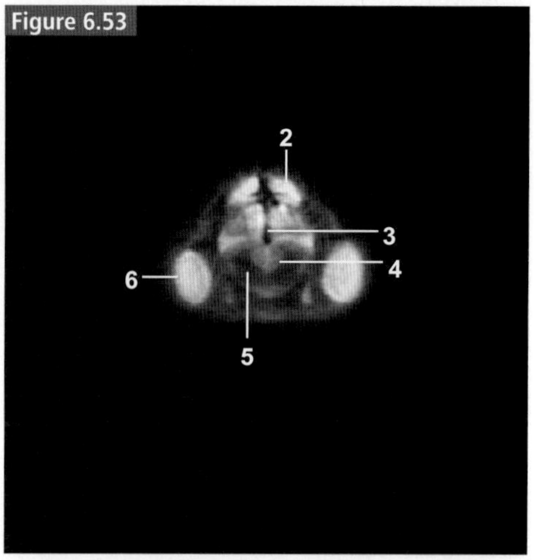

Figure 6.53

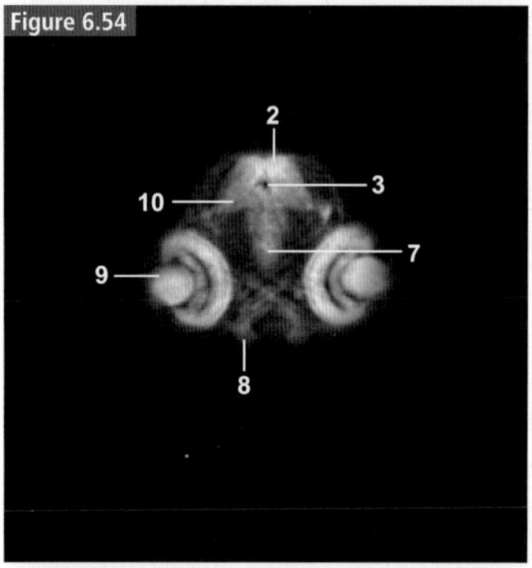

Figure 6.54

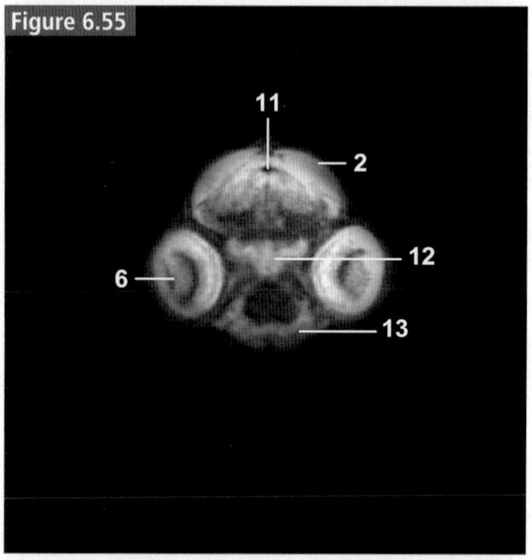

Figure 6.55

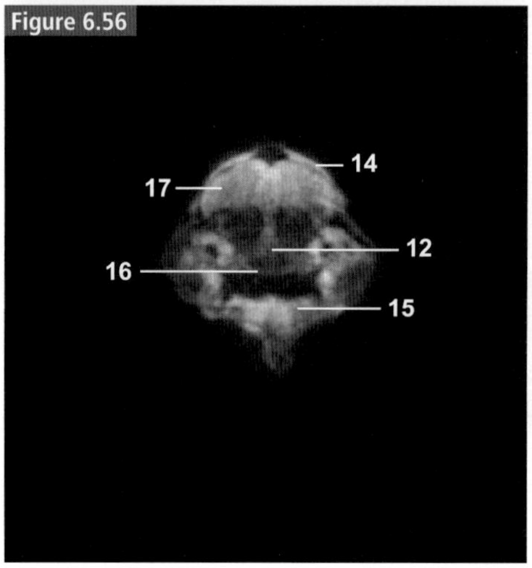

Figure 6.56

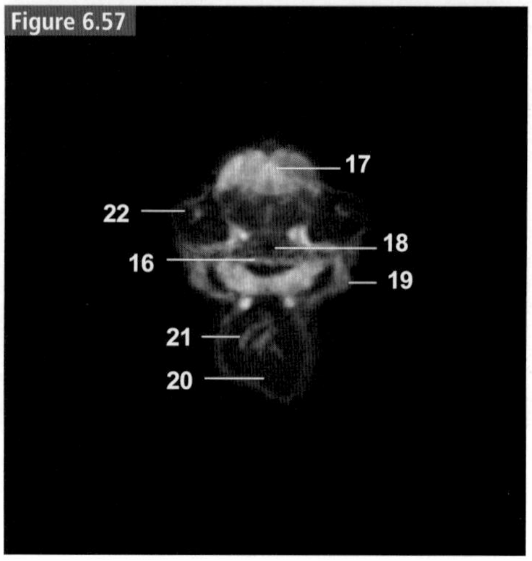

Figure 6.57

1. olfactory bulb
2. optic tectum
3. telencephalic ventricle
4. telencephalon
5. anterior commissure
6. retina
7. subpallium
8. Meckel's cartilage
9. lens
10. pallium
11. tectal ventricle
12. hypothalamus
13. palatoquadrate cartilage
14. cerebellum
15. ceratohyal
16. pharynx
17. medulla oblongata
18. notochord
19. opercle
20. yolk
21. heart
22. immature otolith

Figure 6.58

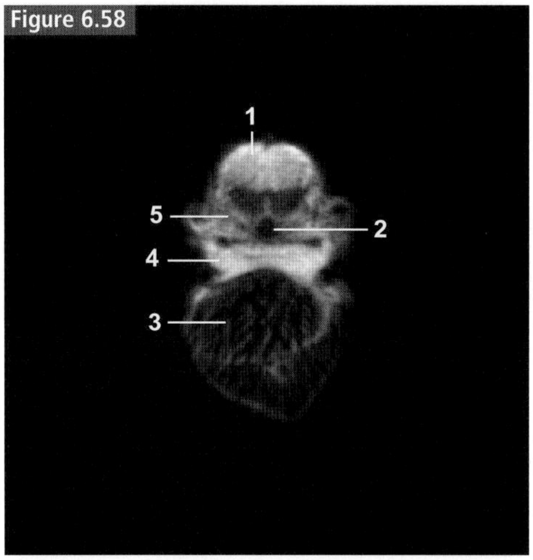

Figure 6.59

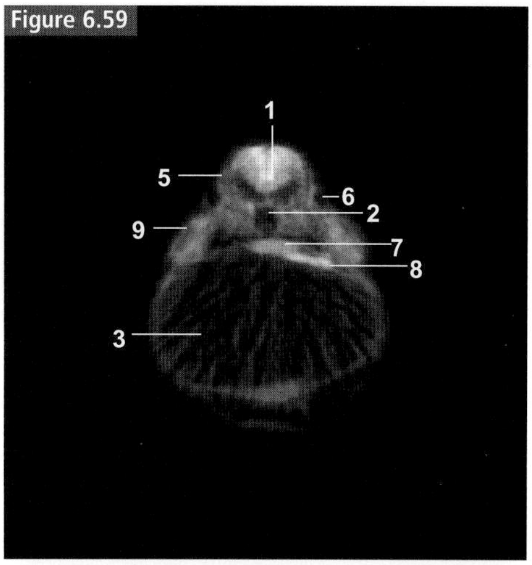

Figure 6.60

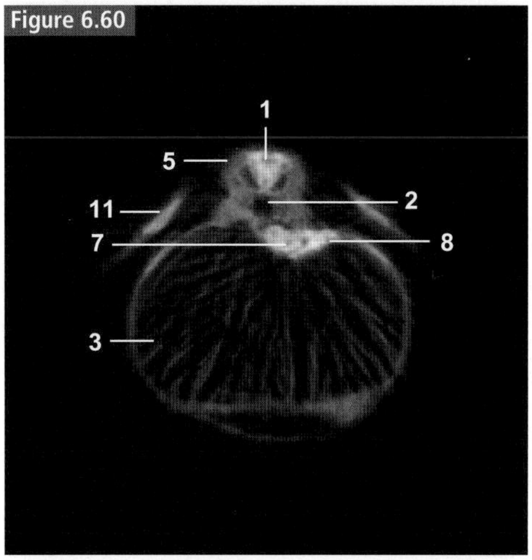

Figure 6.61

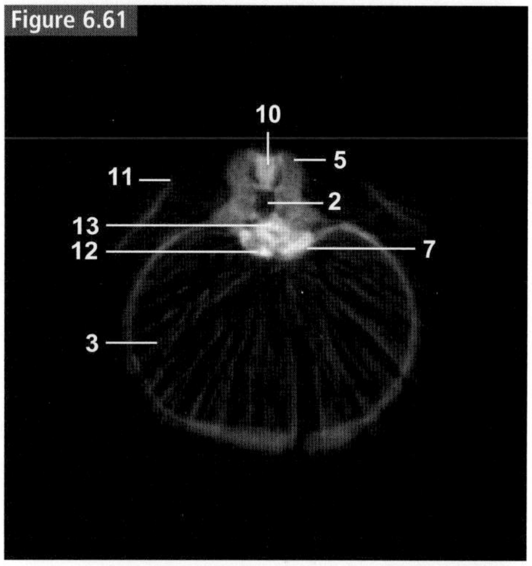

Figure 6.62

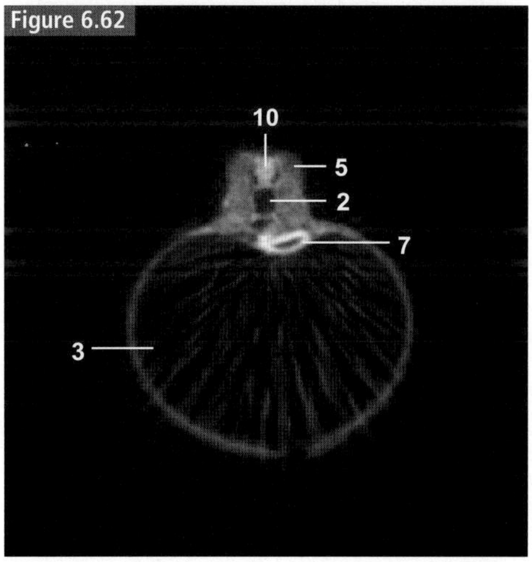

Figure 6.63

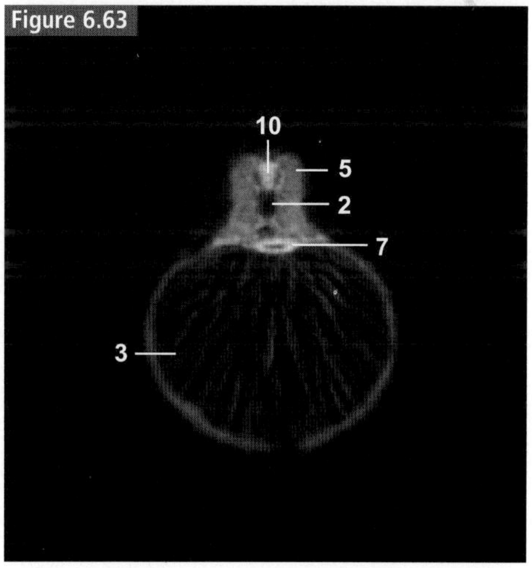

1. medulla oblongata
2. notochord
3. yolk
4. gill arches
5. myotome
6. otic vesicle
7. gut
8. liver
9. pectoral fin musculature
10. spinal cord
11. pectoral fin
12. pancreas
13. swim bladder

Figure 6.64

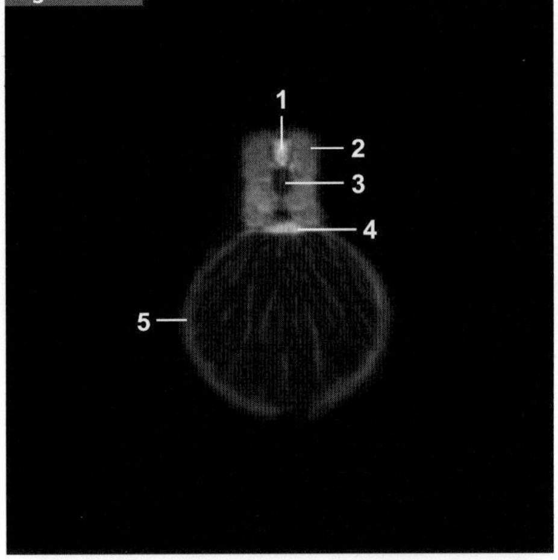

Figure 6.65

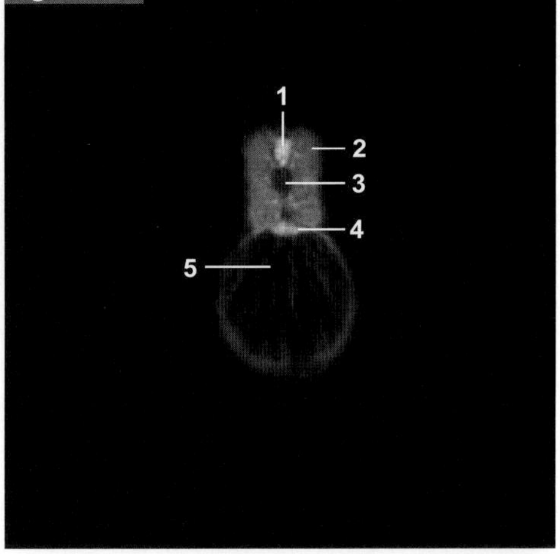

Figure 6.66

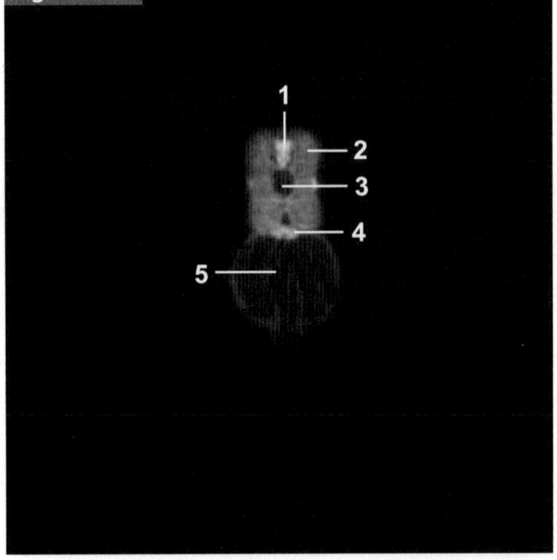

Figure 6.67

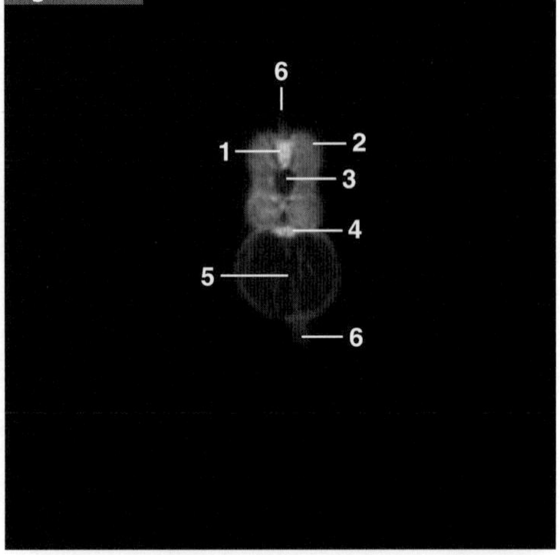

Figure 6.68

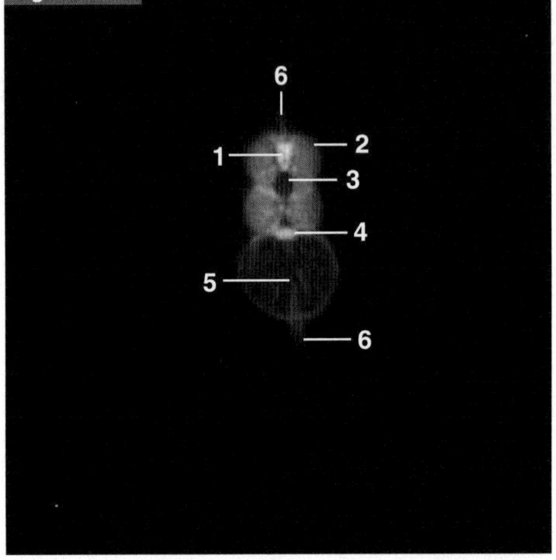

Figure 6.69

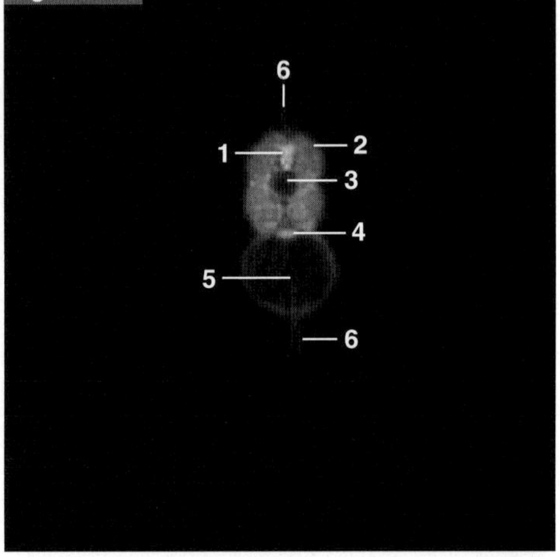

1. spinal cord	**2.** myotome	**3.** notochord	**4.** gut	**5.** yolk extension	**6.** median fin fold

Figure 6.70

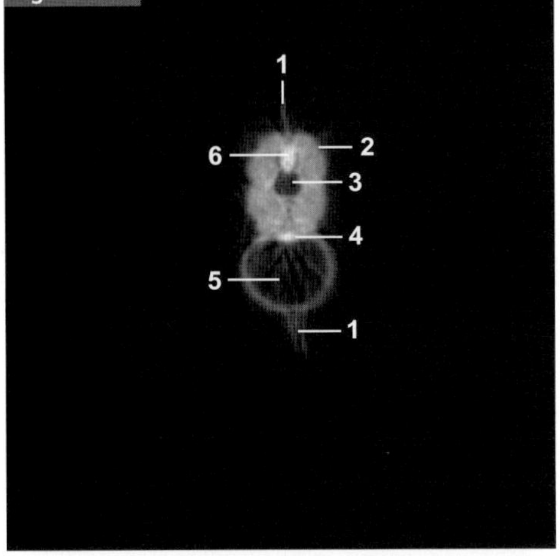

Figure 6.71

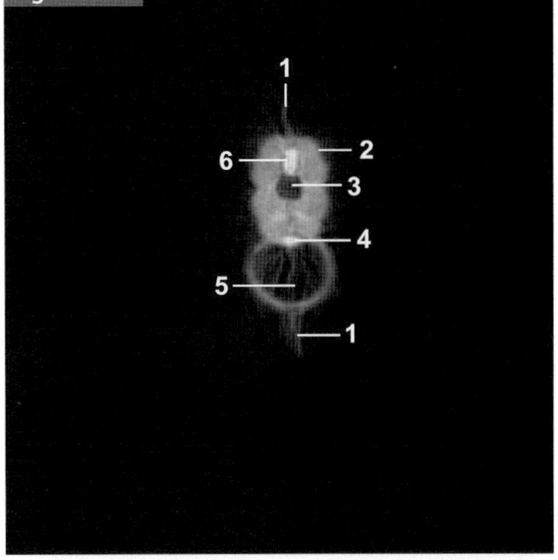

Figure 6.72

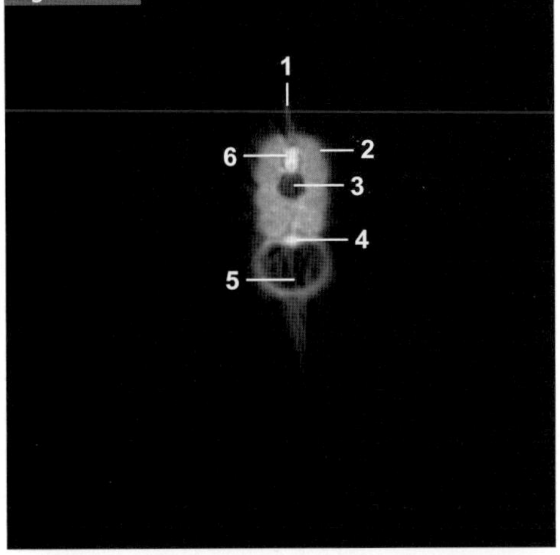

Figure 6.73

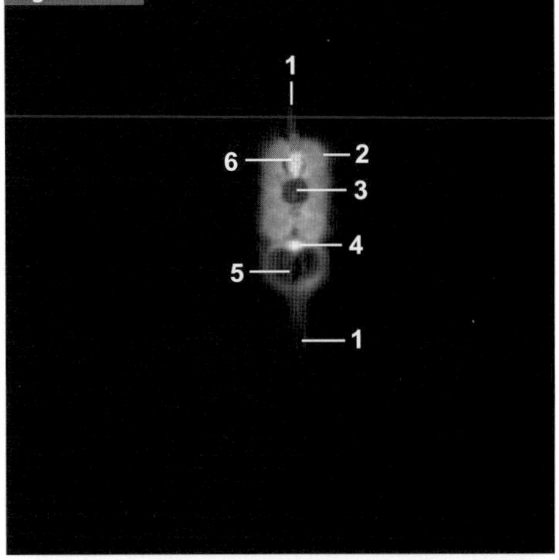

Figure 6.74

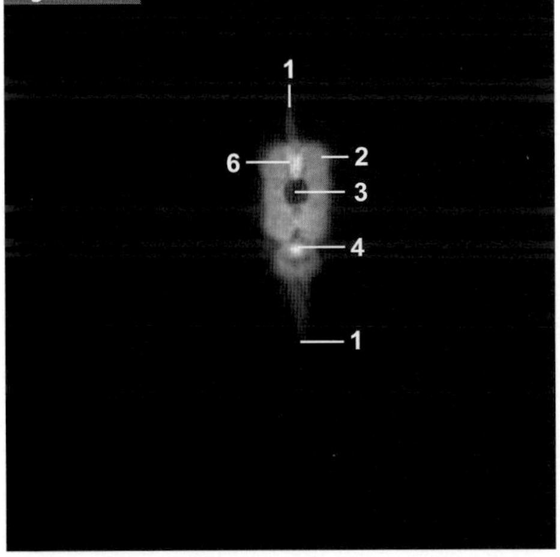

Figure 6.75

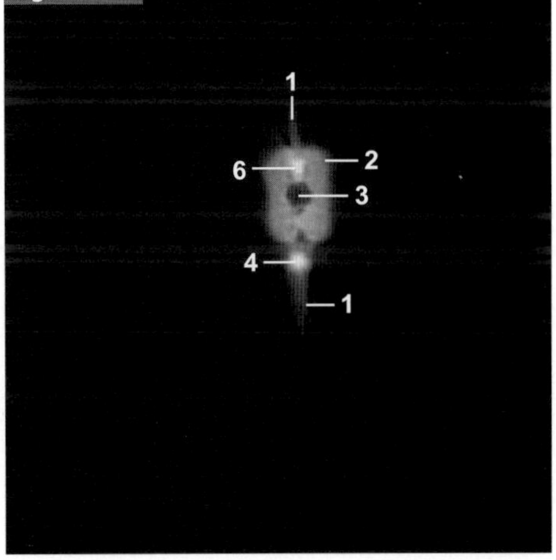

1. median fin fold	**2.** myotome	**3.** notochord	**4.** gut	**5.** yolk extension	**6.** spinal cord

Figure 6.76

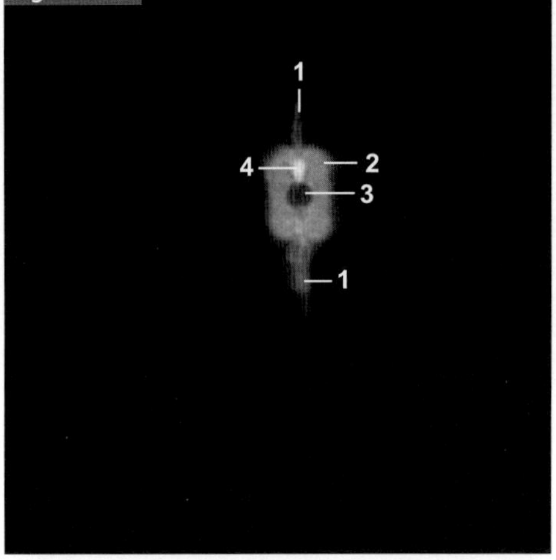

Figure 6.77

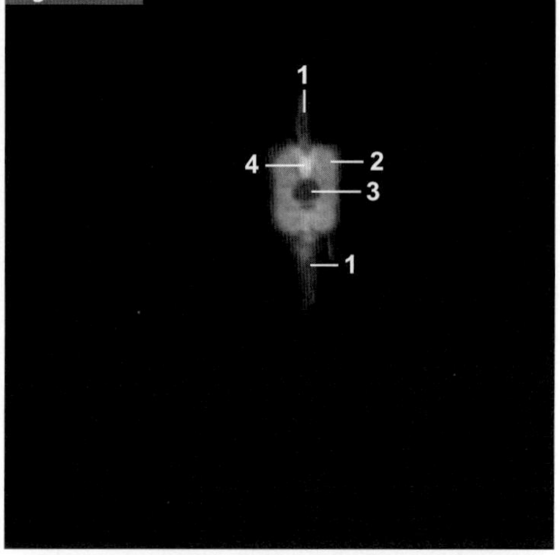

Figure 6.78

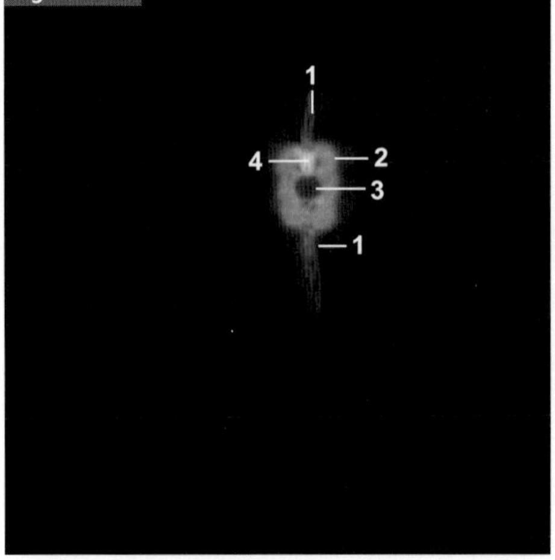

Figure 6.79

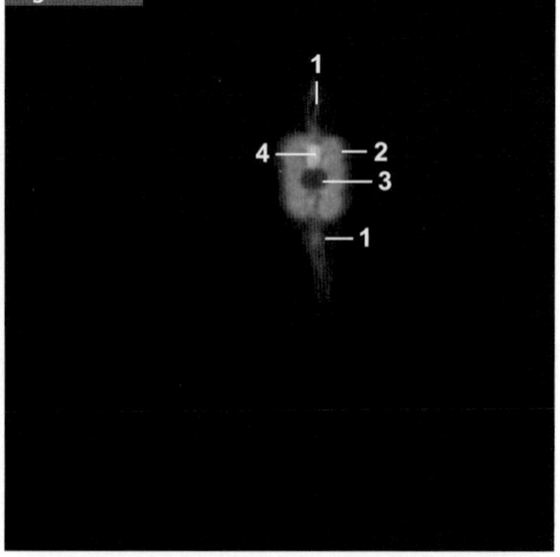

Figure 6.80

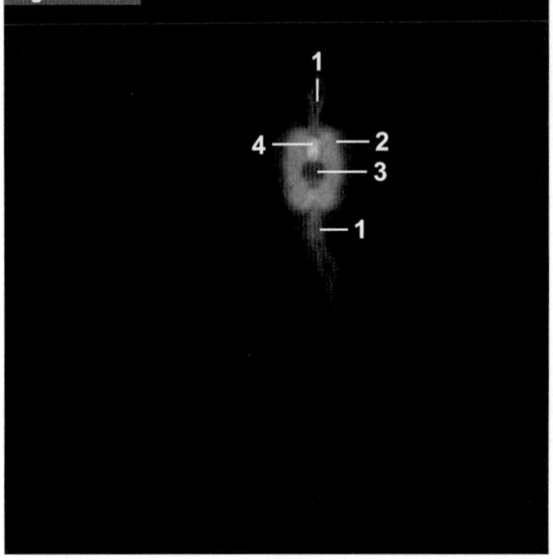

Figure 6.81

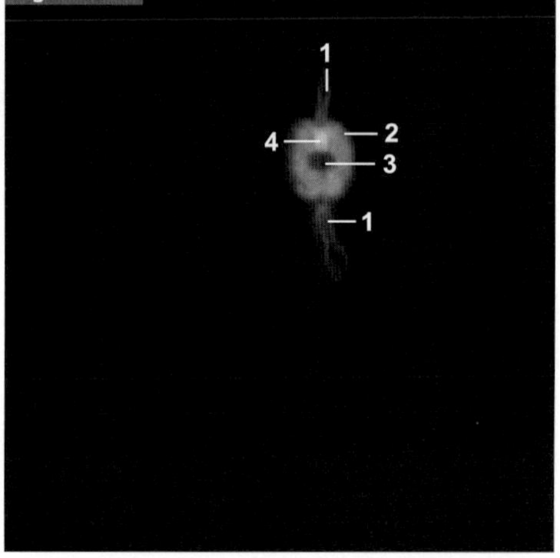

1. median fin fold	**2.** myotome	**3.** notochord	**4.** spinal cord

CHAPTER 7

96 HOURS

Sagittal Sections

Figure 7.1

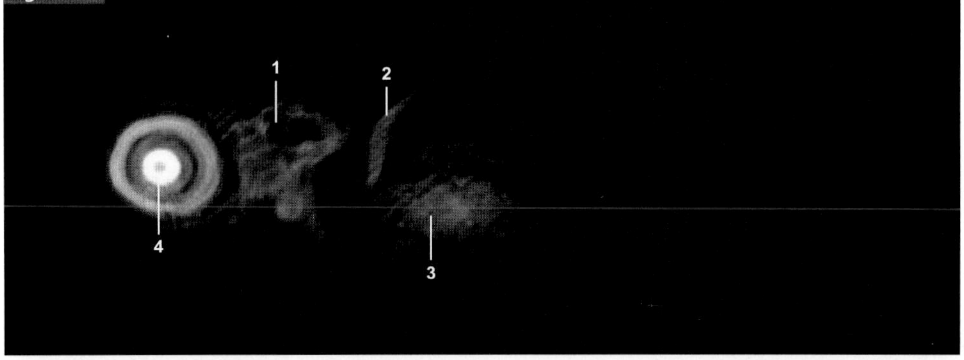

Figure 7.2

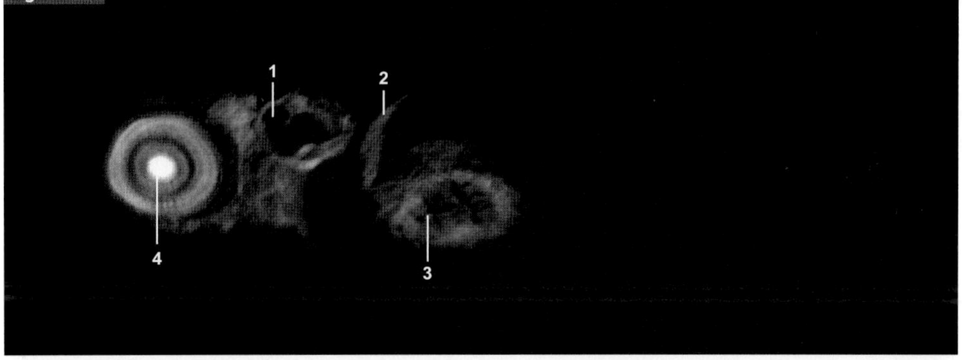

Figure 7.3

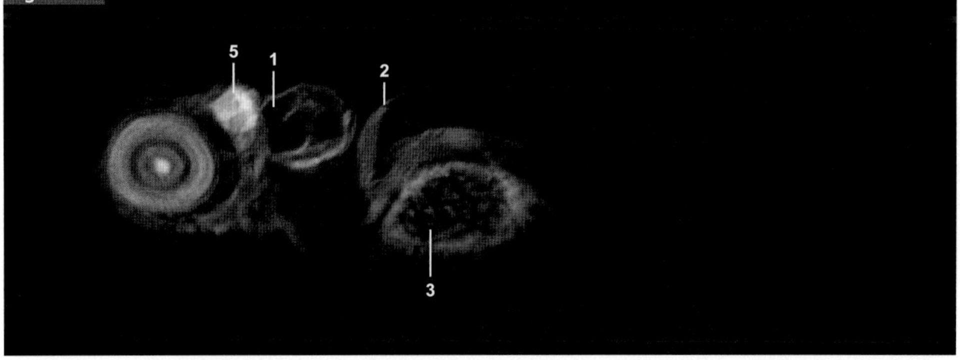

1. otic vesicle	**2.** pectoral fin	**3.** yolk	**4.** lens	**5.** optic tectum

Atlas of Zebrafish Development. 10.1016/B978-0-12-374016-8.00007-8

Figure 7.4

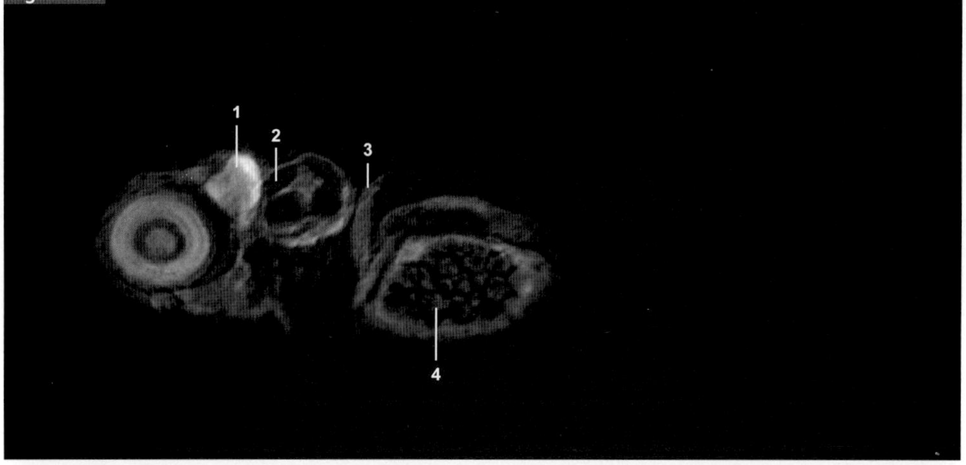

Figure 7.5

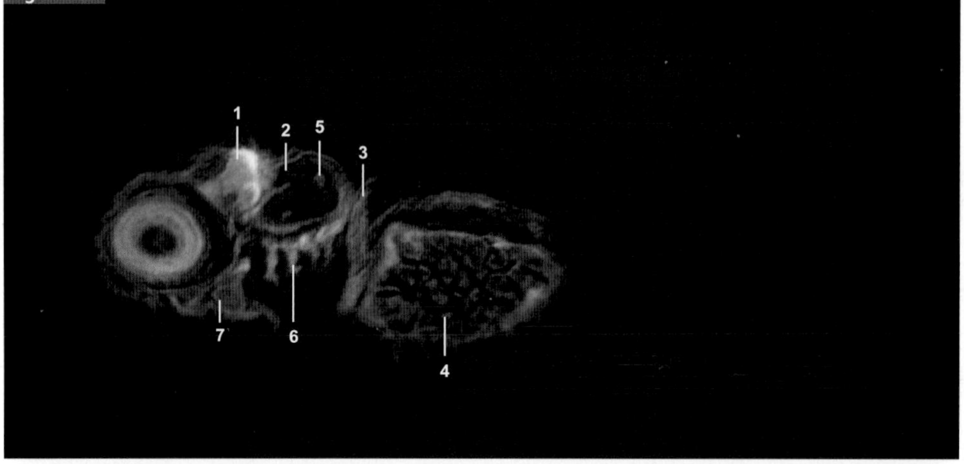

Figure 7.6

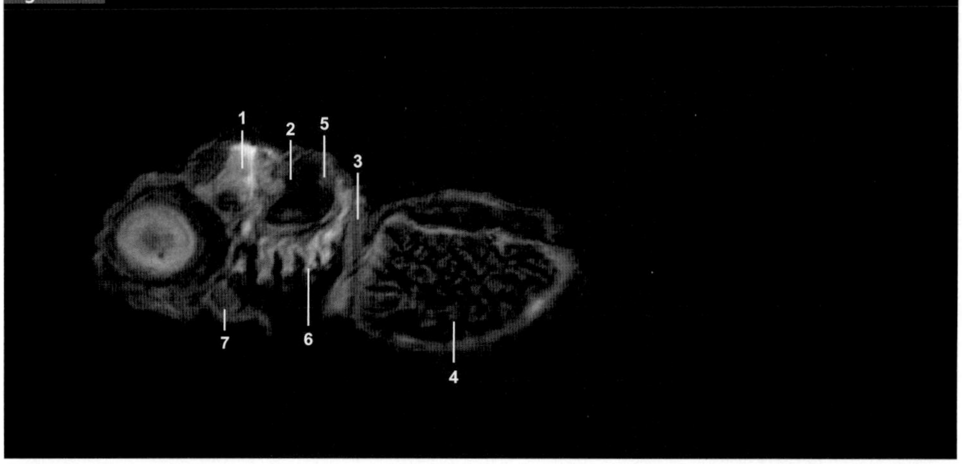

1. optic tectum **2.** otic vesicle **3.** pectoral fin **4.** yolk **5.** immature otolith **6.** gill filaments **7.** hyosymplectic cartilage

Figure 7.7

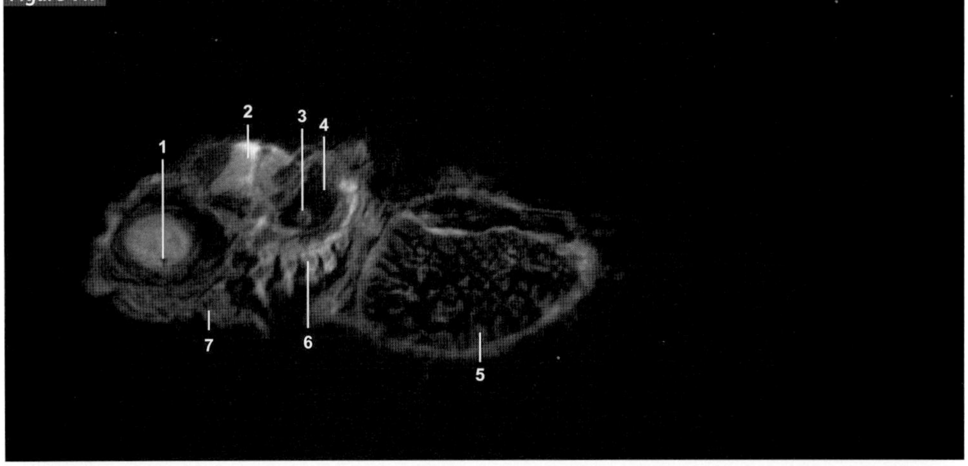

Figure 7.8

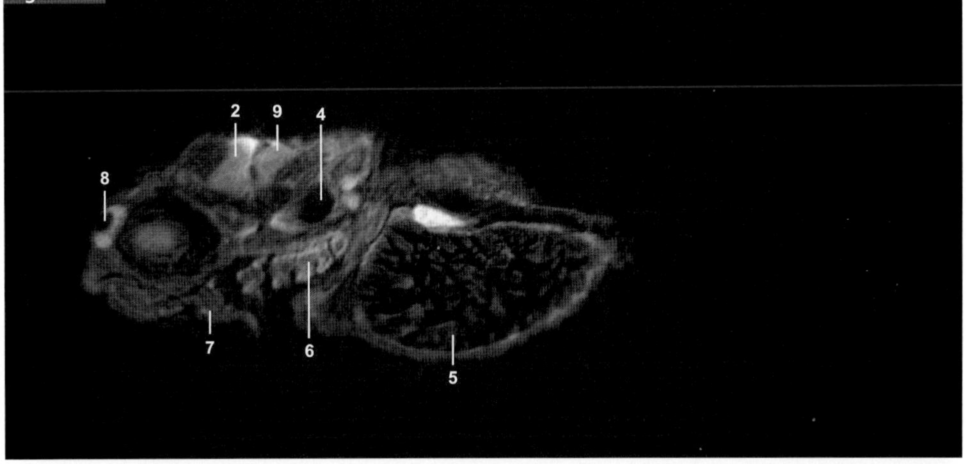

Figure 7.9

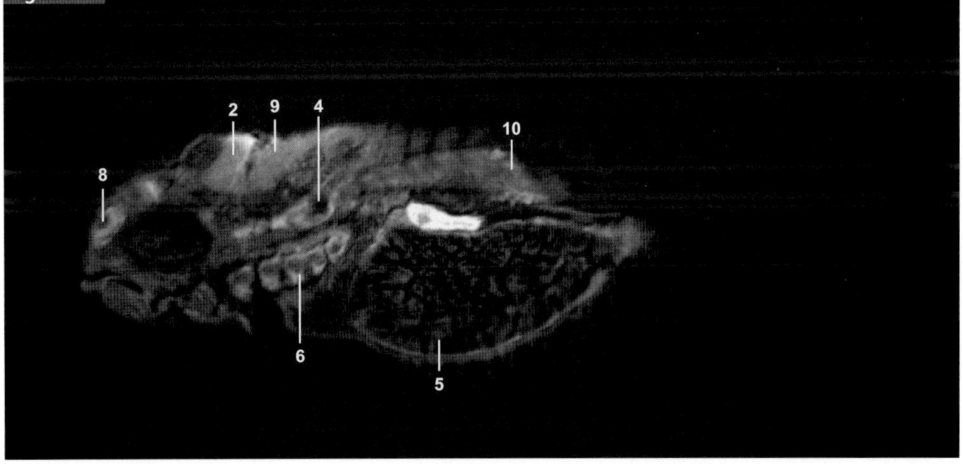

1. cranial nerve II	**3.** immature otolith	**5.** yolk	**7.** hyosymplectic cartilage	**9.** medulla oblongata
2. optic tectum	**4.** otic vesicle	**6.** gill filaments	**8.** olfactory pit	**10.** myotome

Figure 7.10

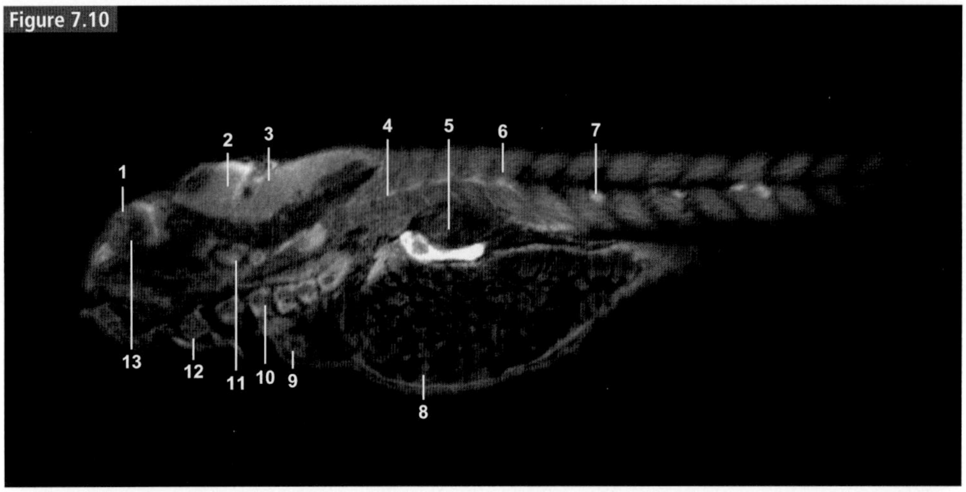

Figure 7.11

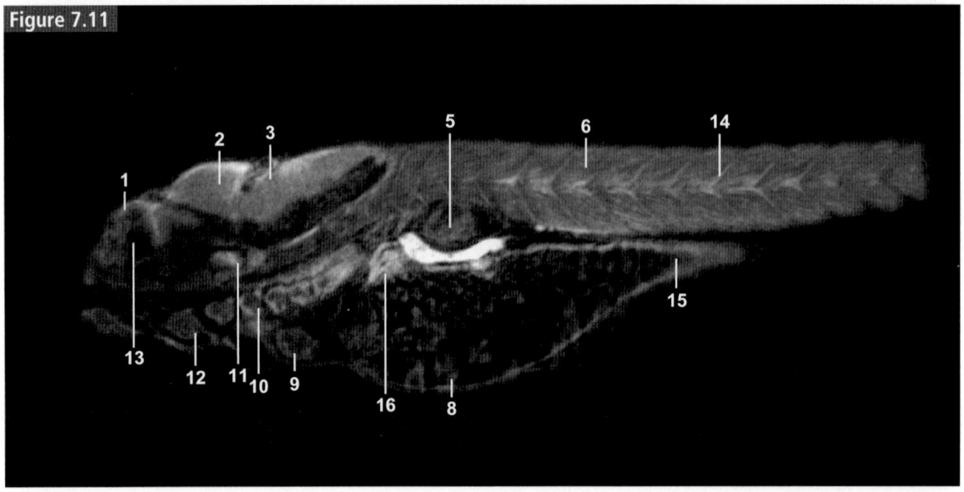

Figure 7.12

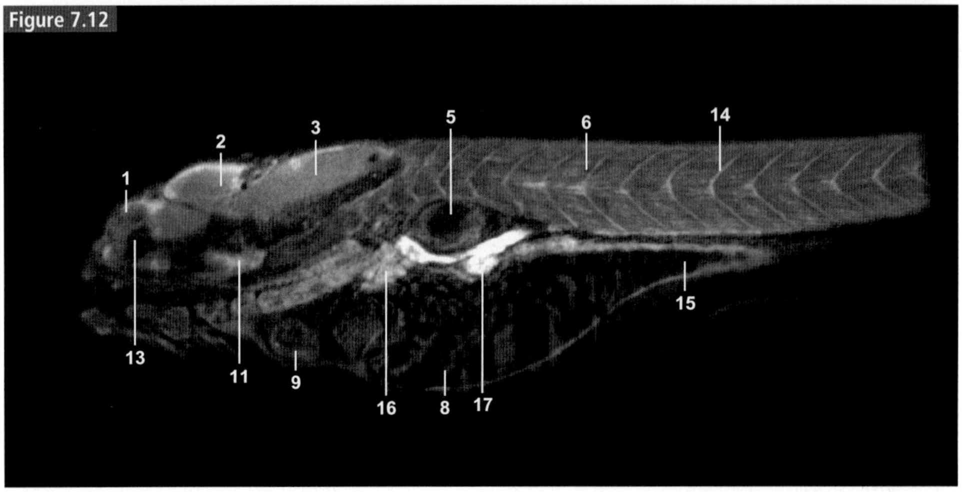

1. telencephalon	**6.** myotome	**11.** hypothalamus	**16.** liver
2. optic tectum	**7.** neuromasts	**12.** palatoquadrate cartilage	**17.** intestine
3. medulla oblongata	**8.** yolk	**13.** anterior commissure	
4. horizontal myoseptum	**9.** cardiac ventricle	**14.** vertical myoseptum	
5. swim bladder	**10.** gill filaments	**15.** yolk extension	

Figure 7.13

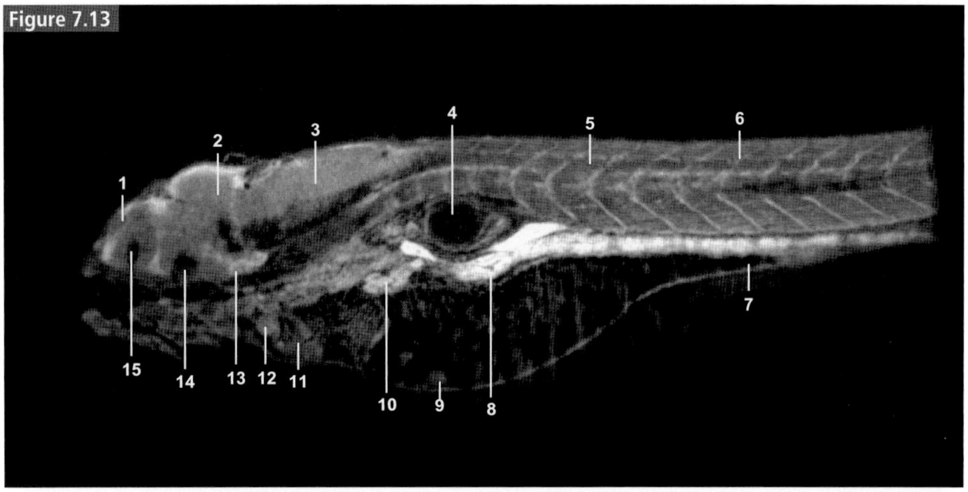

Figure 7.14

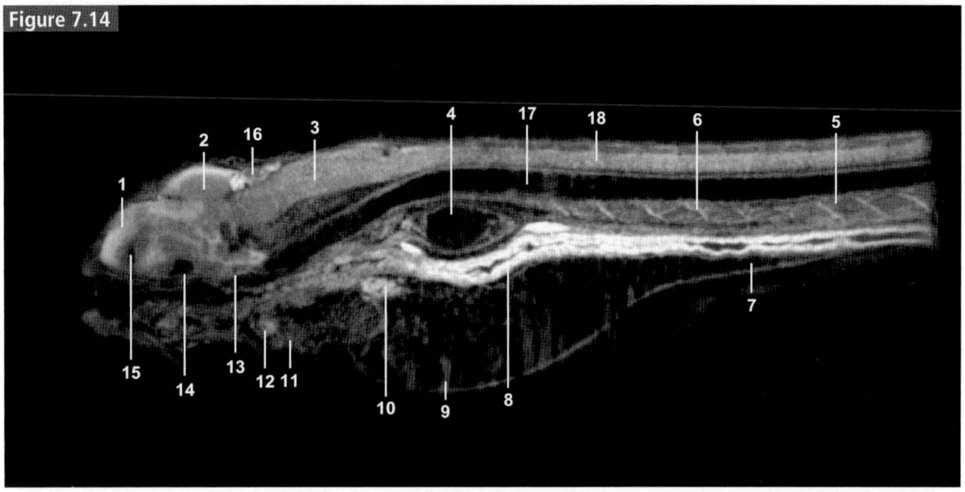

Figure 7.15

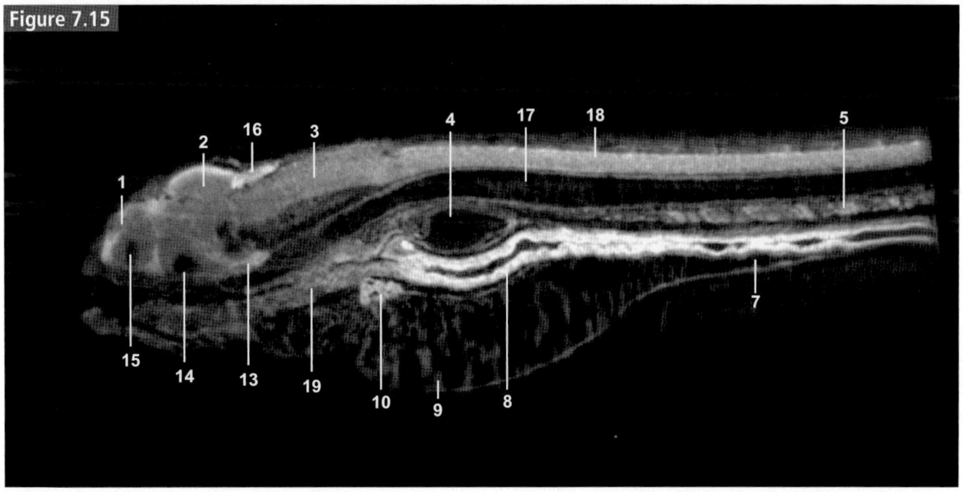

1. telencephalon	**6.** vertical myoseptum	**11.** cardiac ventricle	**16.** cerebellum
2. optic tectum	**7.** yolk extension	**12.** bulbus arteriosus	**17.** notochord
3. medulla oblongata	**8.** intestine	**13.** hypothalamus	**18.** spinal cord
4. swim bladder	**9.** yolk	**14.** optic chiasm	**19.** esophagus
5. myotome	**10.** liver	**15.** anterior commissure	

Figure 7.16

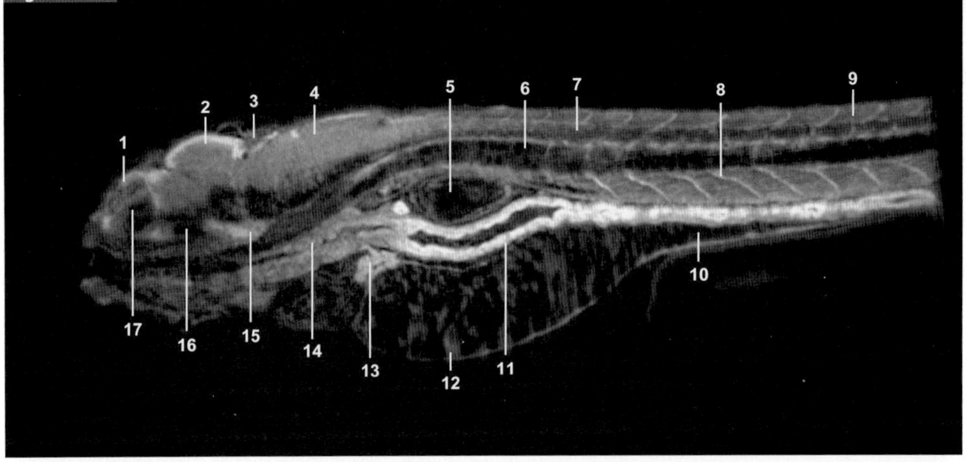

Figure 7.17

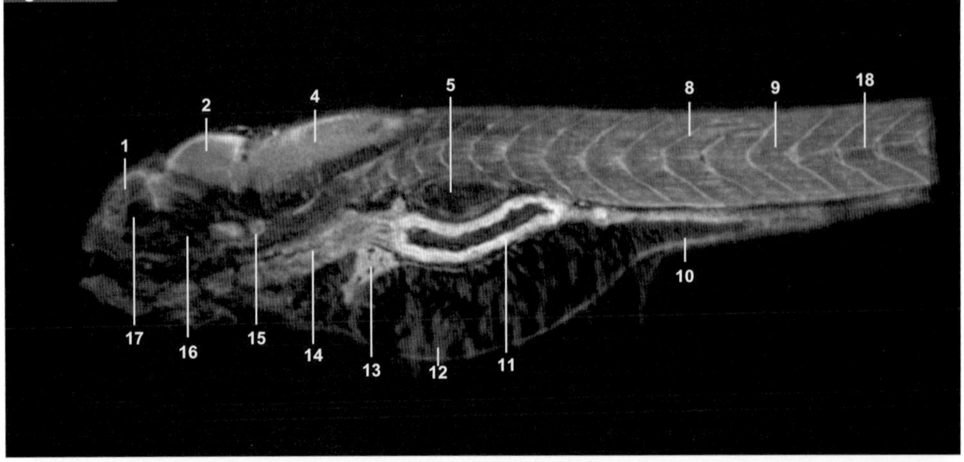

Figure 7.18

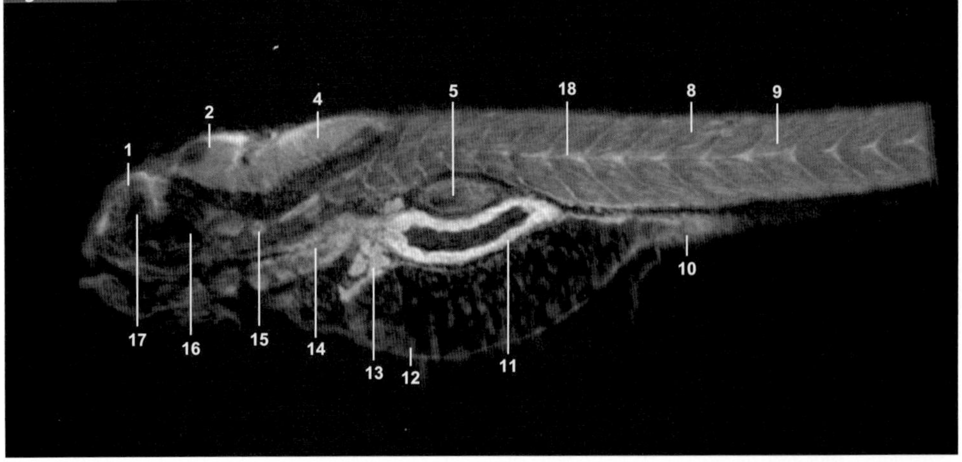

1. telencephalon	**4.** medulla oblongata	**7.** spinal cord	**10.** yolk extension	**13.** liver	**16.** optic chiasm
2. optic tectum	**5.** swim bladder	**8.** vertical myoseptum	**11.** intestine	**14.** esophagus	**17.** anterior commissure
3. cerebellum	**6.** notochord	**9.** myotome	**12.** yolk	**15.** hypothalamus	**18.** horizontal myoseptum

Figure 7.19

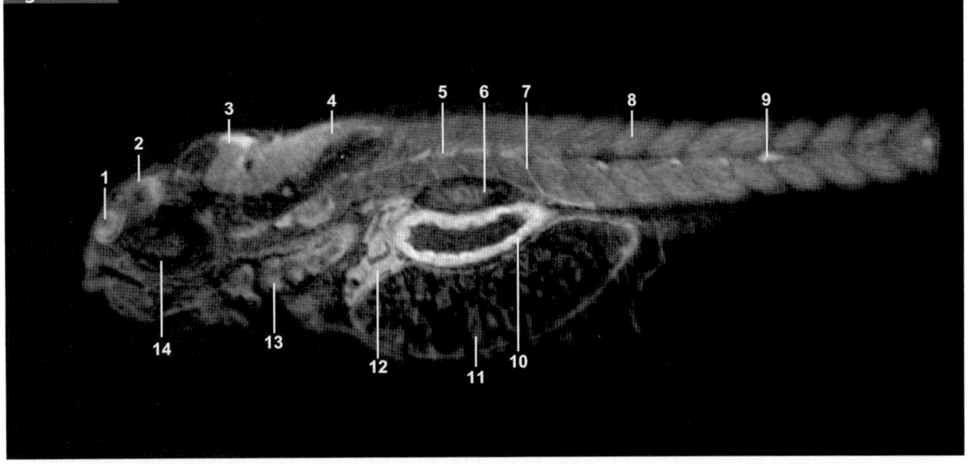

Figure 7.20

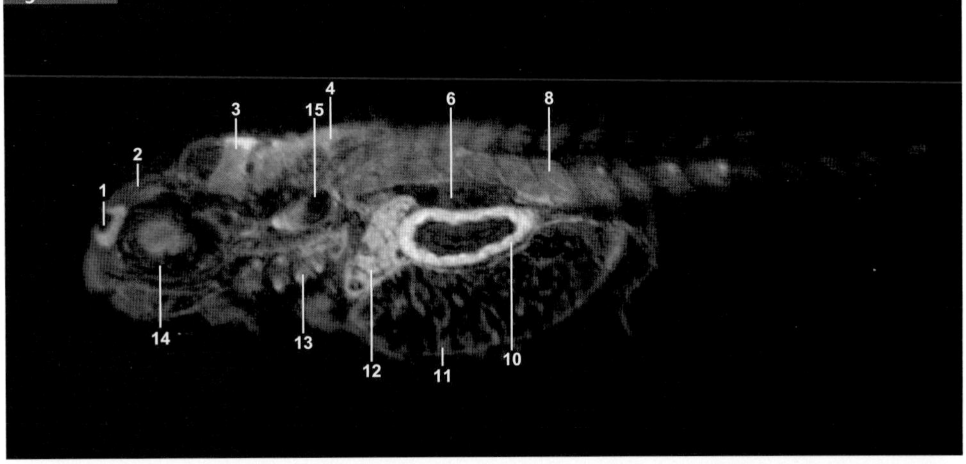

Figure 7.21

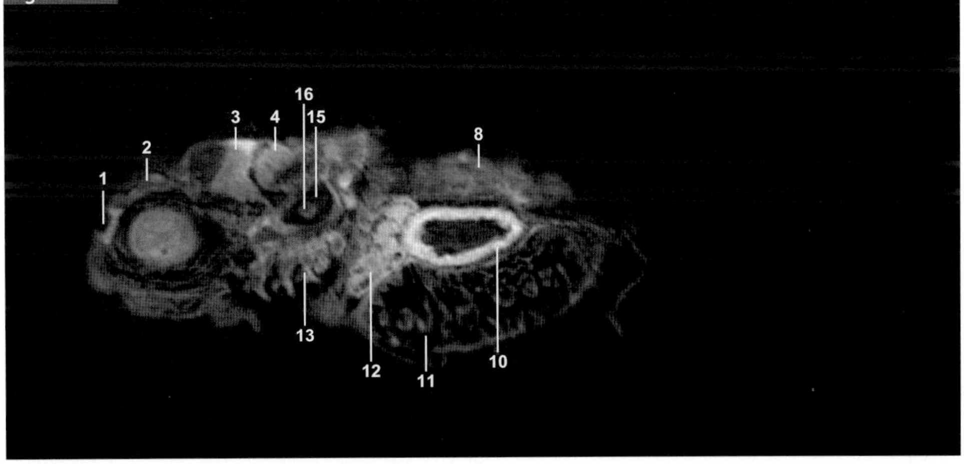

1. olfactory pit	**5.** horizontal myoseptum	**9.** neuromasts	**13.** gill arch
2. telencephalon	**6.** swim bladder	**10.** intestine	**14.** optic chiasm
3. optic tectum	**7.** vertical myoseptum	**11.** yolk	**15.** otic vesicle
4. medulla oblongata	**8.** myotome	**12.** liver	**16.** immature otolith

Figure 7.22

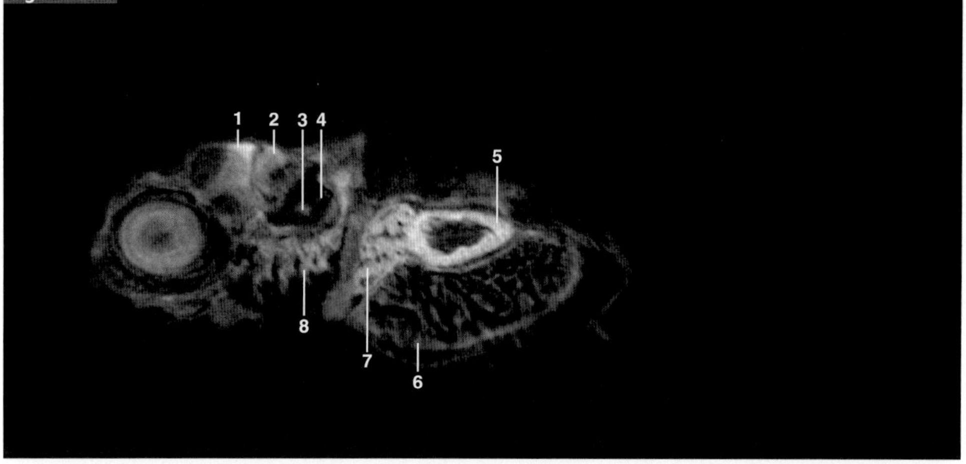

Figure 7.23

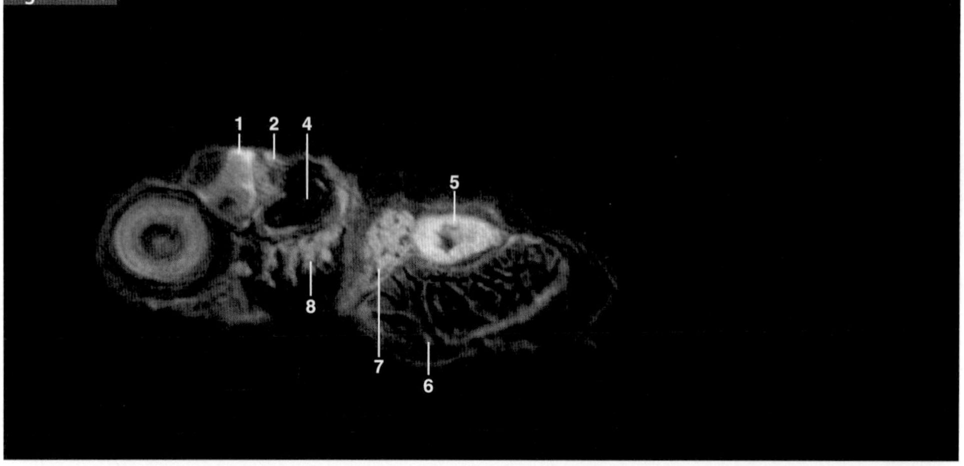

Figure 7.24

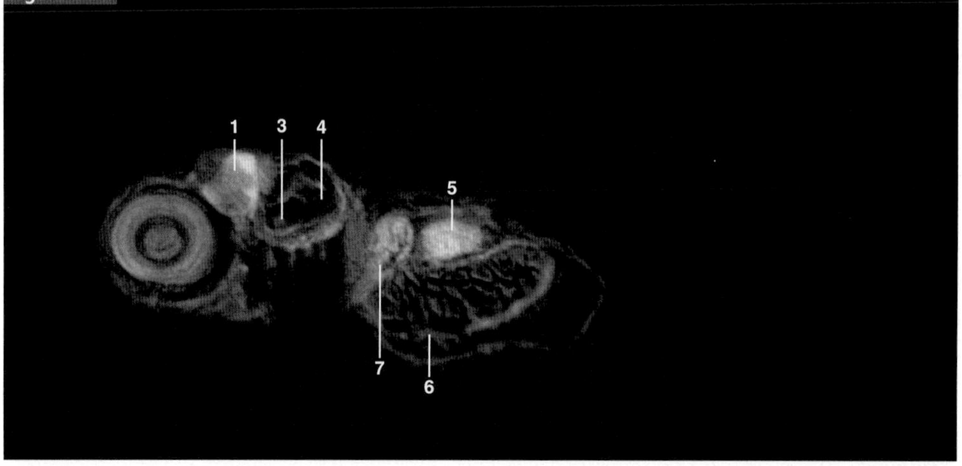

1. optic tectum	**2.** medulla oblongata	**3.** immature otolith	**4.** otic vesicle	**5.** intestine	**6.** yolk	**7.** liver	**8.** gill arch

Figure 7.25

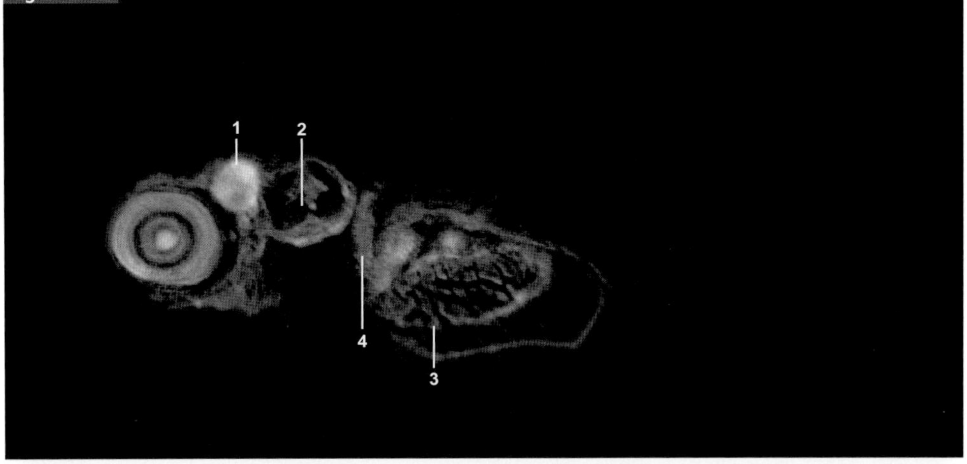

Figure 7.26

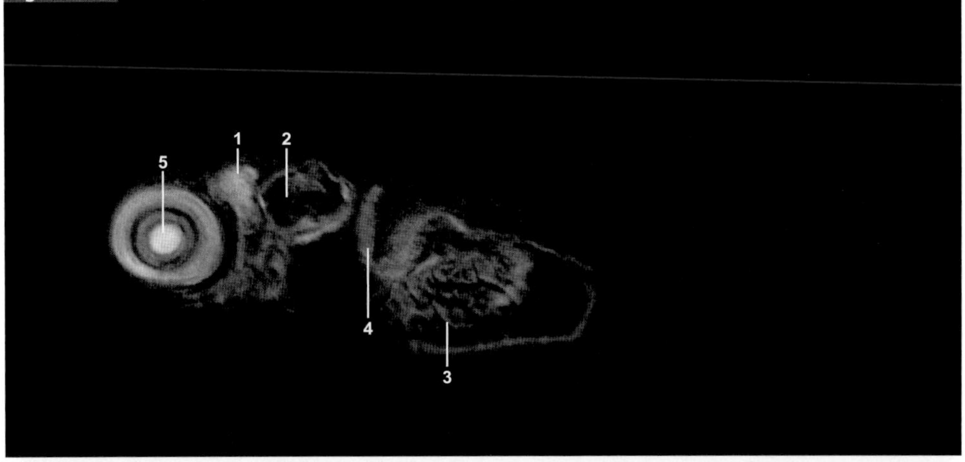

Figure 7.27

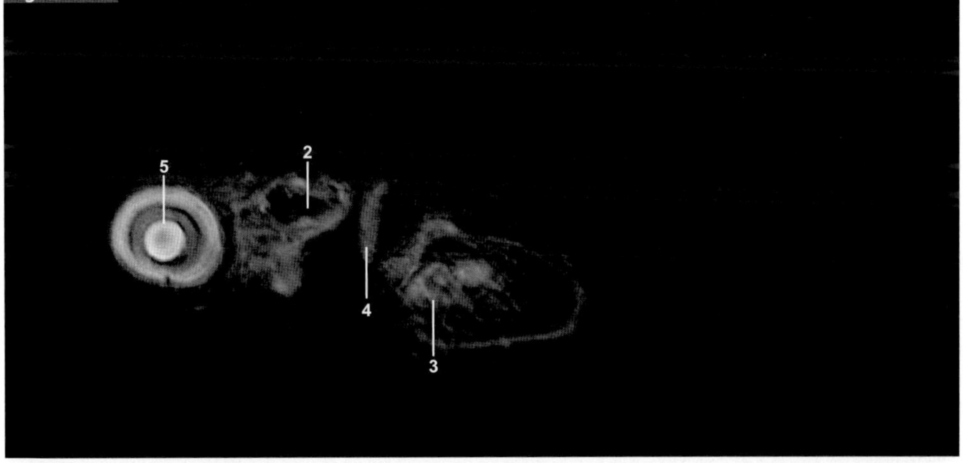

1. optic tectum	**2.** otic vesicle	**3.** yolk	**4.** pectoral fin	**5.** lens

Coronal Sections

Figure 7.28

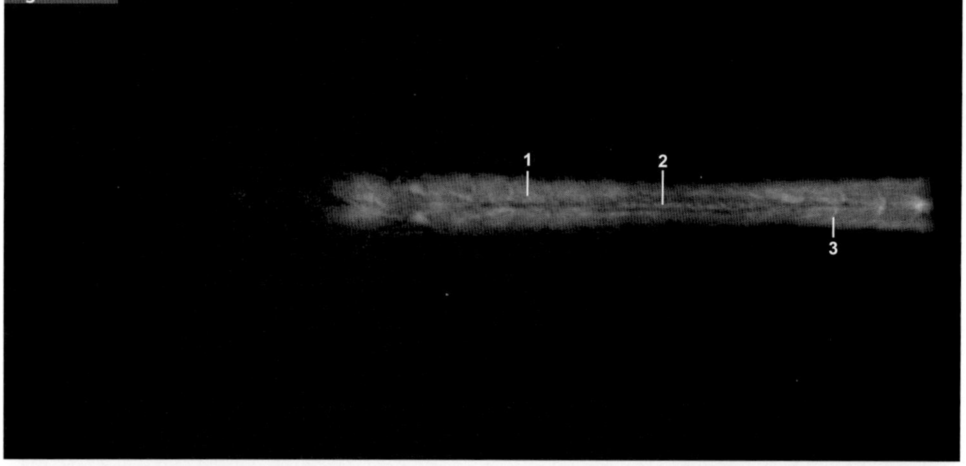

Figure 7.29

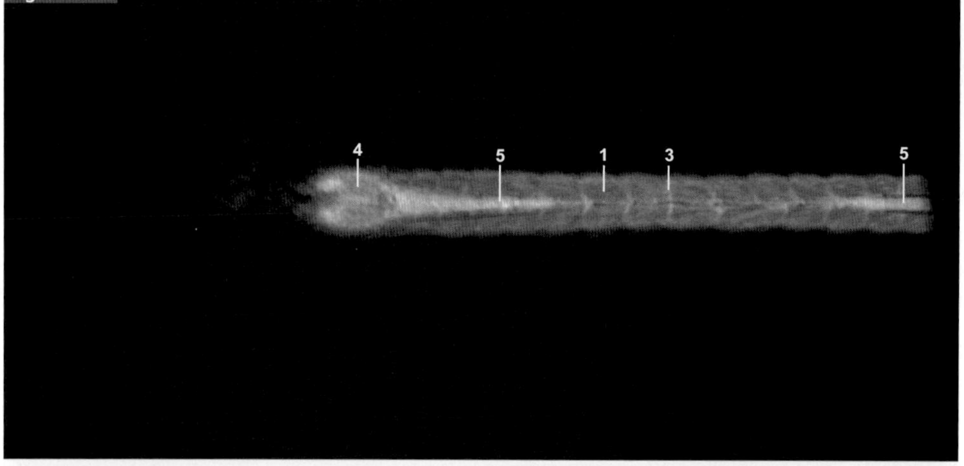

Figure 7.30

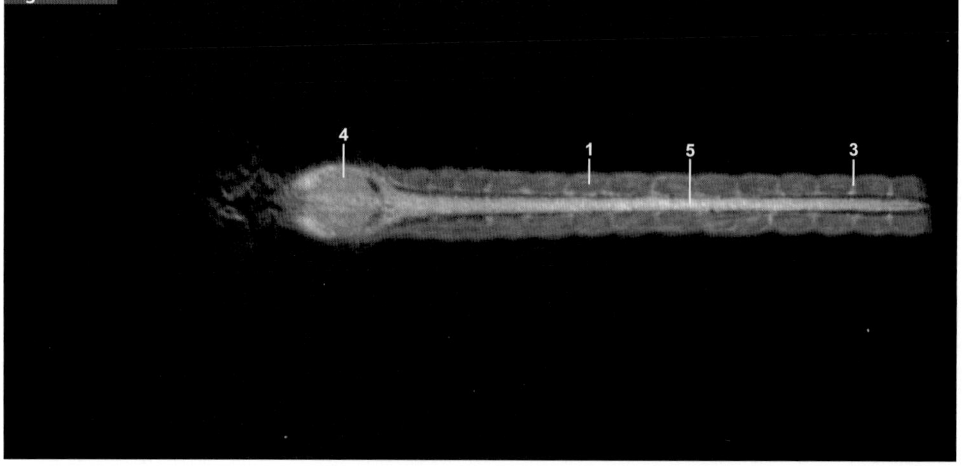

1. myotome	**2.** median fin fold	**3.** vertical myoseptum	**4.** medulla oblongata	**5.** spinal cord

Figure 7.31

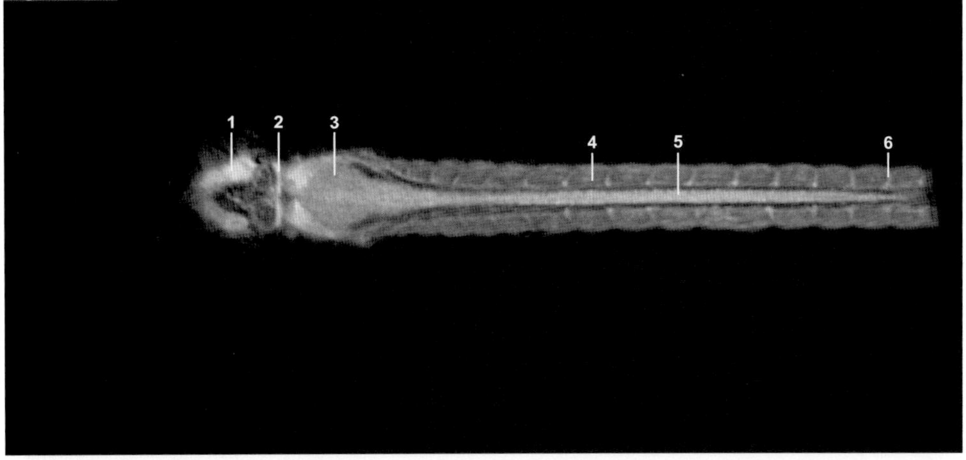

Figure 7.32

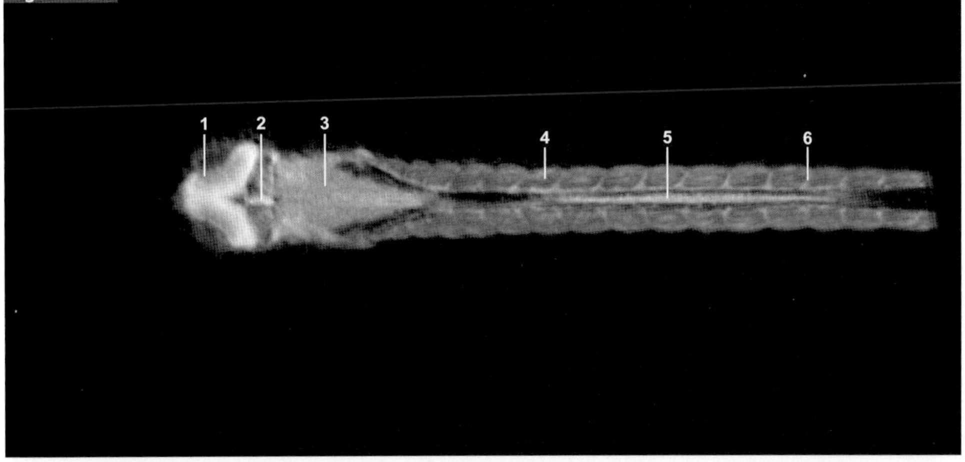

Figure 7.33

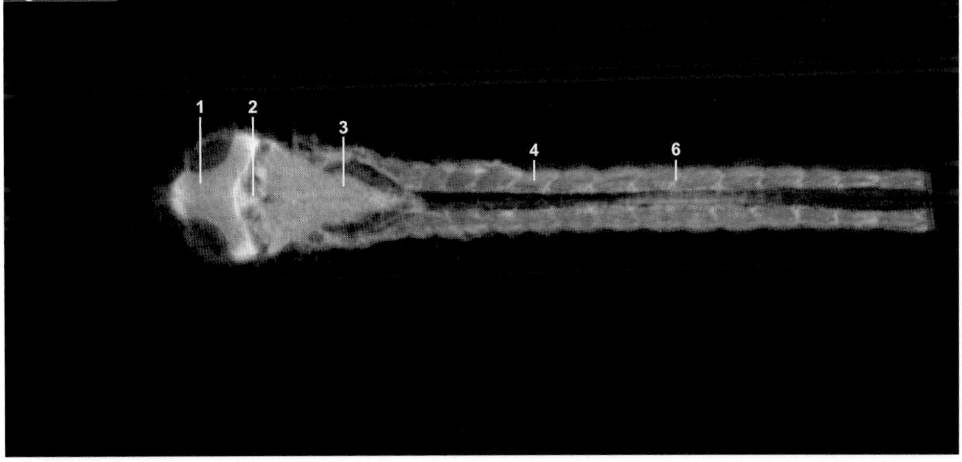

1. optic tectum	**2.** cerebellum	**3.** medulla oblongata	**4.** myotome	**5.** spinal cord	**6.** vertical myoseptum

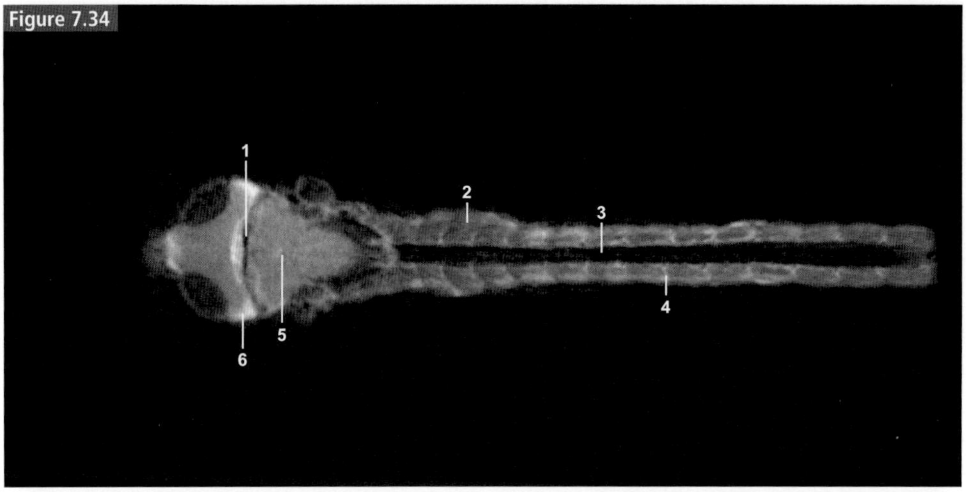

Figure 7.34

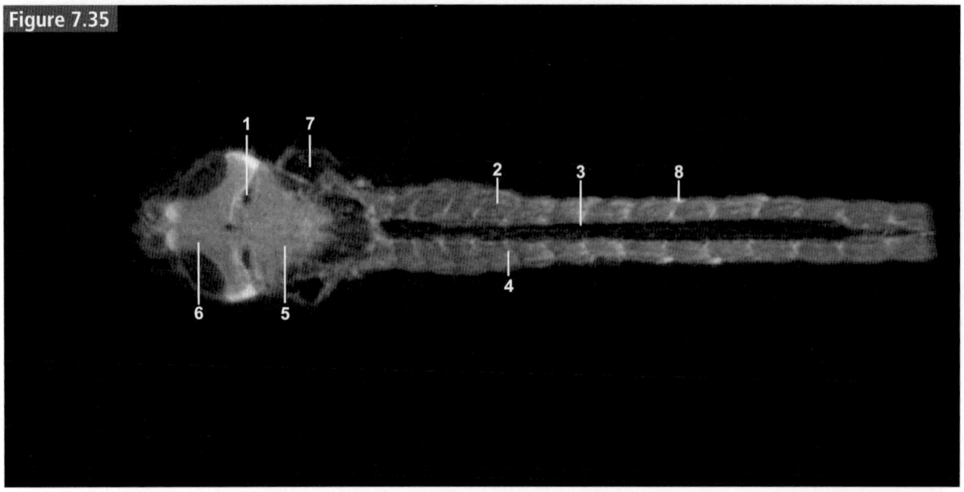

Figure 7.35

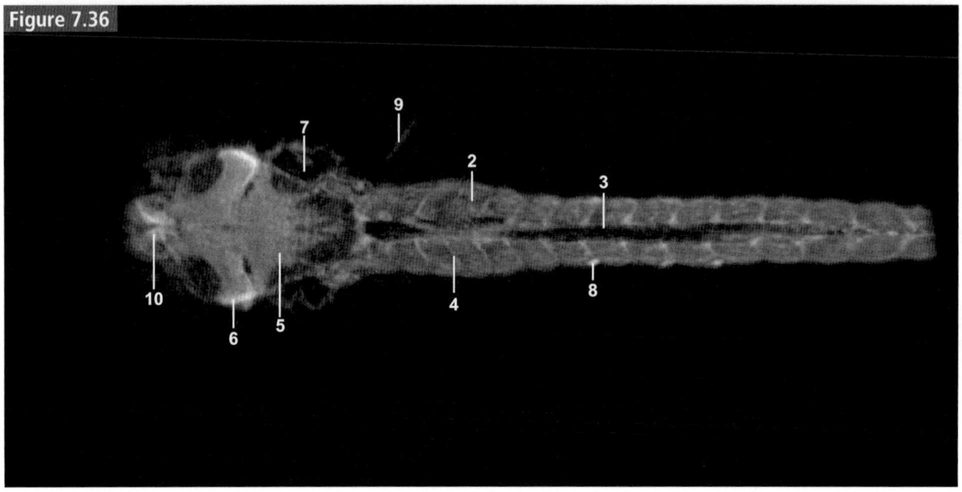

Figure 7.36

1. fourth ventricle	**3.** notochord	**5.** medulla oblongata	**7.** semicircular canal	**9.** pectoral fin
2. myotome	**4.** vertical myoseptum	**6.** optic tectum	**8.** neuromasts	**10.** telencephalon

Figure 7.37

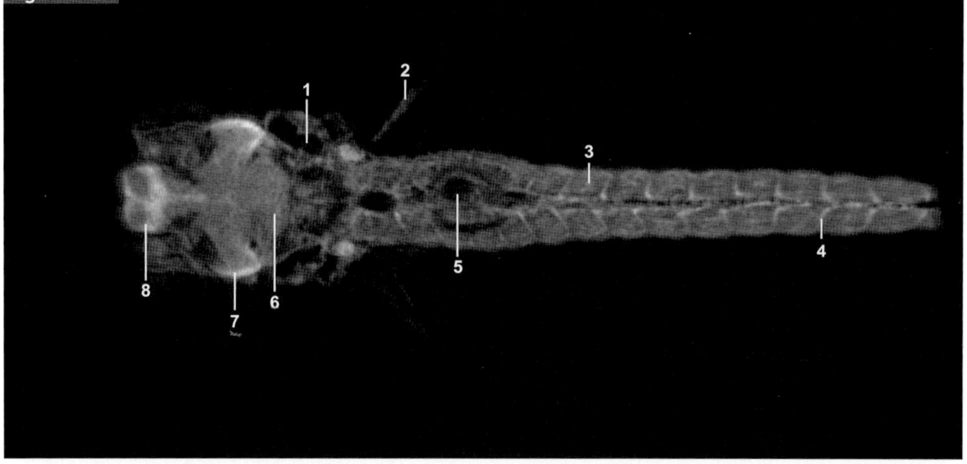

Figure 7.38

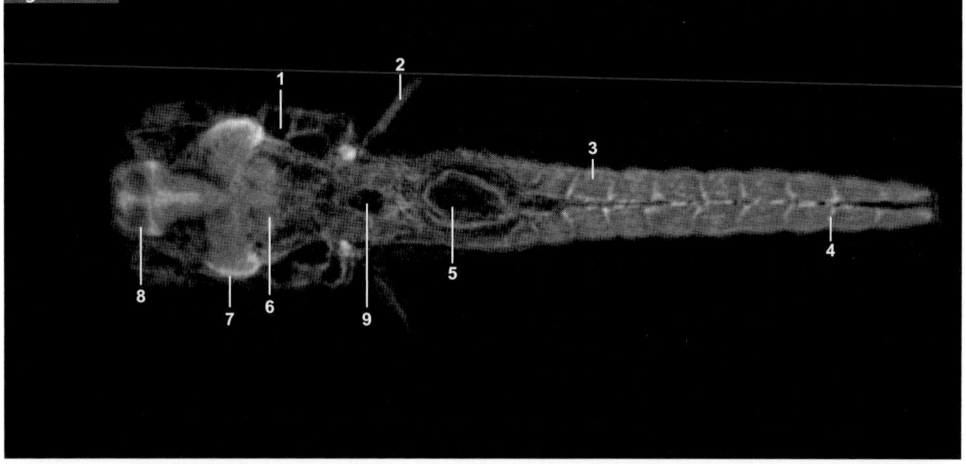

Figure 7.39

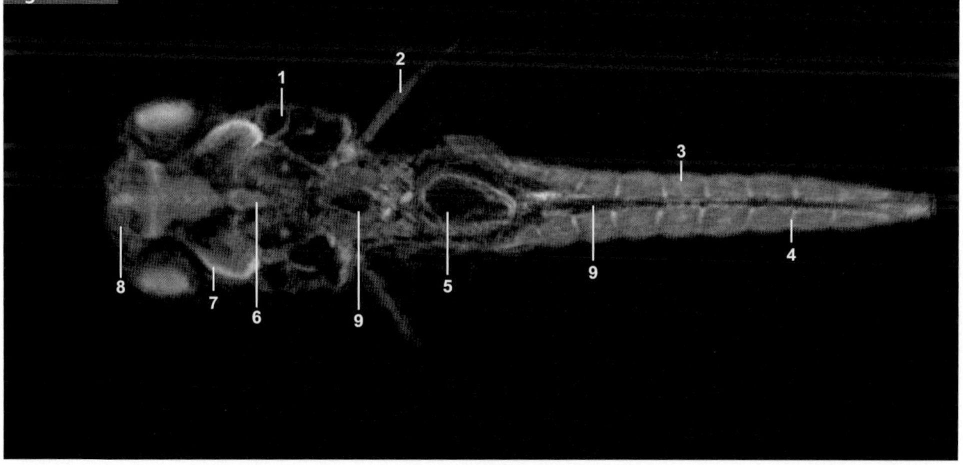

1. semicircular canal	**3.** myotome	**5.** swim bladder	**7.** optic tectum	**9.** notochord
2. pectoral fin	**4.** vertical myoseptum	**6.** medulla oblongata	**8.** telencephalon	

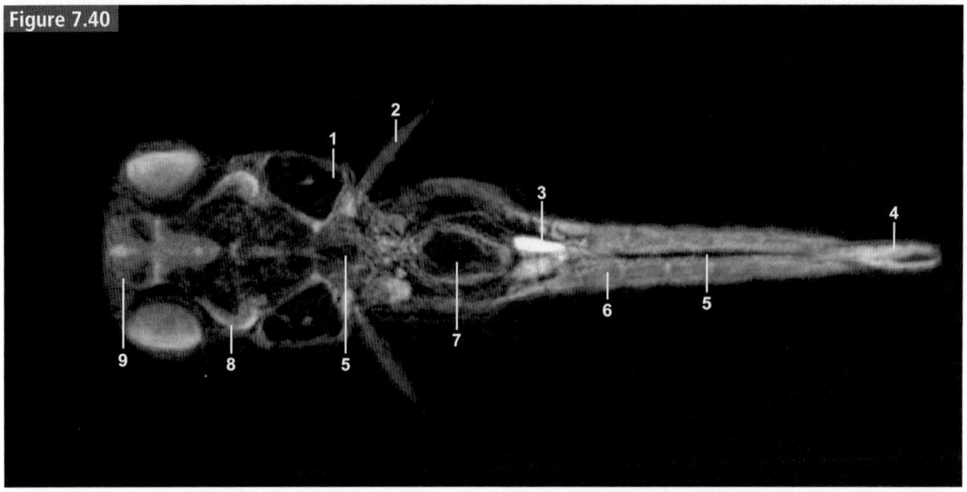

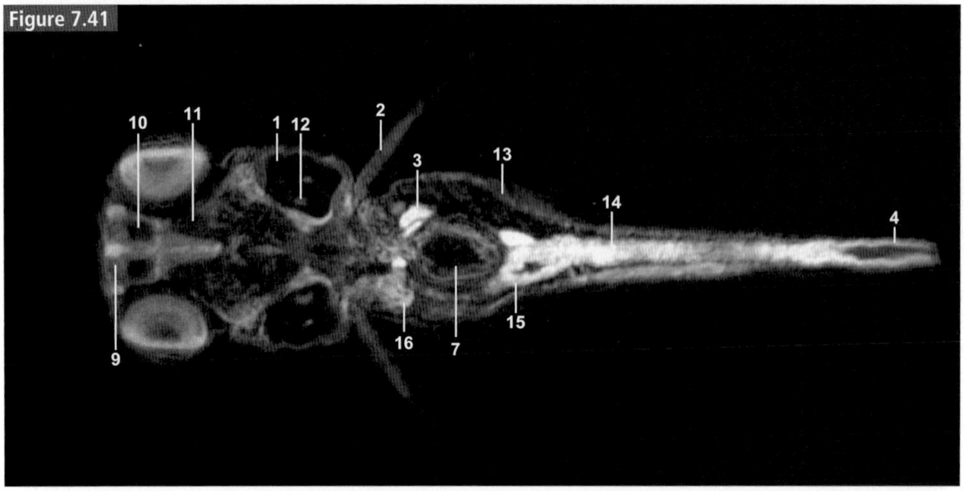

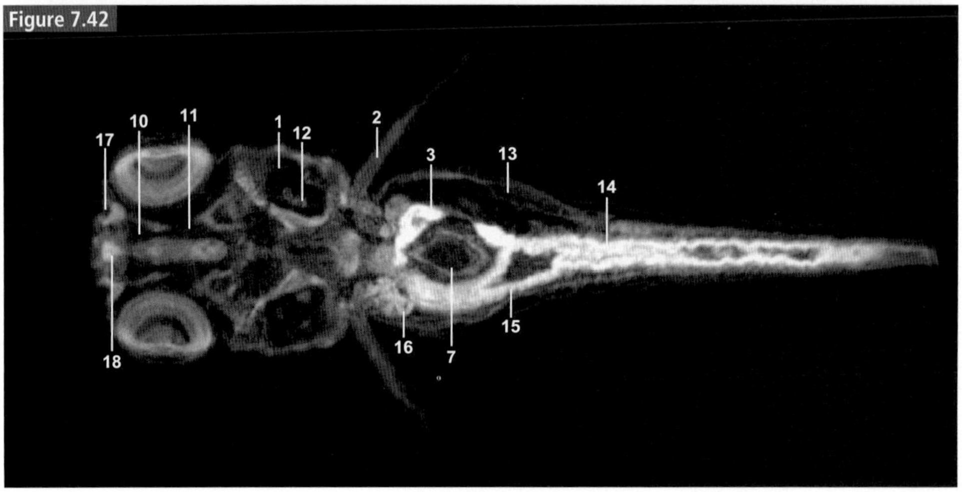

1. semicircular canal
2. pectoral fin
3. pancreas
4. posterior intestine
5. notochord
6. myotome
7. swim bladder
8. optic tectum
9. telencephalon
10. anterior commissure
11. optic chiasm
12. immature otolith
13. yolk
14. mid intestine
15. intestinal bulb
16. liver
17. olfactory pit
18. olfactory bulb

Figure 7.43

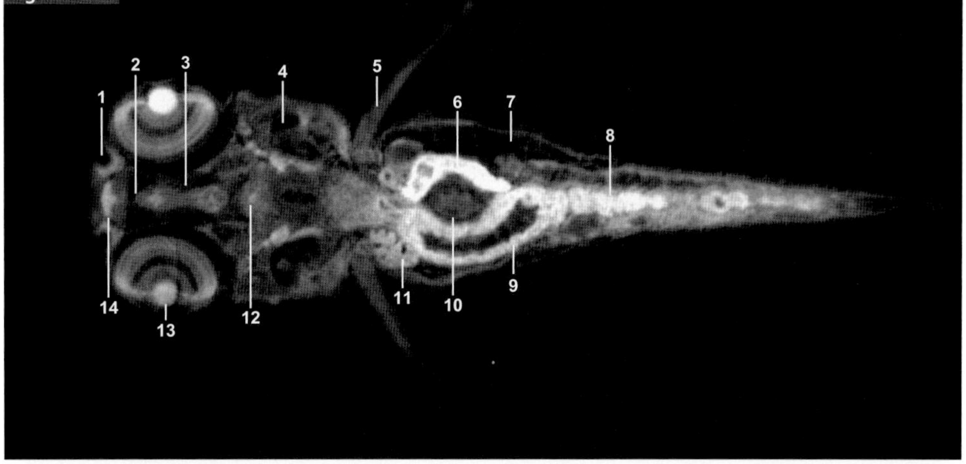

Figure 7.44

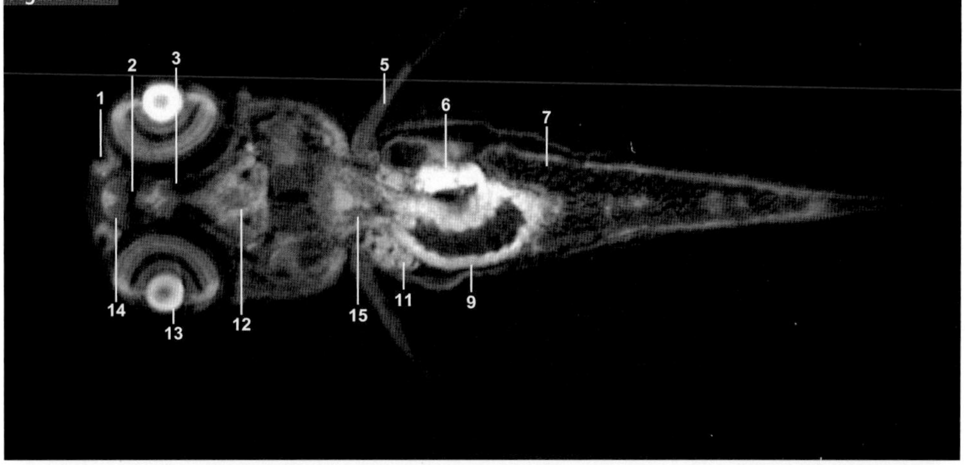

Figure 7.45

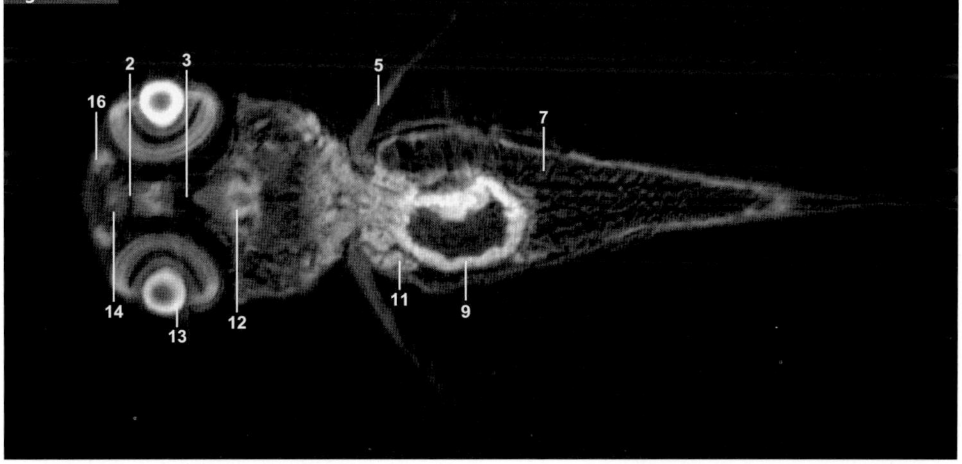

1. olfactory pit	**5.** pectoral fin	**9.** intestinal bulb	**13.** lens
2. anterior commissure	**6.** pancreas	**10.** swim bladder	**14.** olfactory bulb
3. optic chiasm	**7.** yolk	**11.** liver	**15.** esophagus
4. semicircular canal	**8.** mid intestine	**12.** hypothalamus	**16.** olfactory epithelium

Figure 7.46

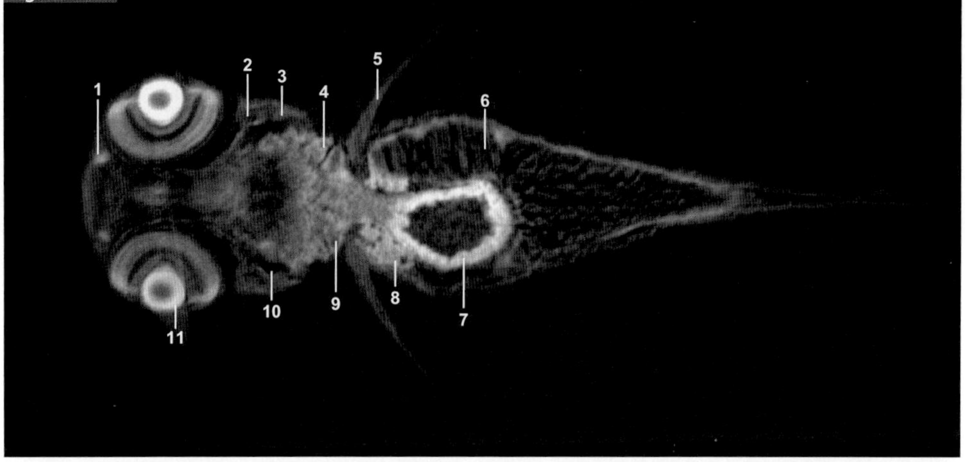

Figure 7.47

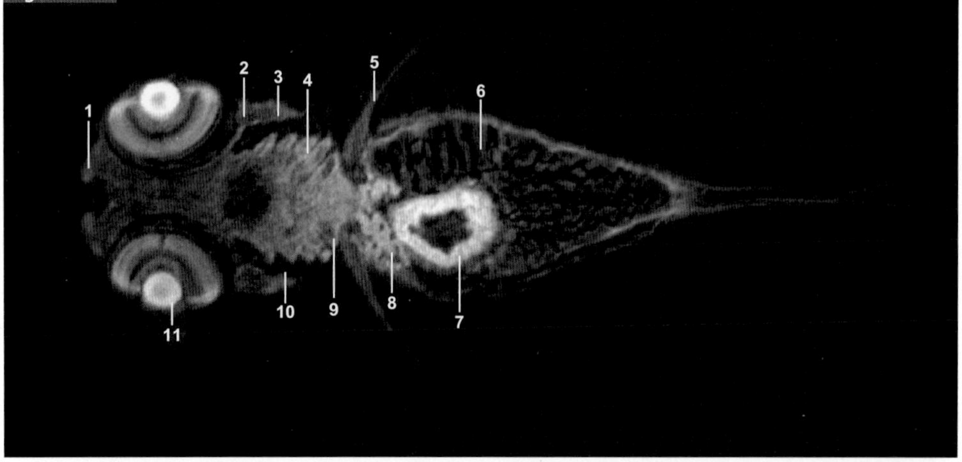

Figure 7.48

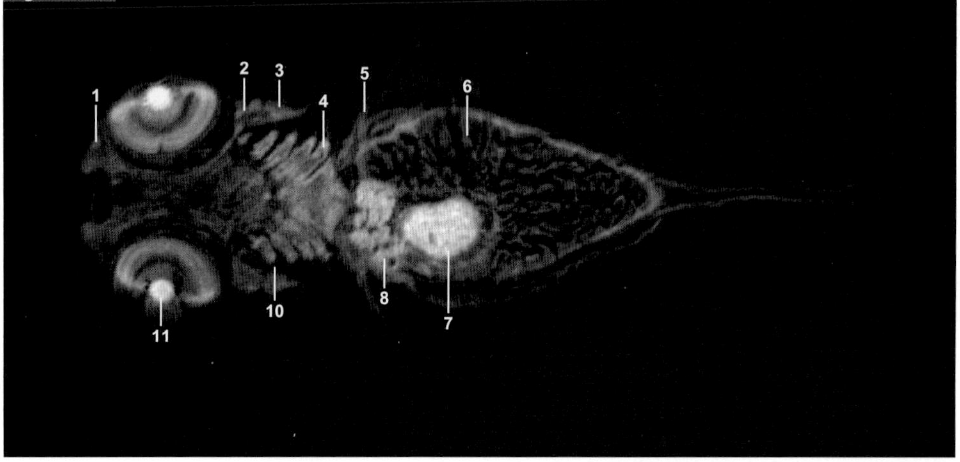

1. dorsal mandibular arch
2. hyomandibula
3. opercle
4. gill arch
5. pectoral fin
6. yolk
7. intestinal bulb
8. liver
9. ceratobranchial 5
10. opercular cavity
11. lens

Figure 7.49

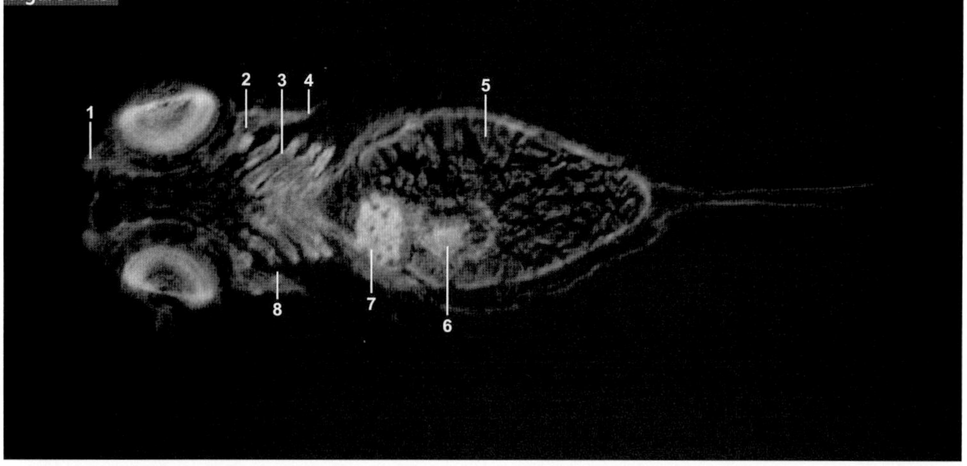

Figure 7.50

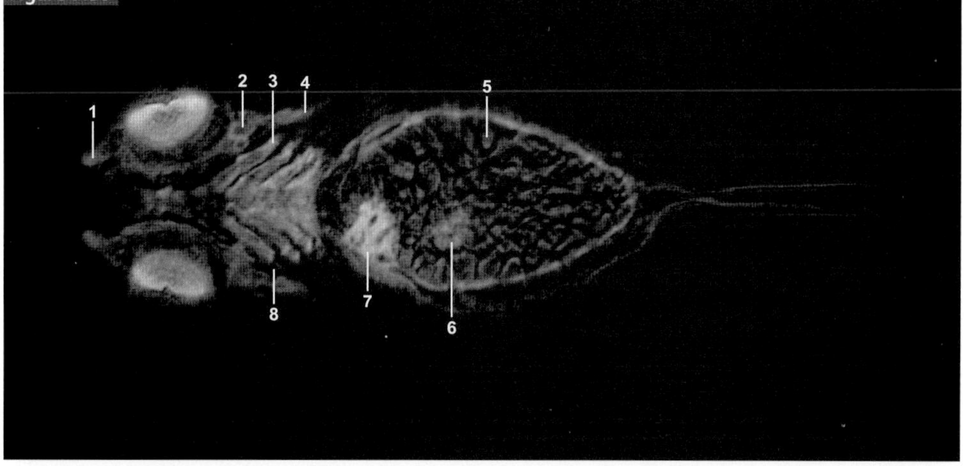

Figure 7.51

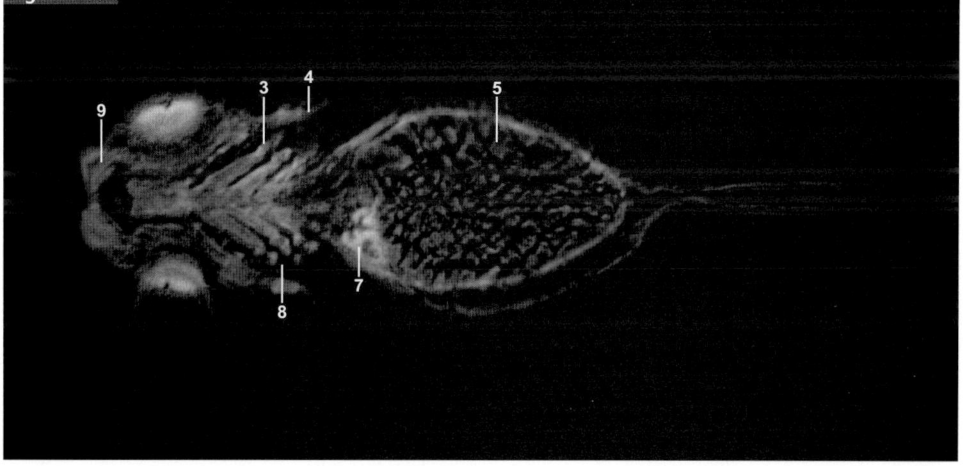

1. dorsal mandibular arch	**3.** gill arch	**5.** yolk	**7.** liver	**9.** ventral mandibular arch
2. hyomandibula	**4.** opercle	**6.** intestinal bulb	**8.** opercular cavity	

Figure 7.52

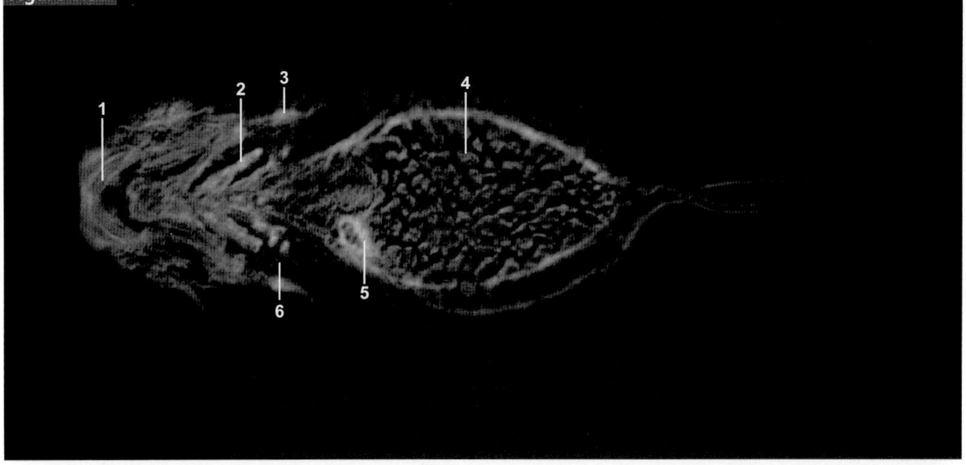

Figure 7.53

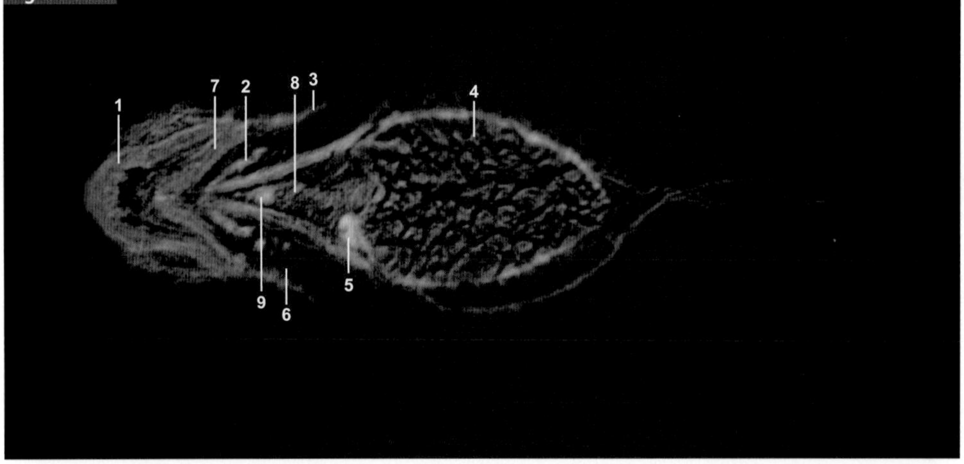

Figure 7.54

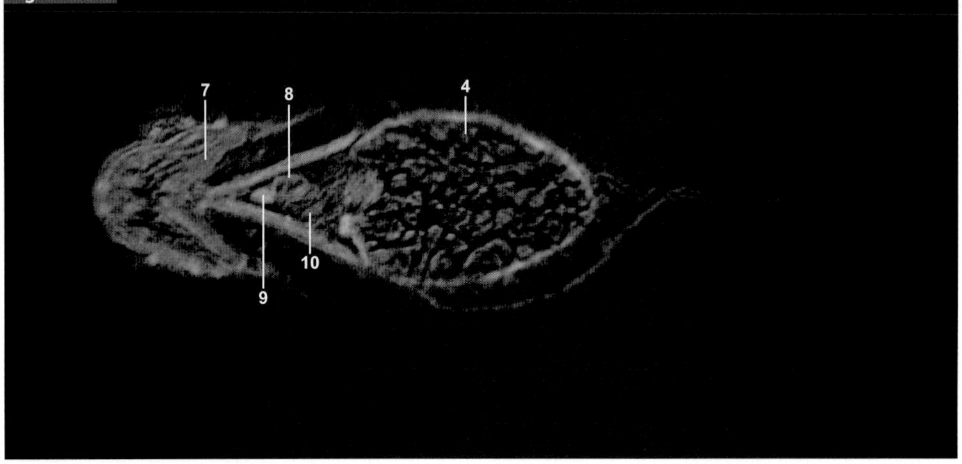

1. ventral mandibular arch	**3.** opercle	**5.** liver	**7.** ceratohyal	**9.** bulbus arteriosus
2. gill arch	**4.** yolk	**6.** opercular cavity	**8.** cardiac ventricle	**10.** atrium

Figure 7.55

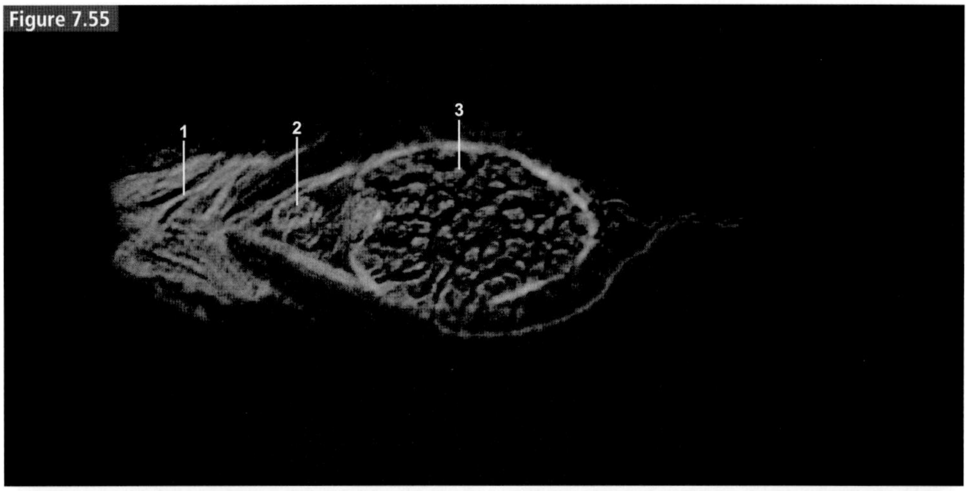

Figure 7.56

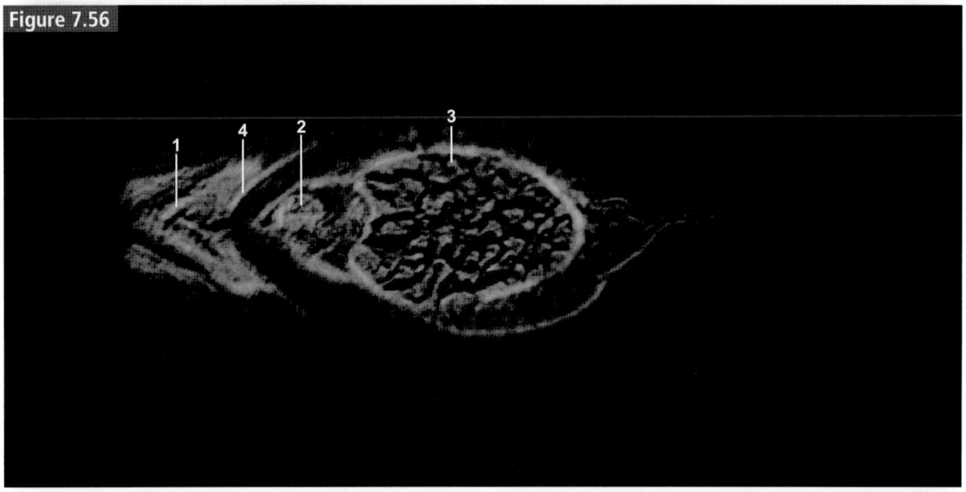

Figure 7.57

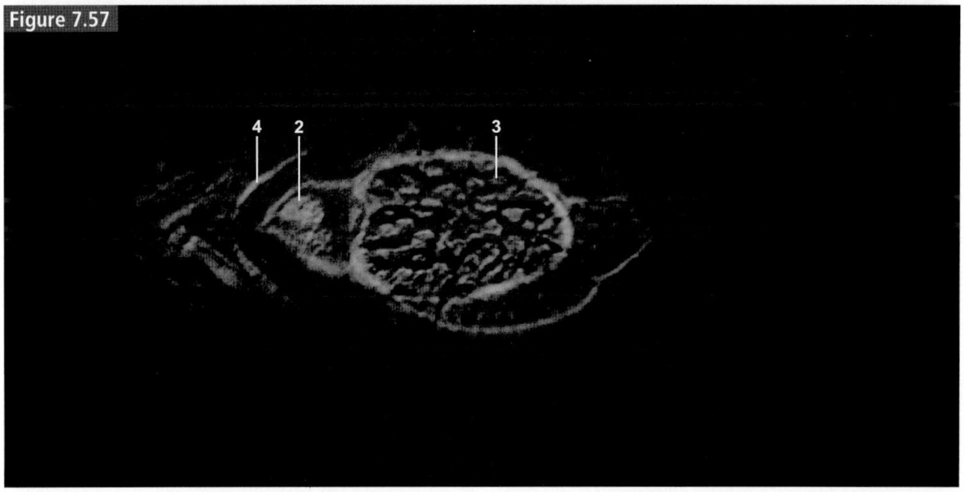

1. ceratohyal	**2.** cardiac ventricle	**3.** yolk	**4.** branchiostegal ray

Transverse Sections

Figure 7.58

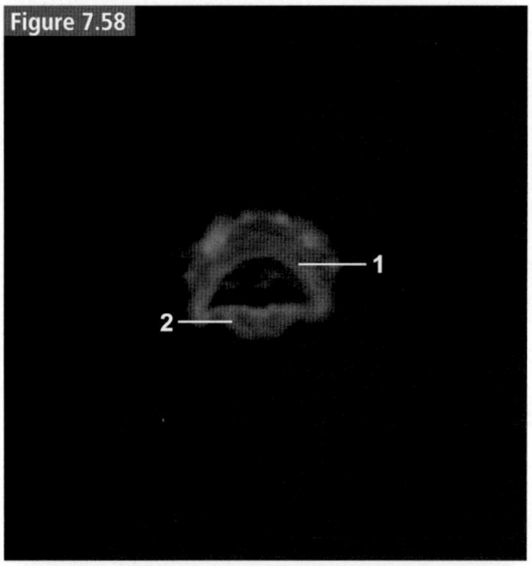

Figure 7.59

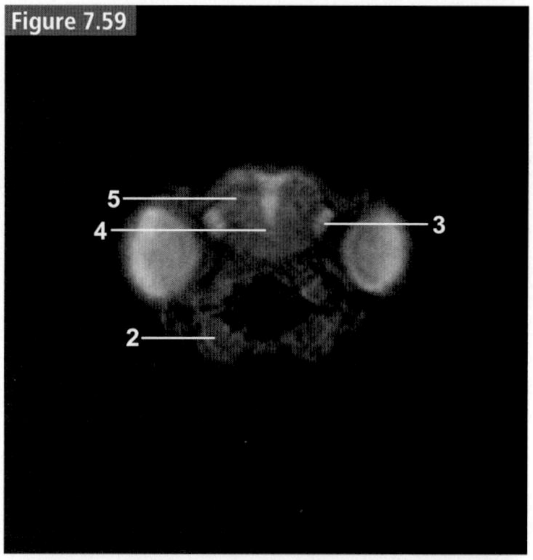

Figure 7.60

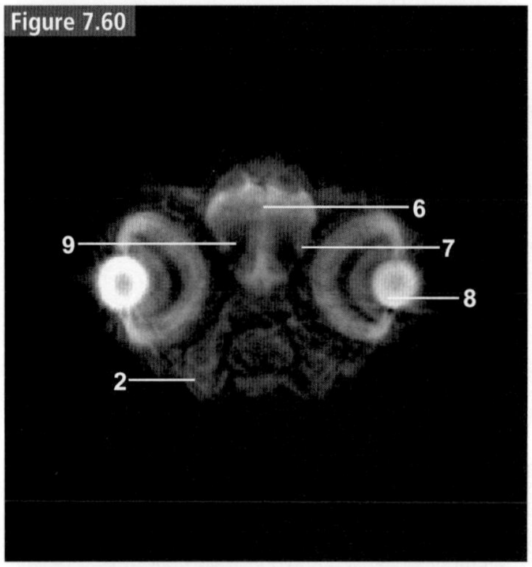

Figure 7.61

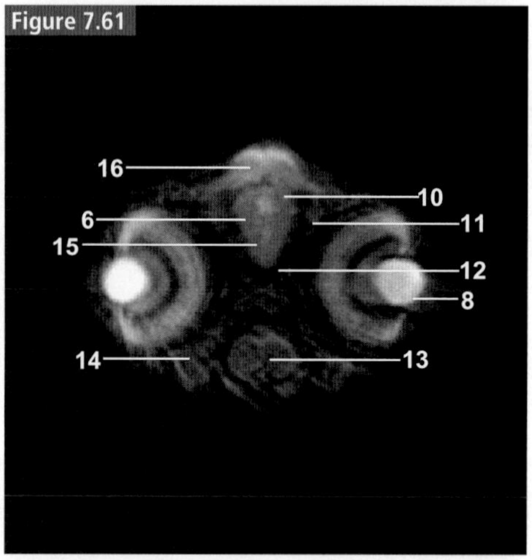

Figure 7.62

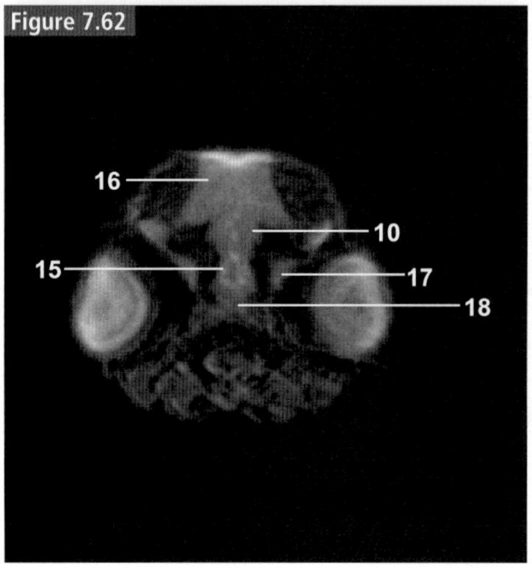

Figure 7.63

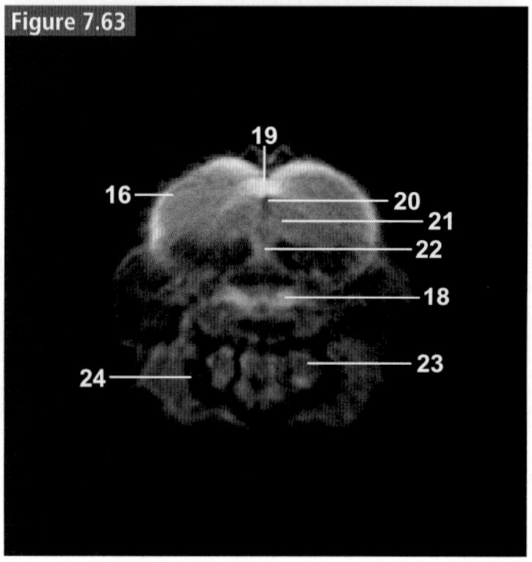

1. dorsal mandibular arch	**7.** migrated area of eminentia thalami	**13.** ceratohyal	**19.** cerebellum
2. ventral mandibular arch	**8.** lens	**14.** palatoquadrate cartilage	**20.** tectal ventricle
3. olfactory epithelium	**9.** anterior commissure	**15.** posterior tuberculum	**21.** medulla oblongata
4. subpallium	**10.** pretectum	**16.** optic tectum	**22.** floor plate
5. pallium	**11.** migrated pretectal area	**17.** migrated posterior tubercular area	**23.** gill arch
6. thalamus	**12.** postoptic commissure	**18.** hypothalamus	**24.** opercular cavity

Figure 7.64

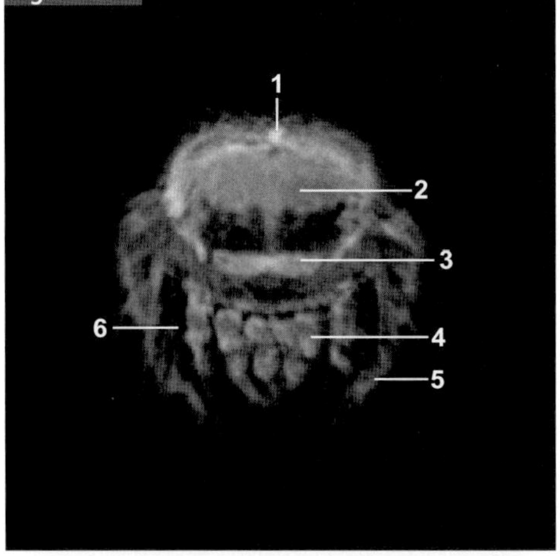

Figure 7.65

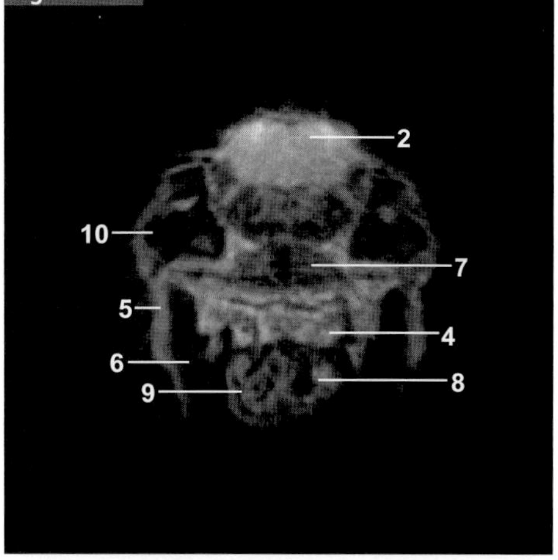

Figure 7.66

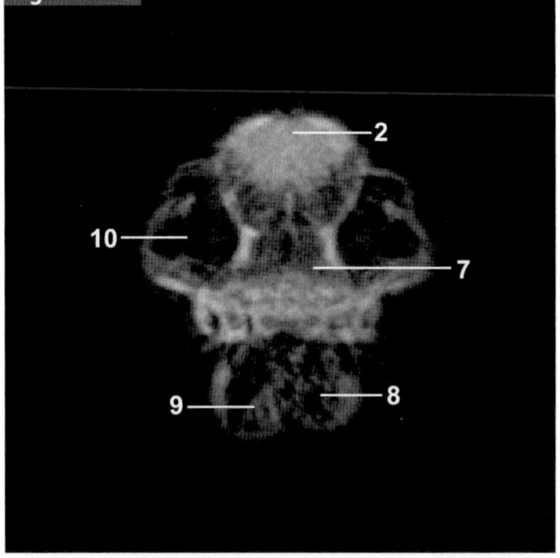

Figure 7.67

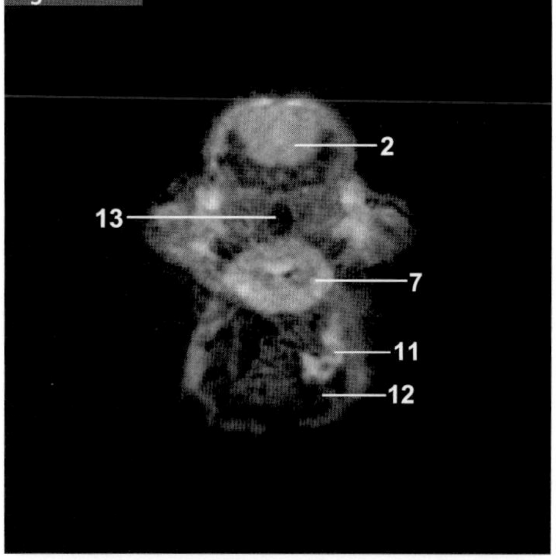

Figure 7.68

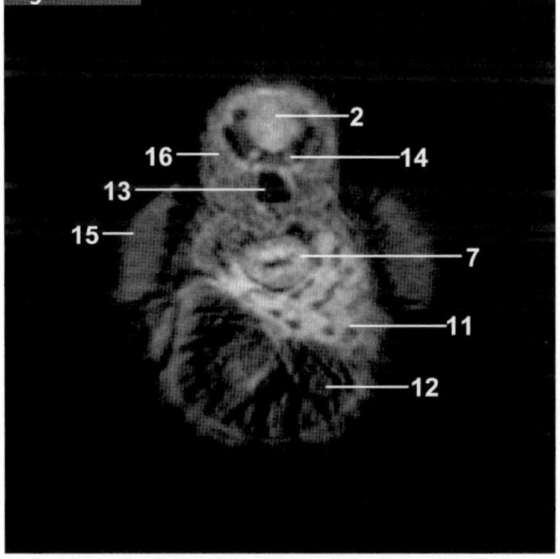

Figure 7.69

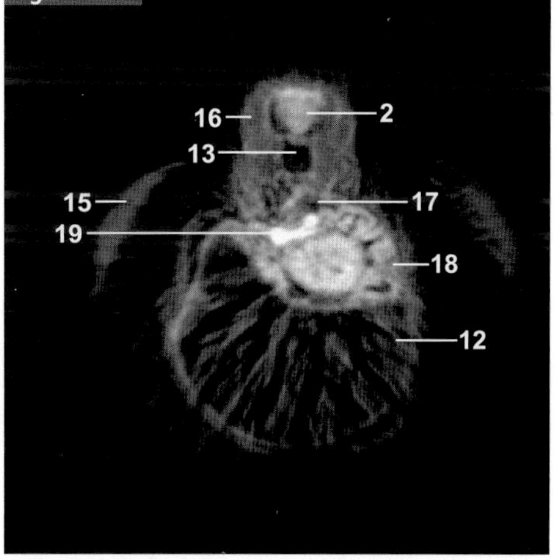

1. cerebellum	**6.** opercular cavity	**11.** liver	**16.** myotome
2. medulla oblongata	**7.** esophagus	**12.** yolk	**17.** swim bladder
3. hypothalamus	**8.** atrium	**13.** notochord	**18.** intestinal bulb
4. gill arch	**9.** cardiac ventricle	**14.** inferior reticular formation	**19.** pancreas
5. opercle	**10.** otic vesicle	**15.** pectoral fin	

Figure 7.70

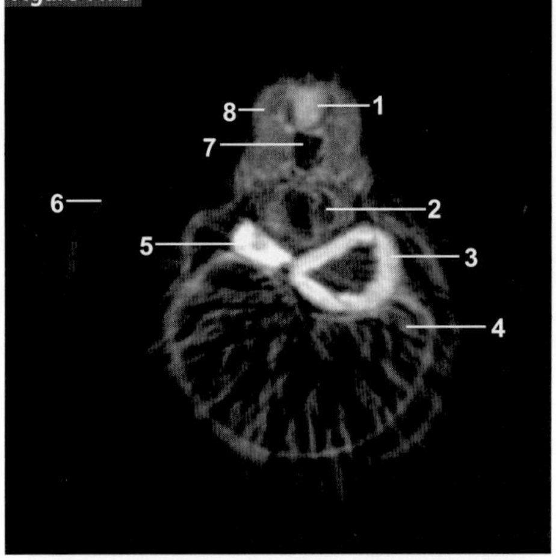

Figure 7.71

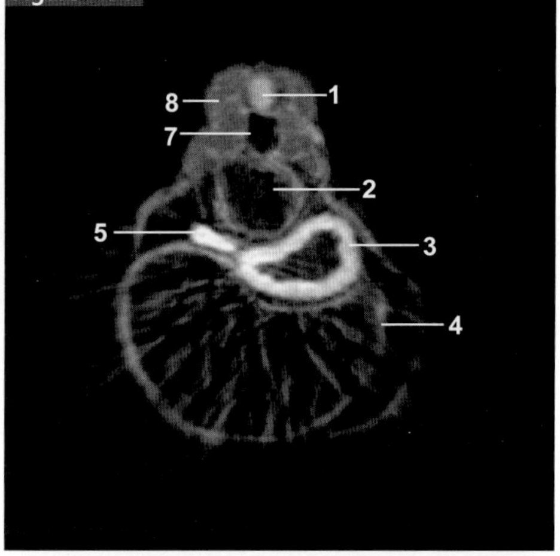

Figure 7.72

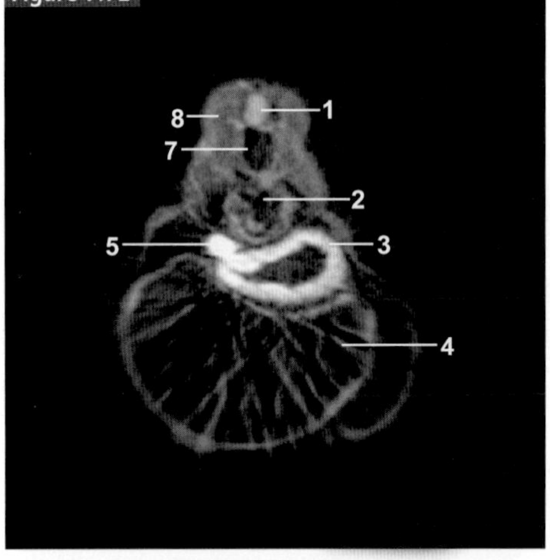

Figure 7.73

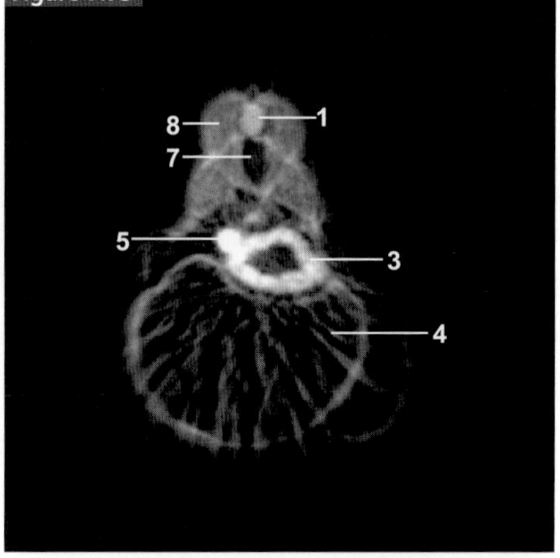

Figure 7.74

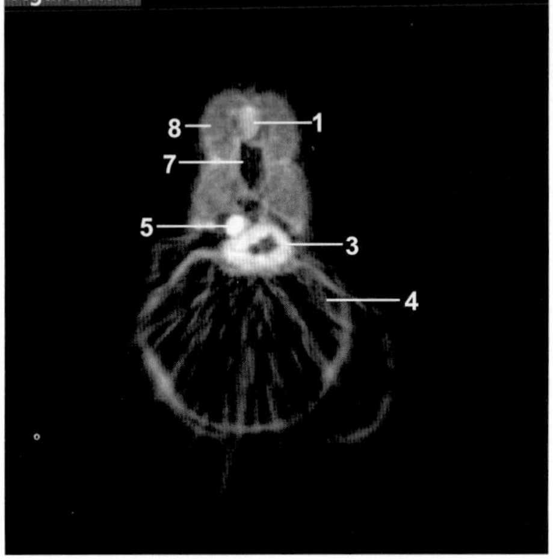

Figure 7.75

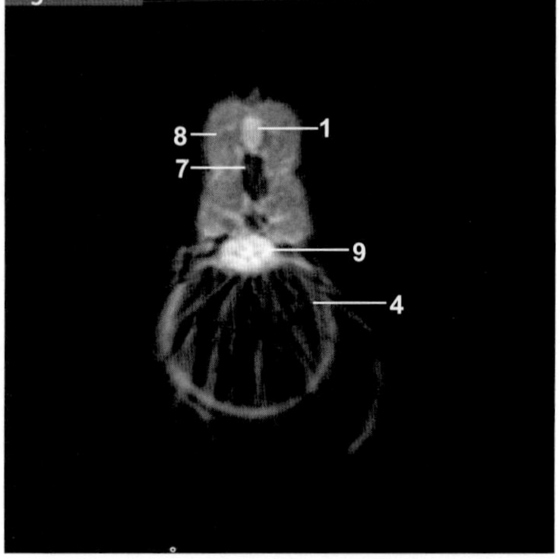

1. spinal cord **2.** swim bladder **3.** intestinal bulb **4.** yolk **5.** pancreas **6.** pectoral fin **7.** notochord **8.** myotome **9.** mid intestine

Figure 7.76

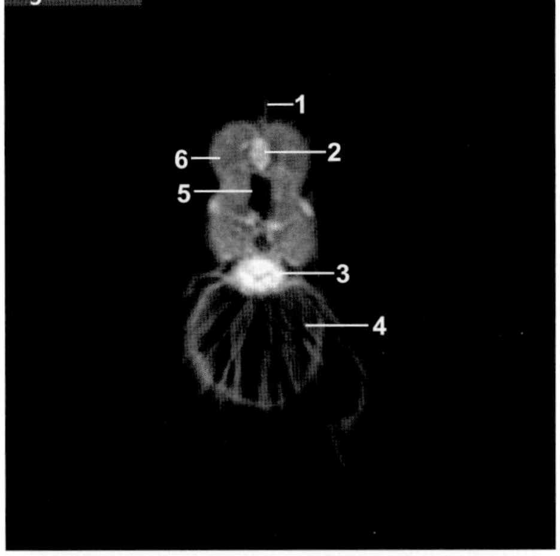

Figure 7.77

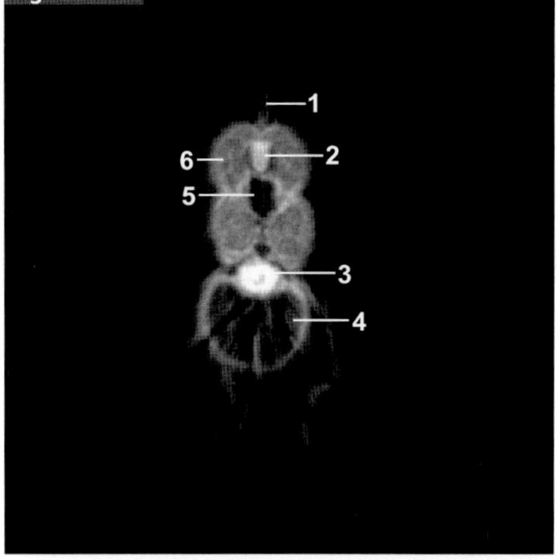

Figure 7.78

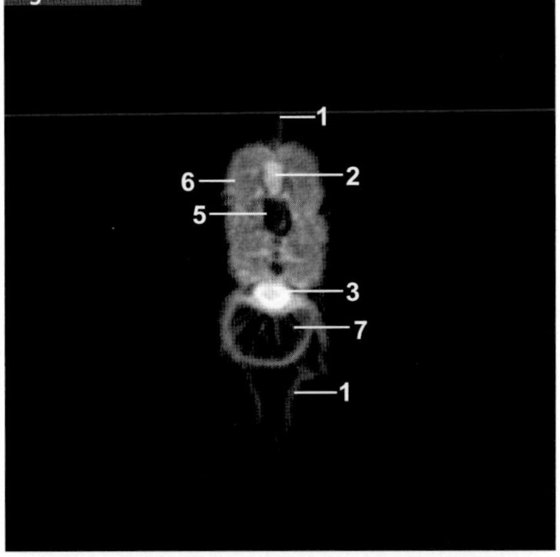

Figure 7.79

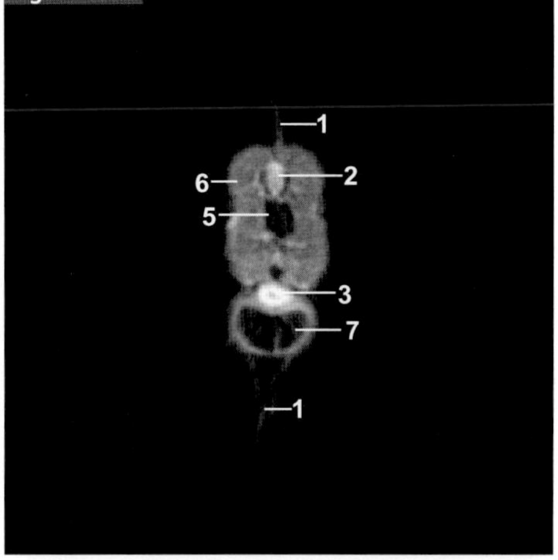

Figure 7.80

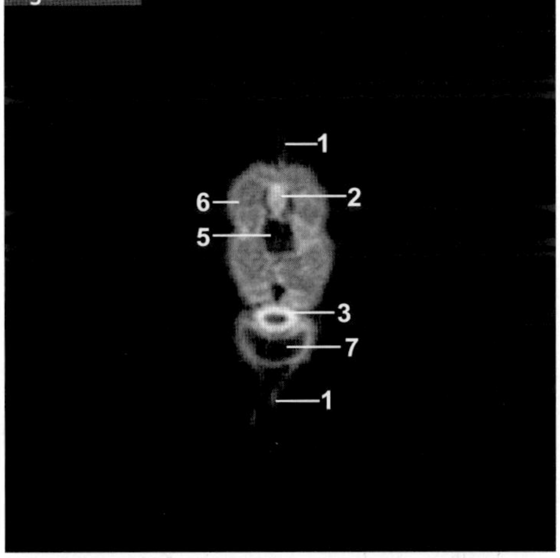

Figure 7.81

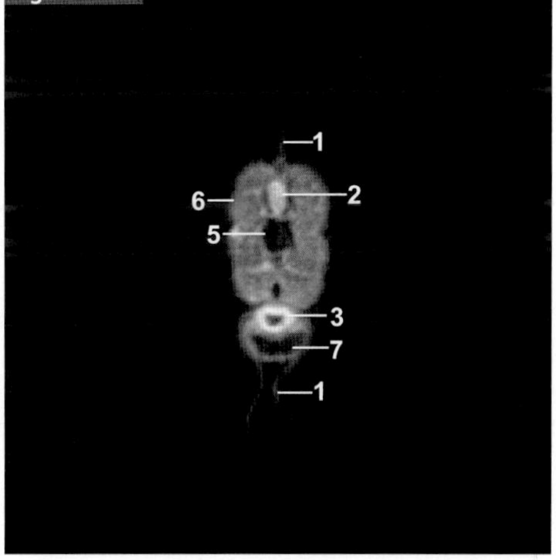

1. median fin fold	**2.** spinal cord	**3.** posterior intestine	**4.** yolk	**5.** notochord	**6.** myotome	**7.** yolk extension

CHAPTER 8

120 HOURS

Sagittal Sections

Figure 8.1

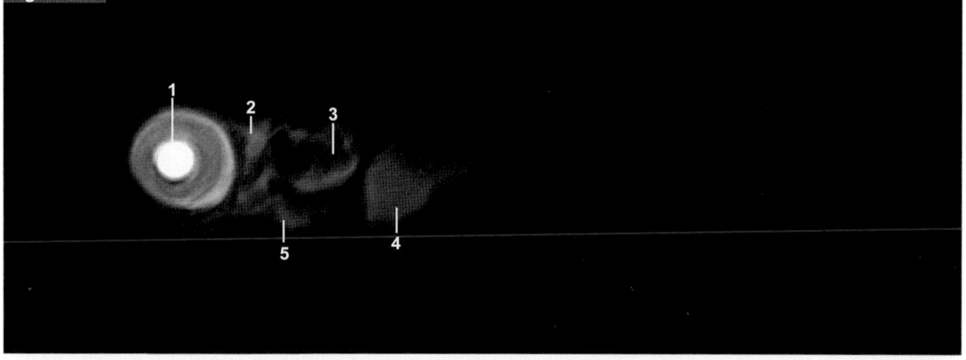

Figure 8.2

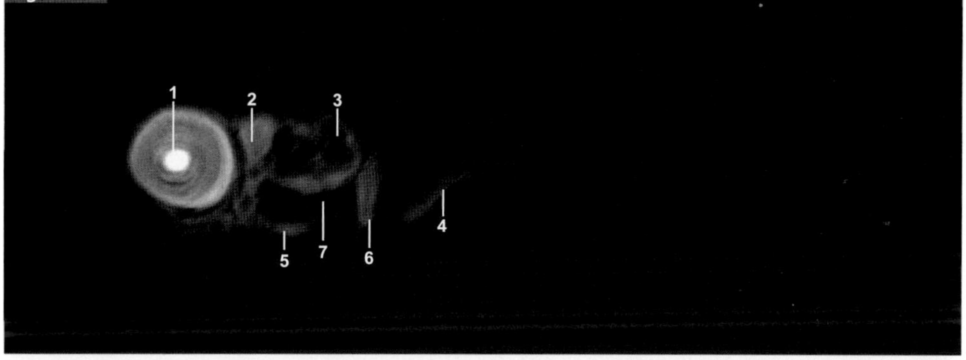

Figure 8.3

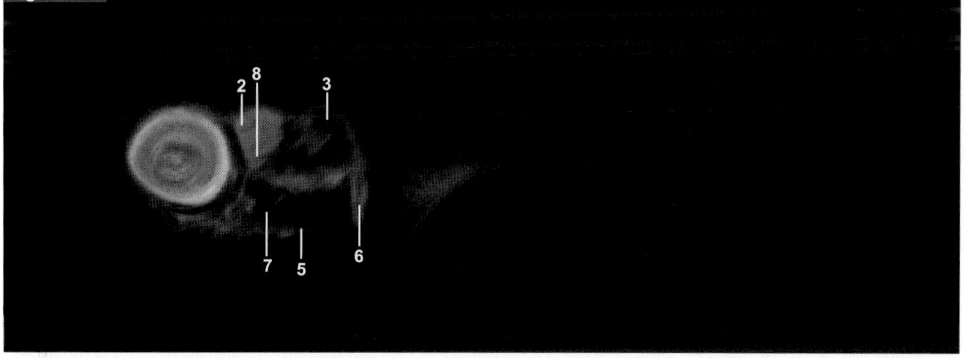

1. lens **2.** optic tectum **3.** semicircular canal **4.** pectoral fin **5.** opercle **6.** pectoral fin musculature **7.** opercular cavity **8.** torus lateralis

Atlas of Zebrafish Development. 10.1016/B978-0-12-374016-8.00008-X

Figure 8.4

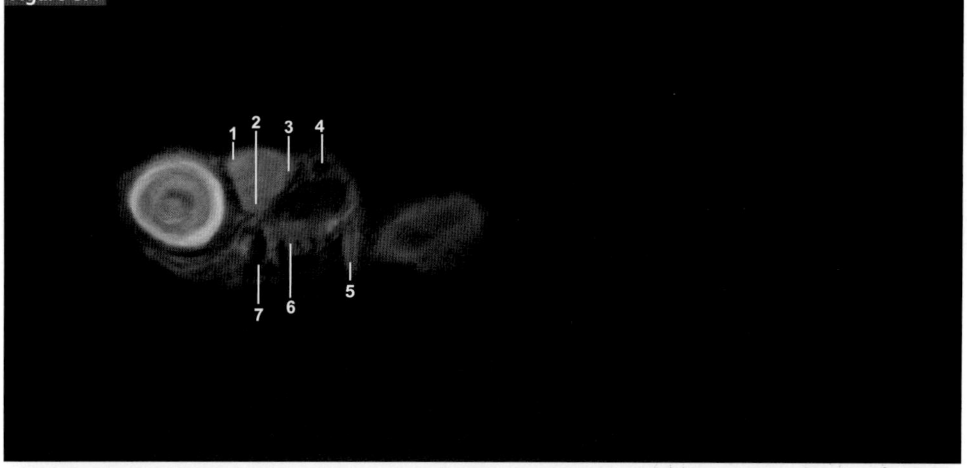

Figure 8.5

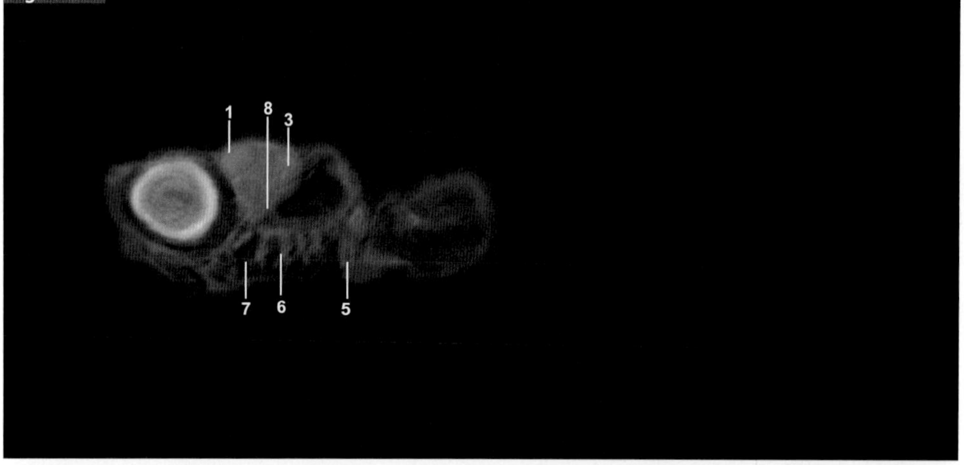

Figure 8.6

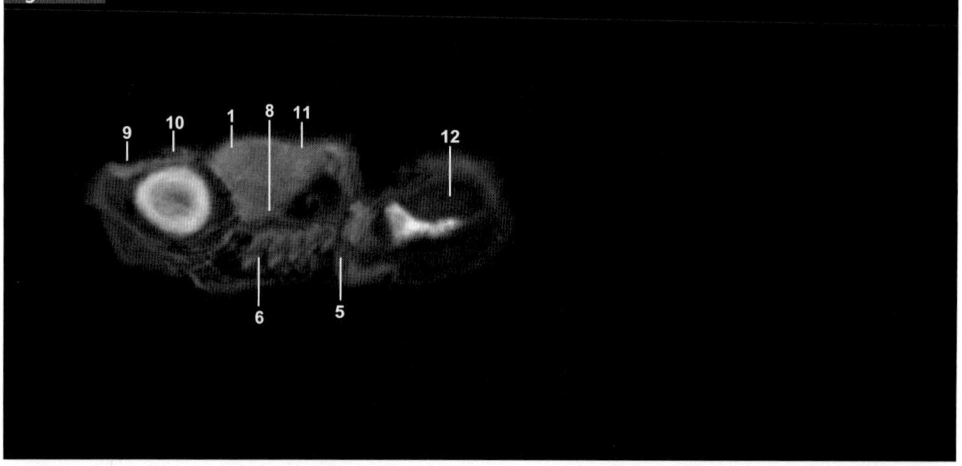

1. optic tectum	**3.** granular eminence	**5.** pectoral fin musculature	**7.** opercular cavity	**9.** olfactory pit	**11.** cerebellum
2. torus lateralis	**4.** semicircular canal	**6.** gill filaments	**8.** inferior lobe	**10.** telencephalon	**12.** swim bladder

Figure 8.7

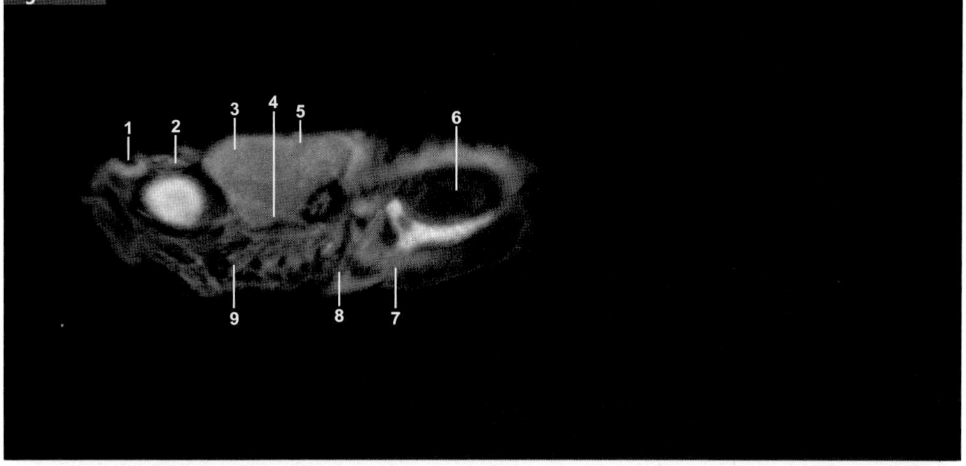

Figure 8.8

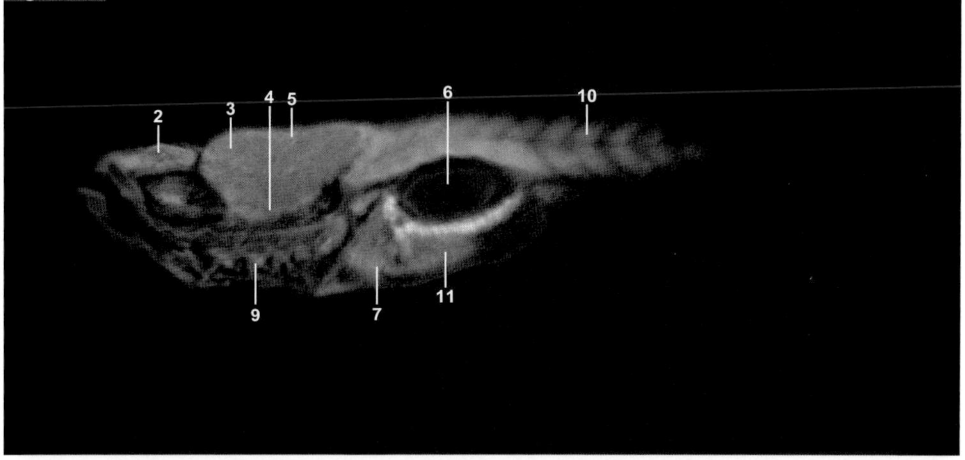

Figure 8.9

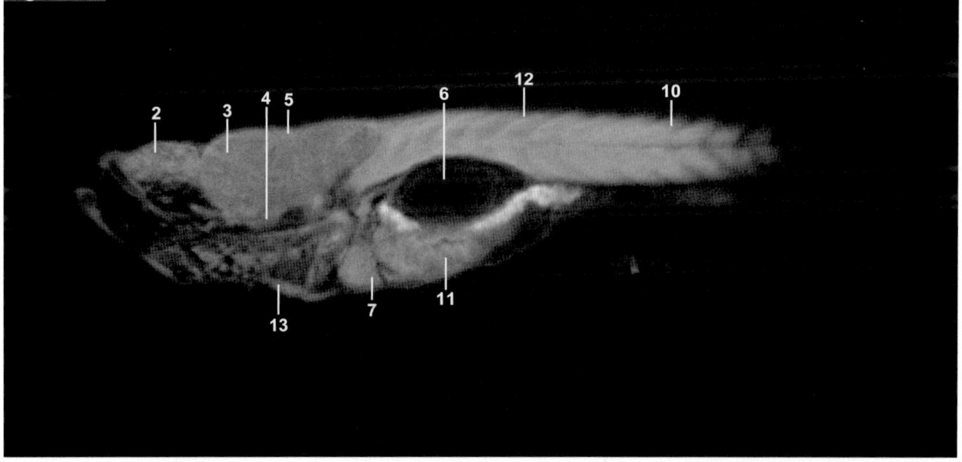

1. olfactory pit	**4.** inferior lobe	**7.** liver	**10.** myotome	**13.** sternohyoid
2. telencephalon	**5.** cerebellum	**8.** pectoral fin musculature	**11.** intestinal bulb	
3. optic tectum	**6.** swim bladder	**9.** gill filaments	**12.** vertical myoseptum	

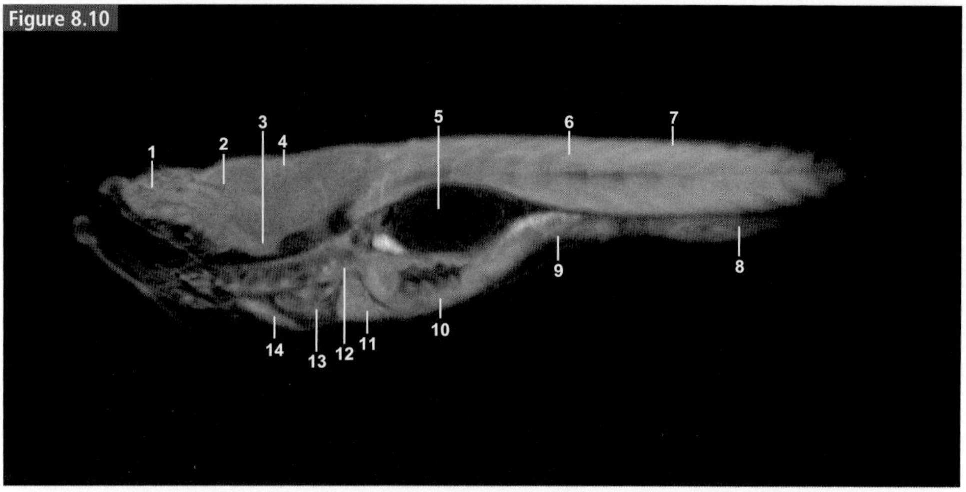

Figure 8.10

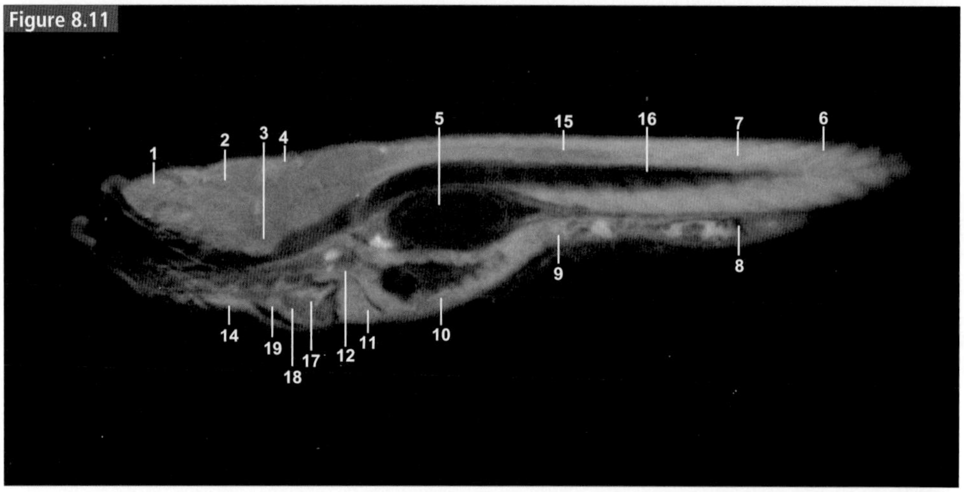

Figure 8.11

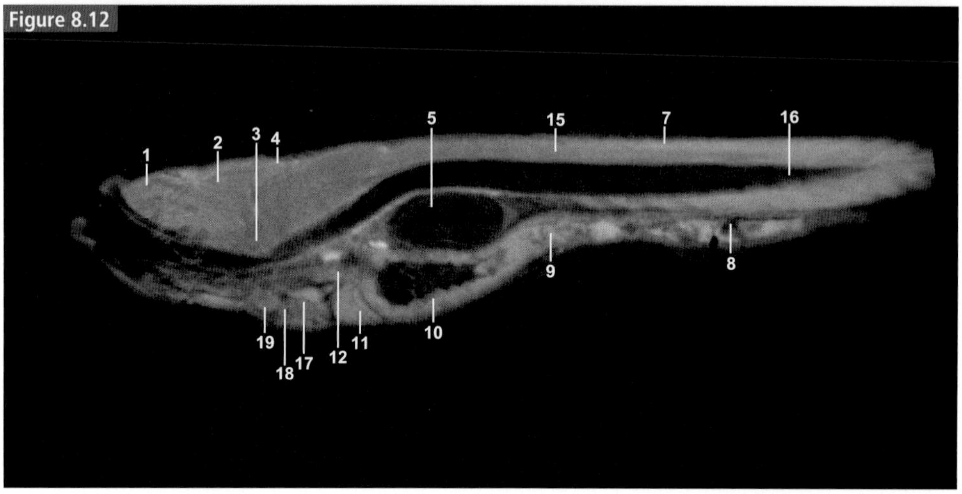

Figure 8.12

1. telencephalon	**5.** swim bladder	**9.** mid intestine	**13.** cardiac ventricle	**17.** atrium
2. optic tectum	**6.** vertical myoseptum	**10.** intestinal bulb	**14.** sternohyoid	**18.** cardiac ventricle
3. inferior lobe	**7.** myotome	**11.** liver	**15.** spinal cord	**19.** bulbus arteriosus
4. cerebellum	**8.** posterior intestine	**12.** esophagus	**16.** notochord	

Figure 8.13

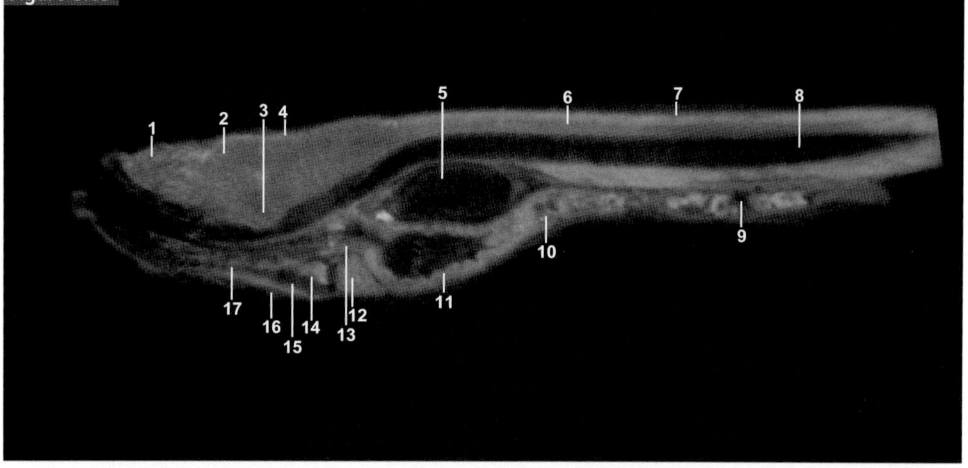

Figure 8.14

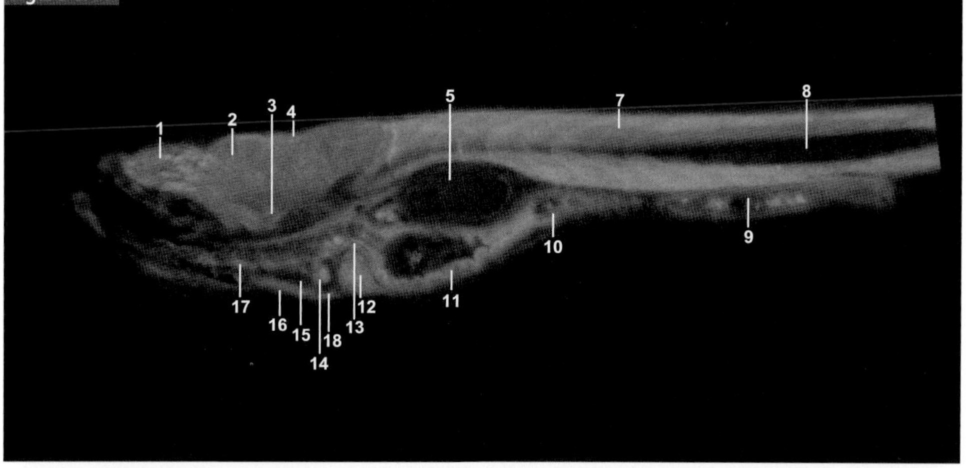

Figure 8.15

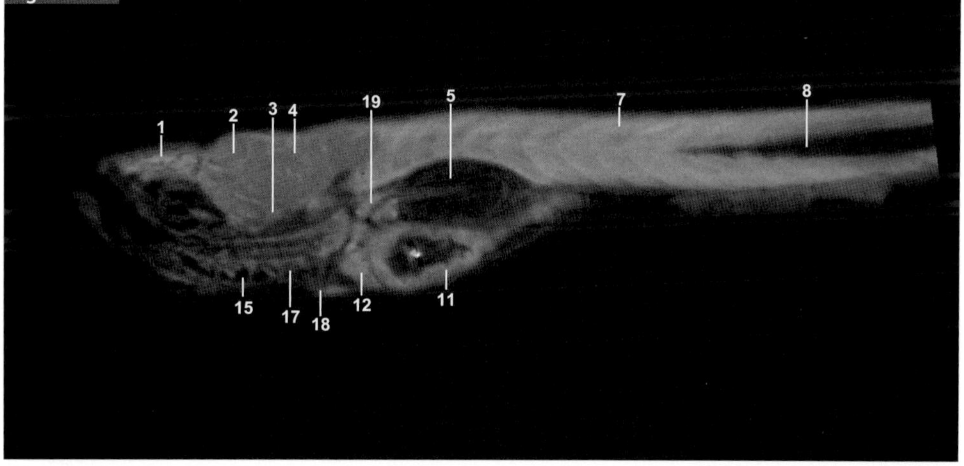

1. telencephalon	**5.** swim bladder	**9.** posterior intestine	**13.** esophagus	**17.** gill filaments
2. optic tectum	**6.** spinal cord	**10.** mid intestine	**14.** atrium	**18.** pectoral fin musculature
3. inferior lobe	**7.** myotome	**11.** intestinal bulb	**15.** opercular cavity	**19.** pronephros
4. cerebellum	**8.** notochord	**12.** liver	**16.** sternohyoid	

Figure 8.16

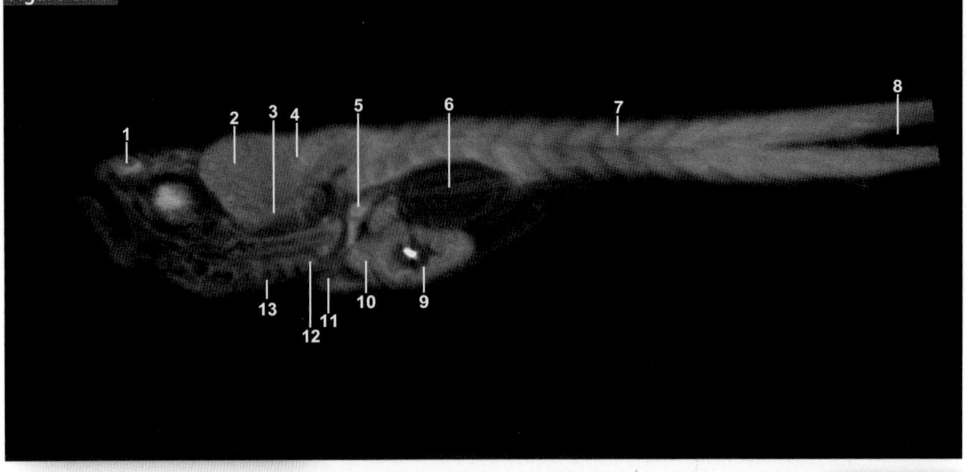

Figure 8.17

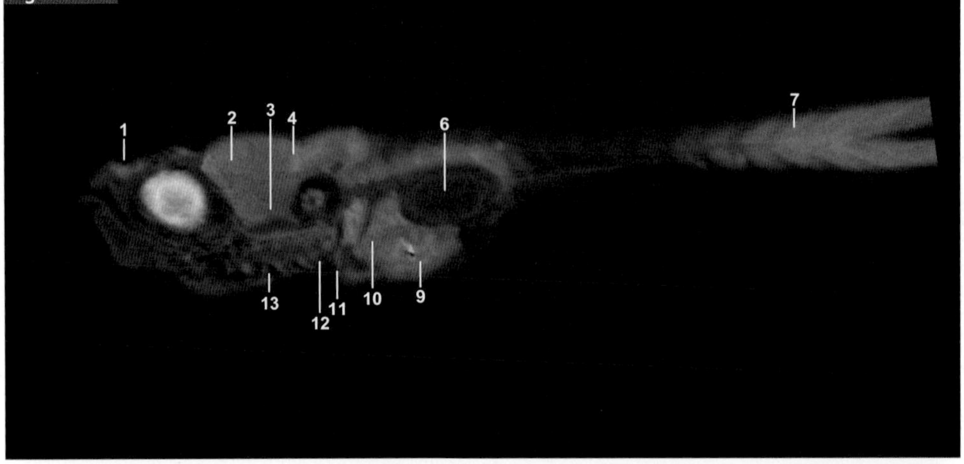

Figure 8.18

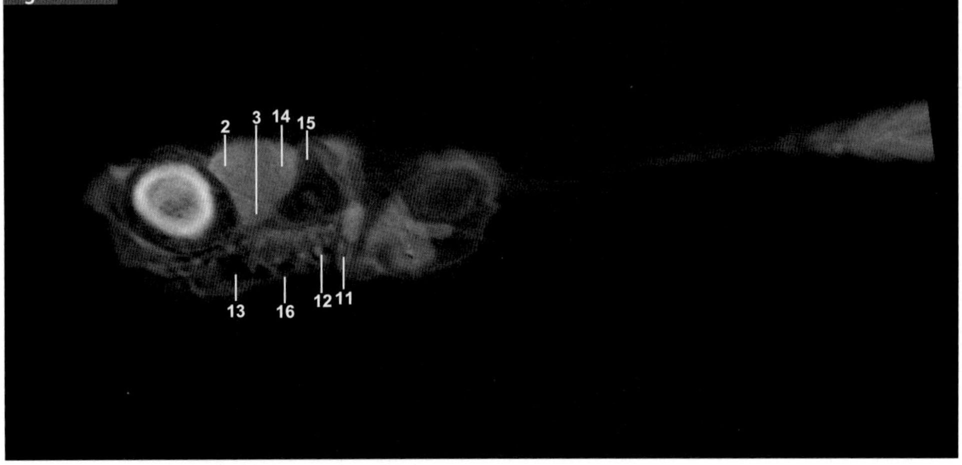

1. olfactory pit	**5.** pronephros	**9.** intestinal bulb	**13.** opercular cavity
2. optic tectum	**6.** swim bladder	**10.** liver	**14.** granular eminence
3. inferior lobe	**7.** myotome	**11.** pectoral fin musculature	**15.** semicircular canal
4. cerebellum	**8.** vertebral column	**12.** gill filaments	**16.** opercle

Figure 8.19

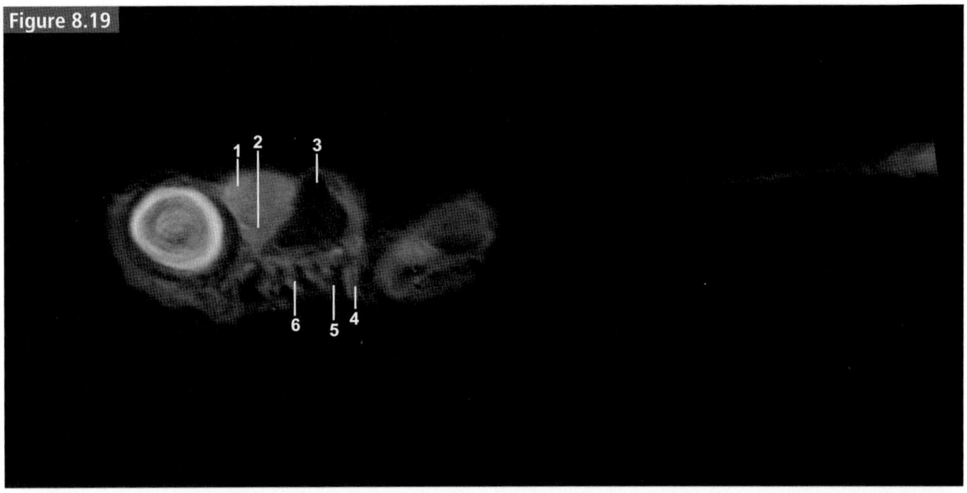

Figure 8.20

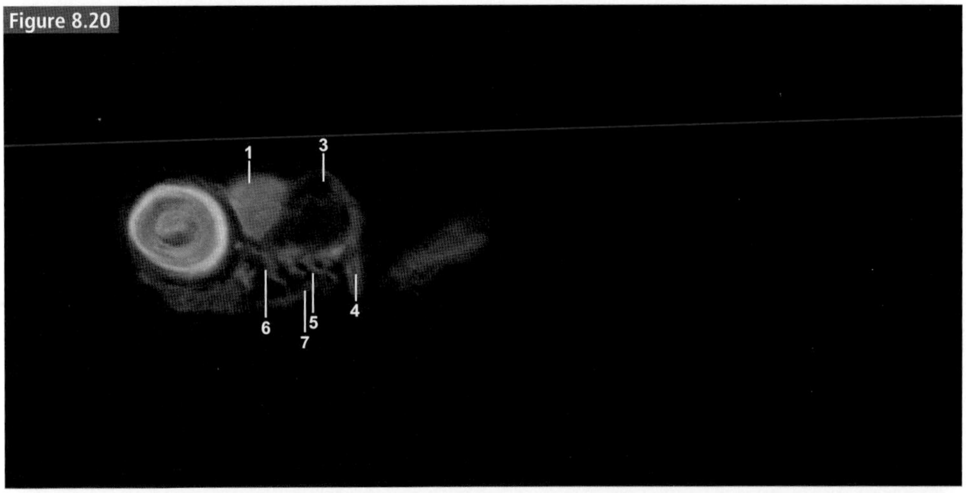

Figure 8.21

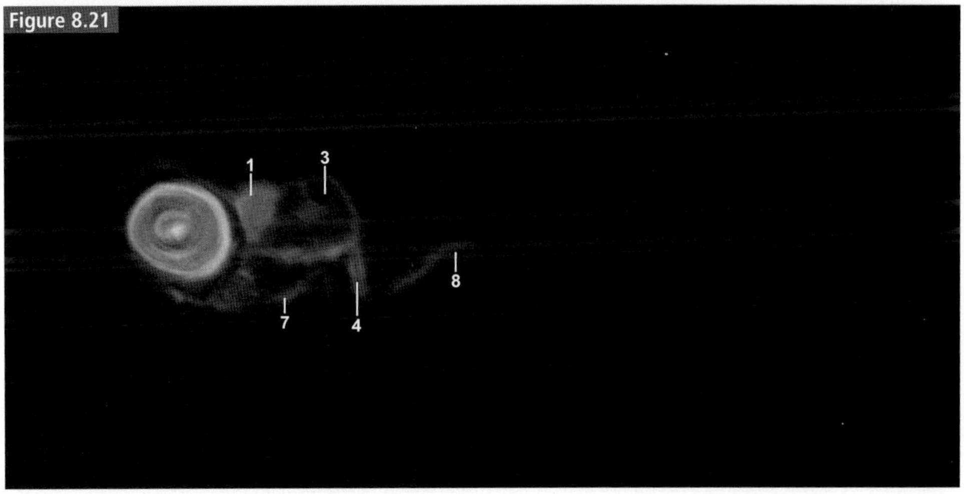

1. optic tectum	**3.** semicircular canal	**5.** opercular cavity	**7.** opercle
2. torus lateralis	**4.** pectoral fin musculature	**6.** gill filaments	**8.** pectoral fin

Figure 8.22

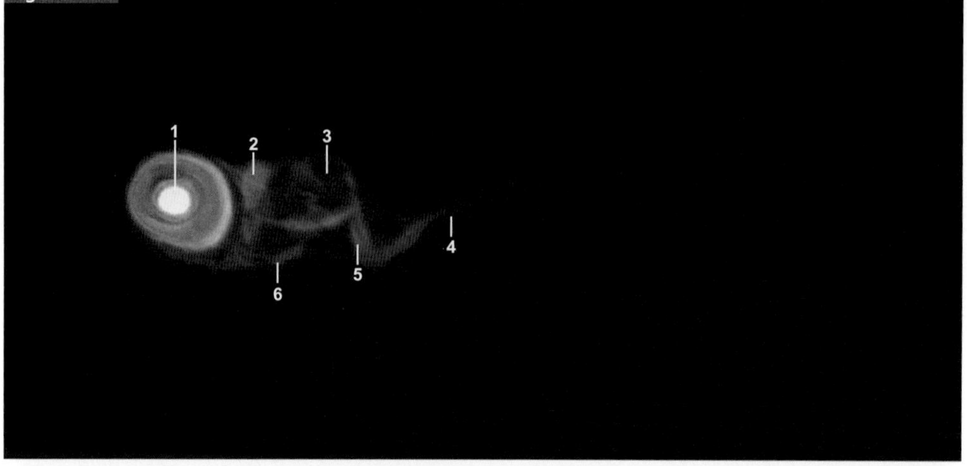

Figure 8.23

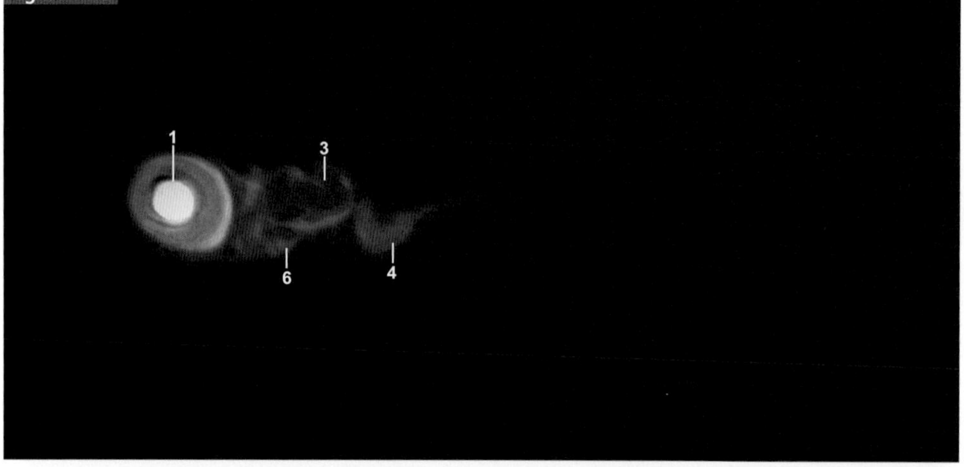

Figure 8.24

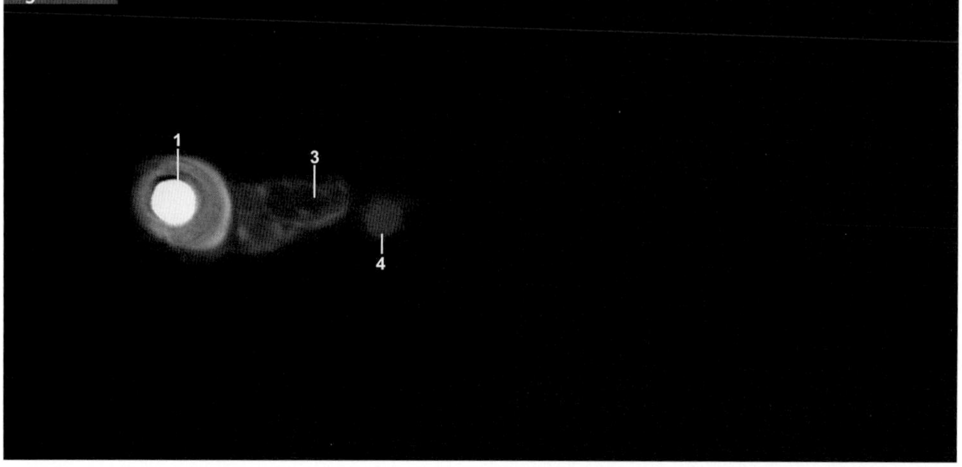

1. lens	**2.** optic tectum	**3.** semicircular canal	**4.** pectoral fin	**5.** pectoral fin musculature	**6.** opercle

Coronal Sections

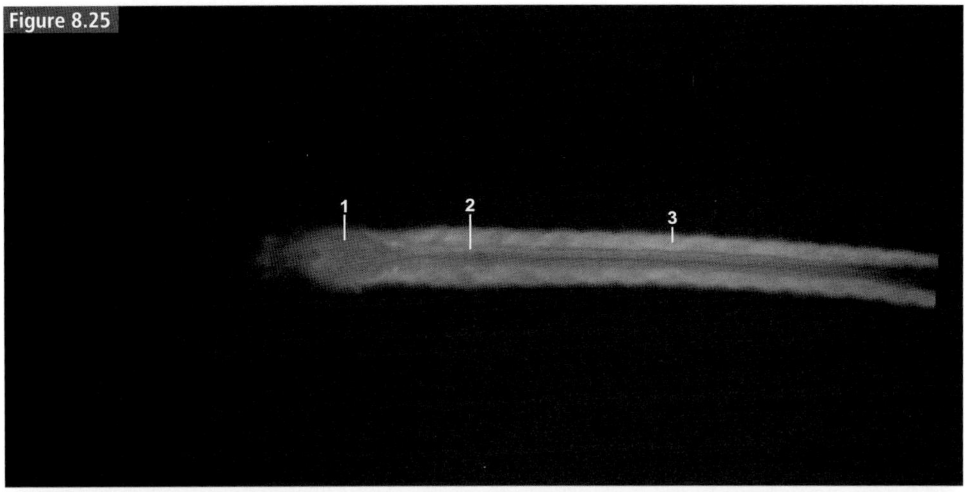

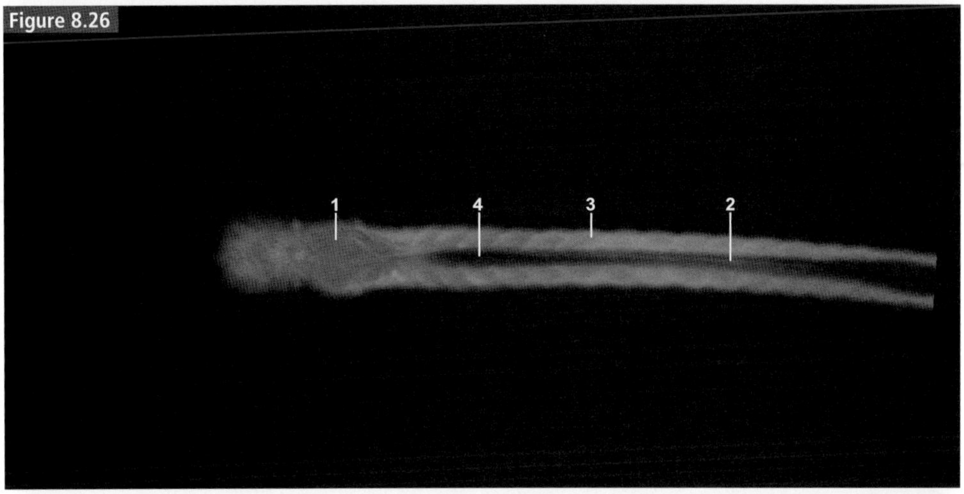

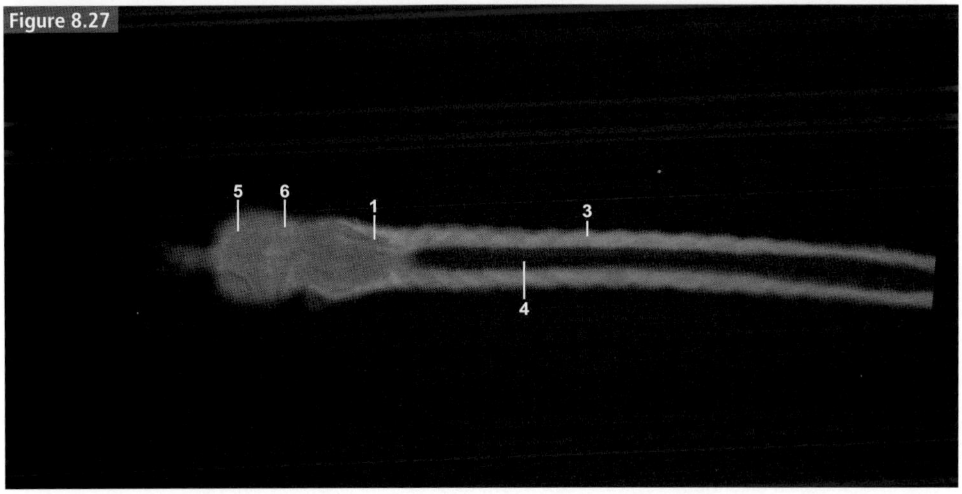

1. medulla oblongata	**2.** spinal cord	**3.** myotome	**4.** notochord	**5.** optic tectum	**6.** cerebellum

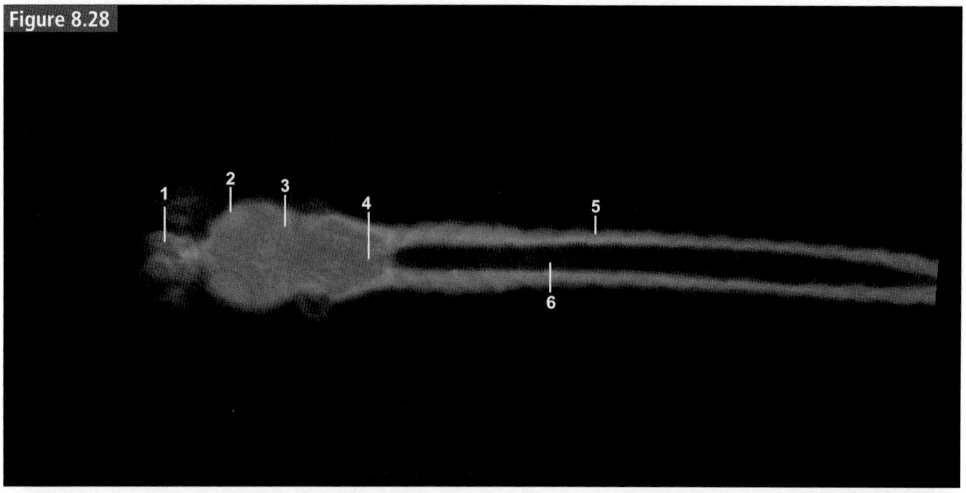

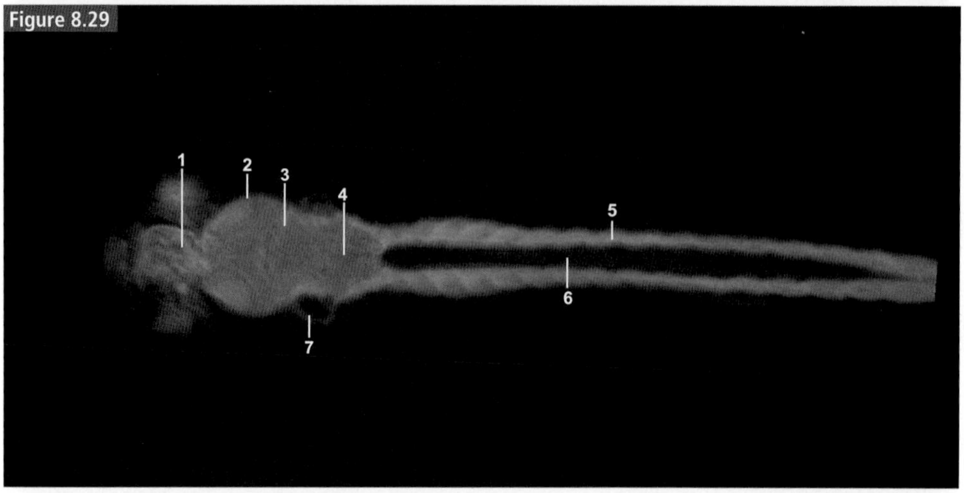

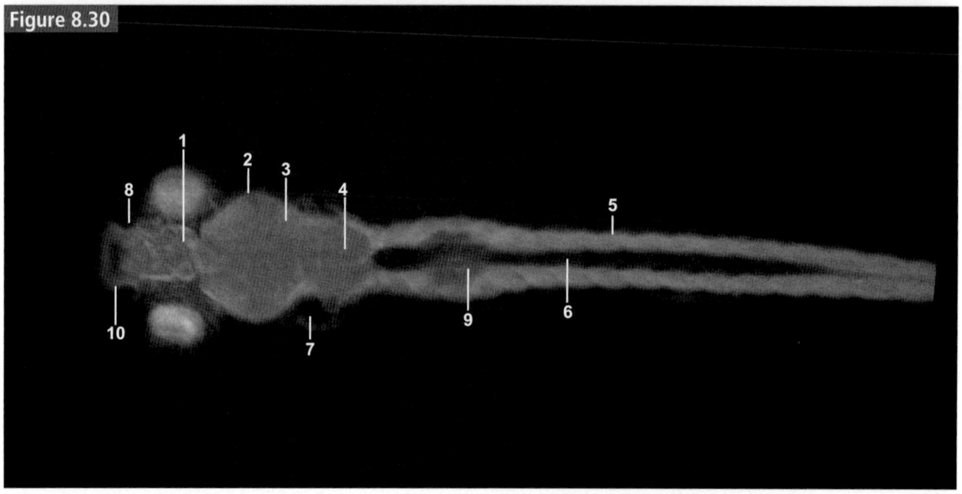

1. telencephalon	**3.** cerebellum	**5.** myotome	**7.** semicircular canal	**9.** swim bladder
2. optic tectum	**4.** medulla oblongata	**6.** notochord	**8.** olfactory pit	**10.** olfactory epithelium

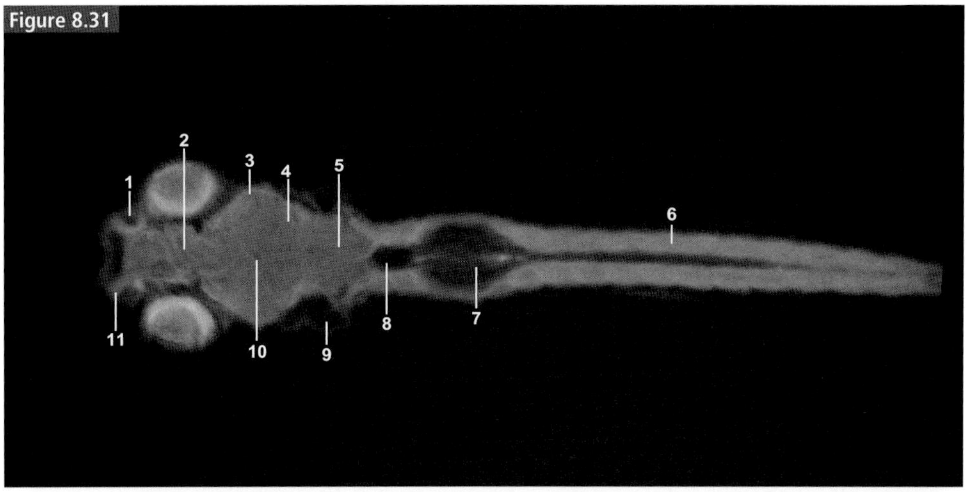

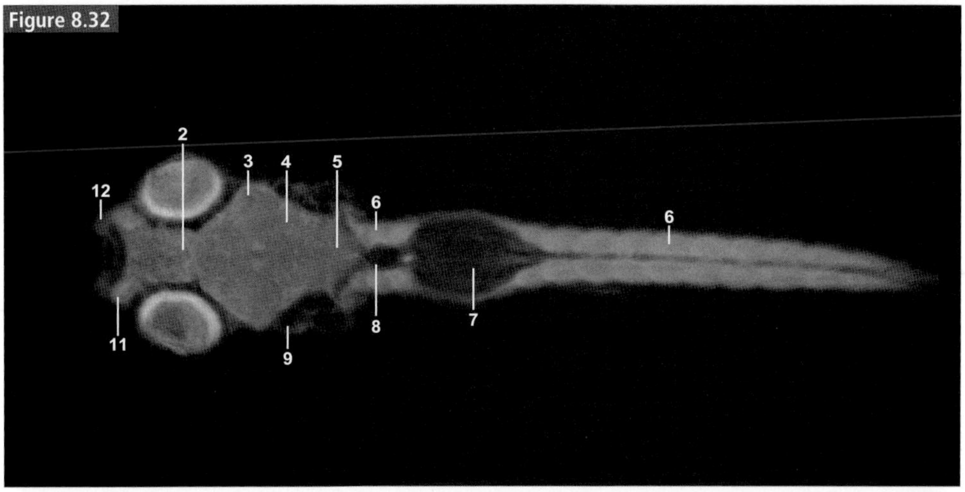

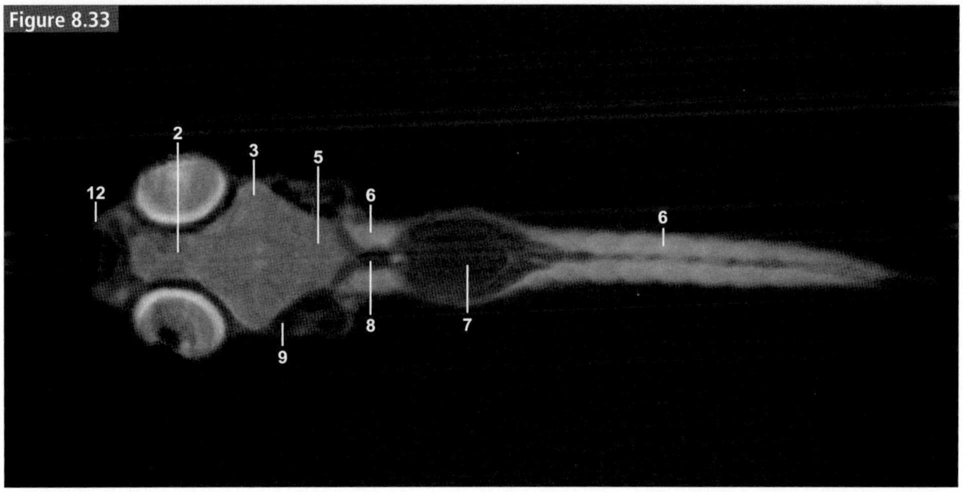

1. olfactory pit	**3.** optic tectum	**5.** medulla oblongata	**7.** swim bladder	**9.** semicircular canal	**11.** olfactory epithelium
2. telencephalon	**4.** cerebellum	**6.** myotome	**8.** notochord	**10.** tectal ventricle	**12.** dorsal mandibular arch

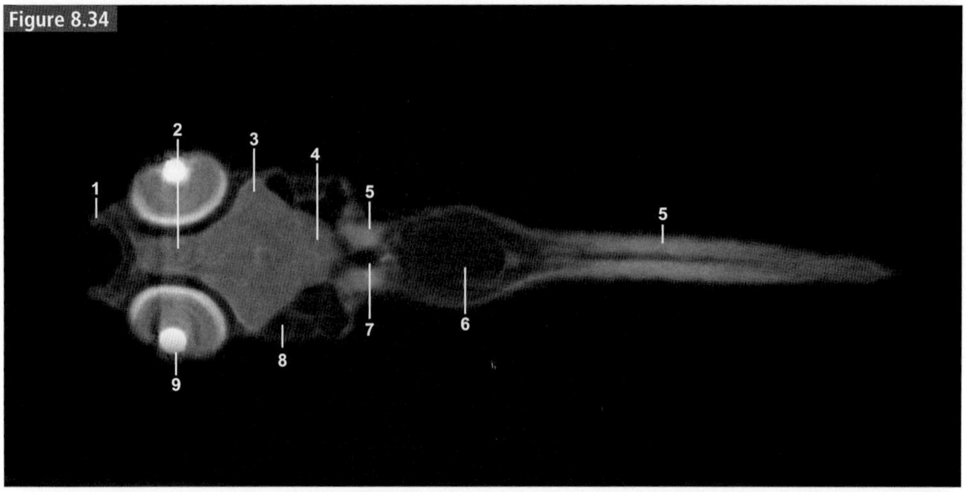

Figure 8.34

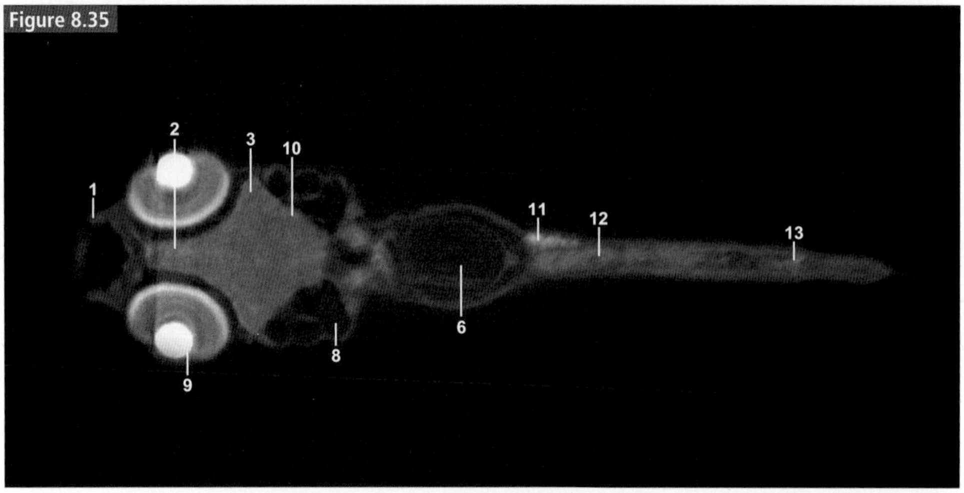

Figure 8.35

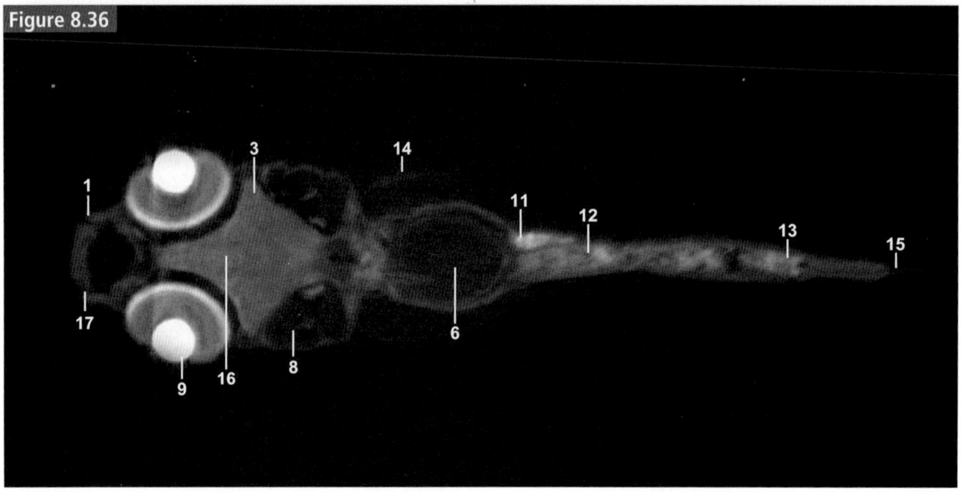

Figure 8.36

1. dorsal mandibular arch	**5.** myotome	**9.** lens	**13.** posterior intestine	**17.** ventral mandibular arch
2. telencephalon	**6.** swim bladder	**10.** granular eminence	**14.** pectoral fin	
3. optic tectum	**7.** notochord	**11.** pancreas	**15.** median fin fold	
4. medulla oblongata	**8.** semicircular canal	**12.** mid intestine	**16.** hypothalamus	

Figure 8.37

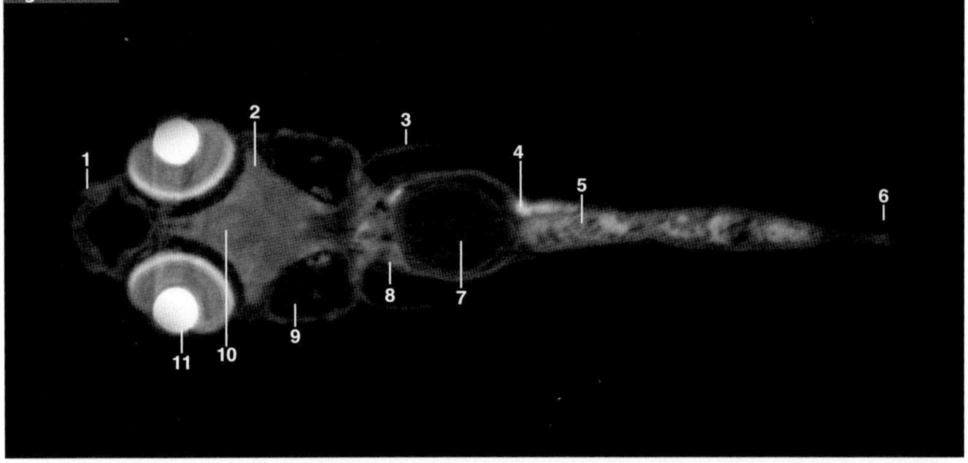

Figure 8.38

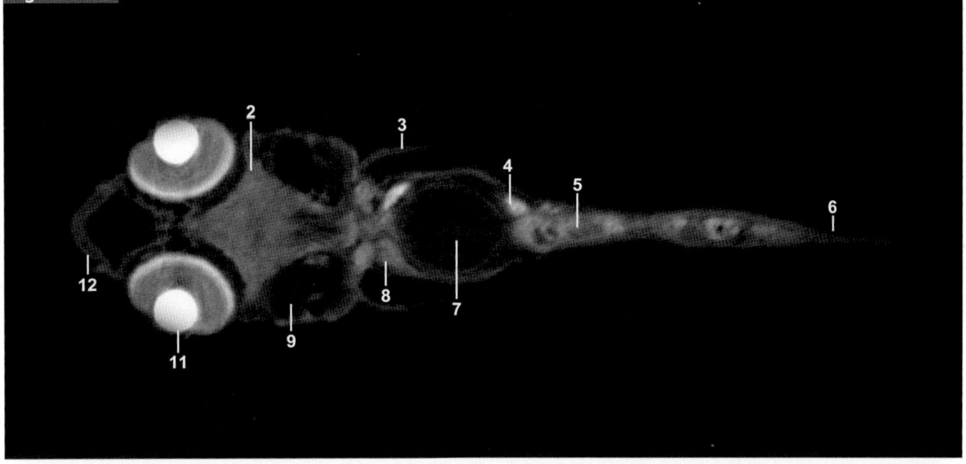

Figure 8.39

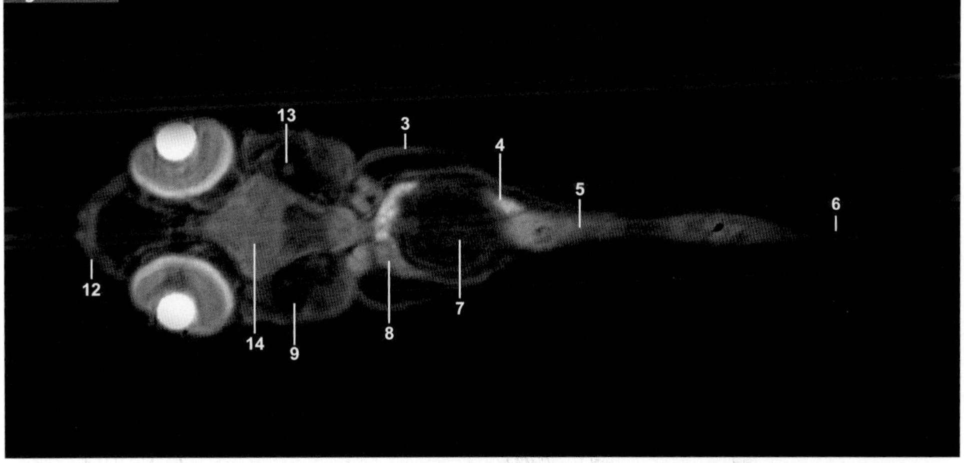

1. dorsal mandibular arch	**4.** pancreas	**7.** swim bladder	**10.** hypothalamus	**13.** otolith
2. optic tectum	**5.** mid intestine	**8.** liver	**11.** lens	**14.** inferior lobe
3. pectoral fin	**6.** median fin fold	**9.** semicircular canal	**12.** ventral mandibular arch	

Figure 8.40

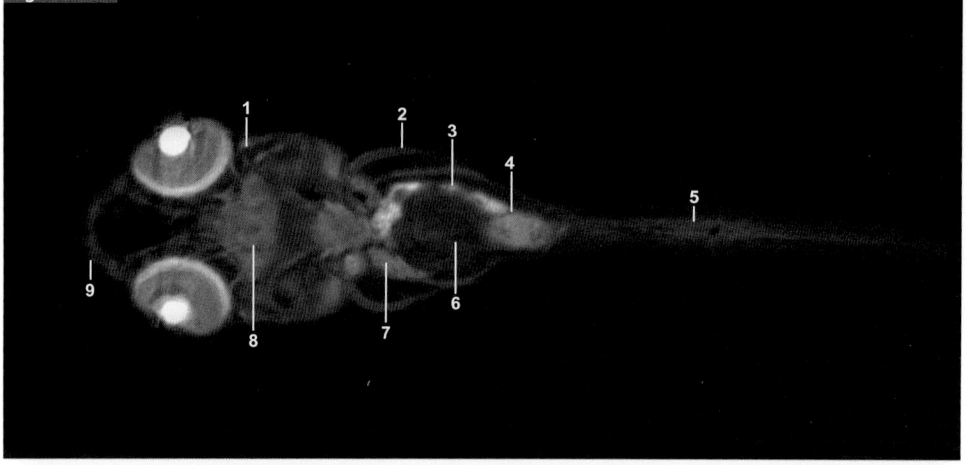

Figure 8.41

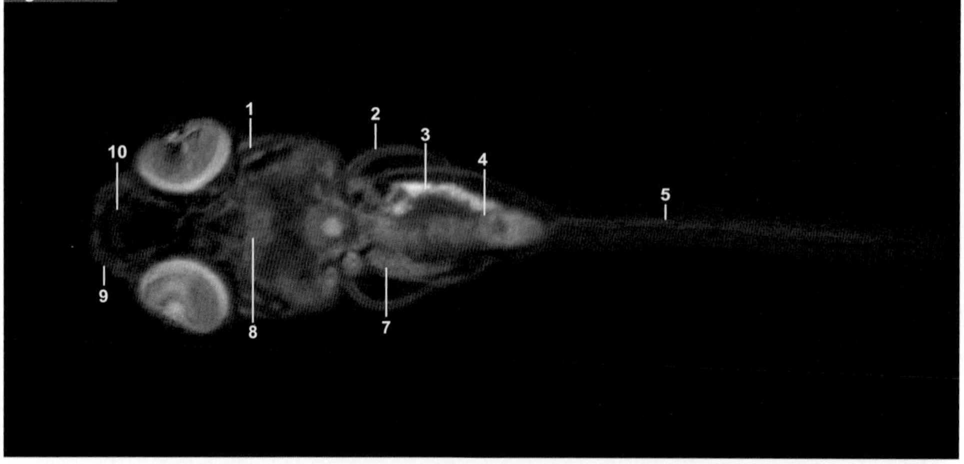

Figure 8.42

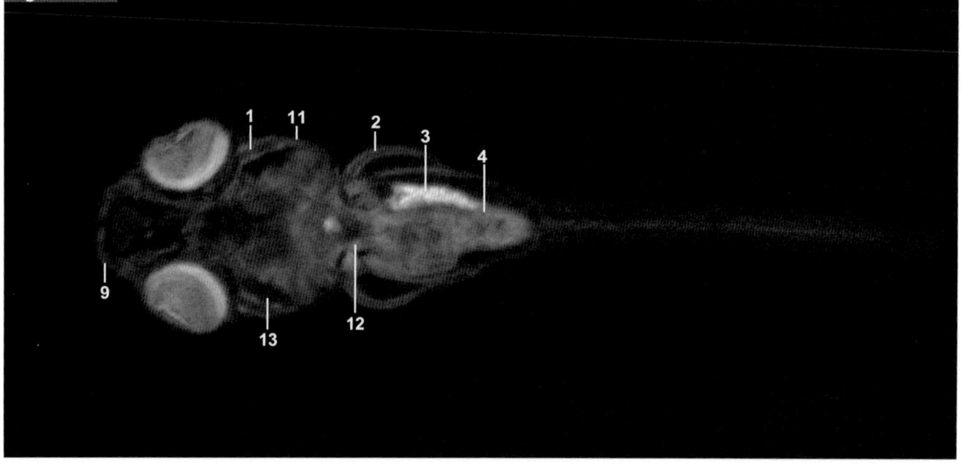

1. hyomandibula	**4.** intestinal bulb	**7.** liver	**10.** intermandibularis	**13.** opercular cavity
2. pectoral fin	**5.** median fin fold	**8.** inferior lobe	**11.** opercle	
3. pancreas	**6.** swim bladder	**9.** ventral mandibular arch	**12.** esophagus	

Figure 8.43

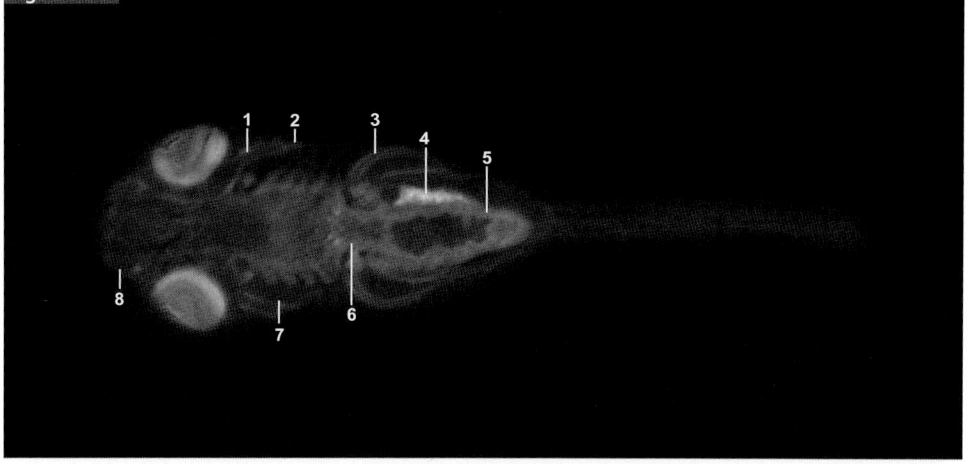

Figure 8.44

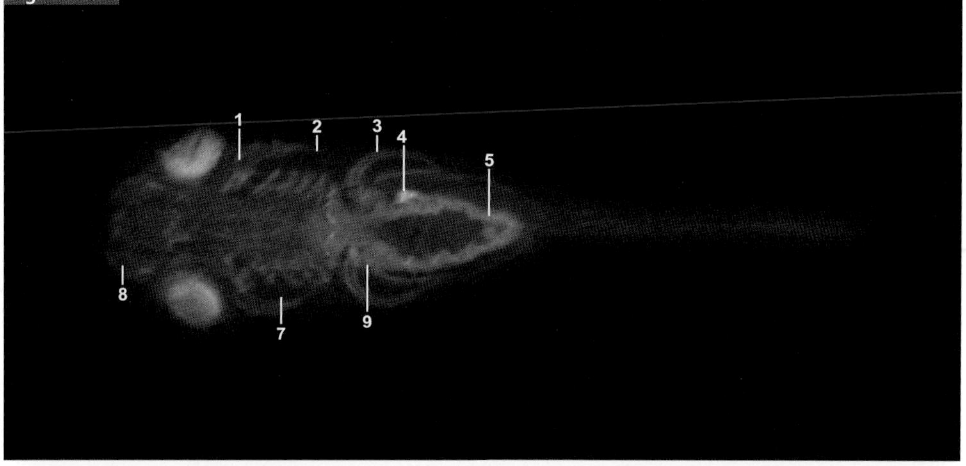

Figure 8.45

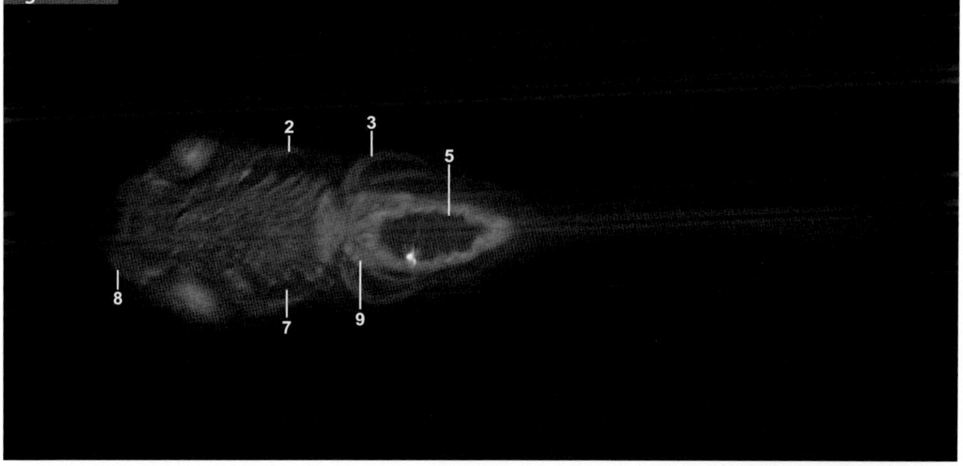

1. hyomandibula	**3.** pectoral fin	**5.** intestinal bulb	**7.** opercular cavity	**9.** liver
2. opercle	**4.** pancreas	**6.** esophagus	**8.** ventral mandibular arch	

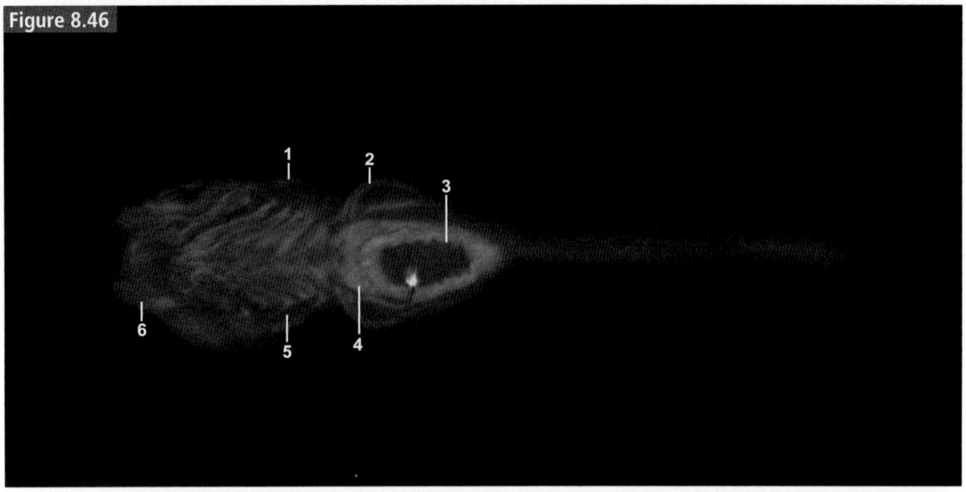

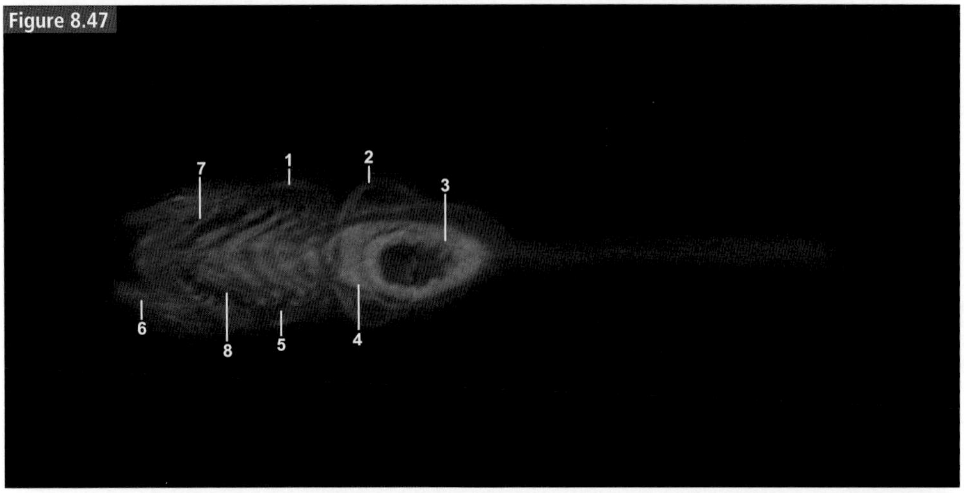

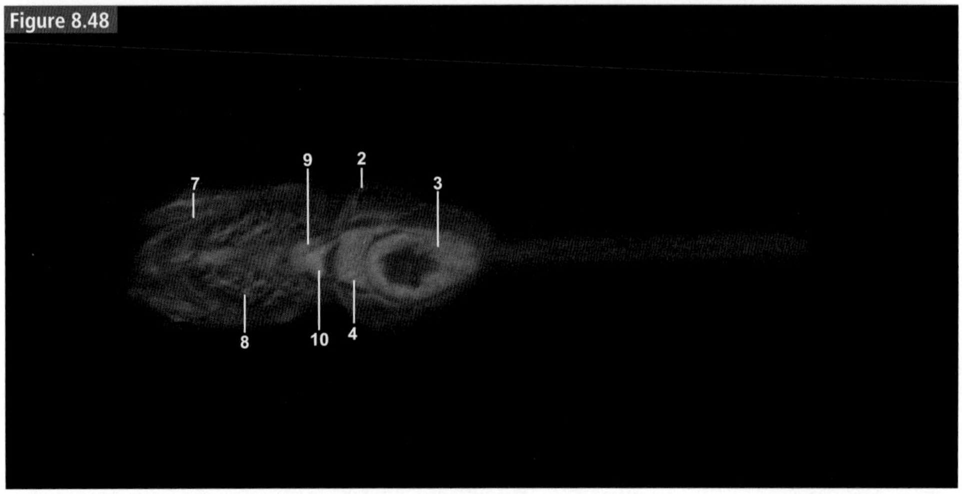

1. opercle	**3.** intestinal bulb	**5.** opercular cavity	**7.** ceratohyal	**9.** cardiac ventricle
2. pectoral fin	**4.** liver	**6.** adductor mandibulae complex	**8.** gill filaments	**10.** atrium

Figure 8.49

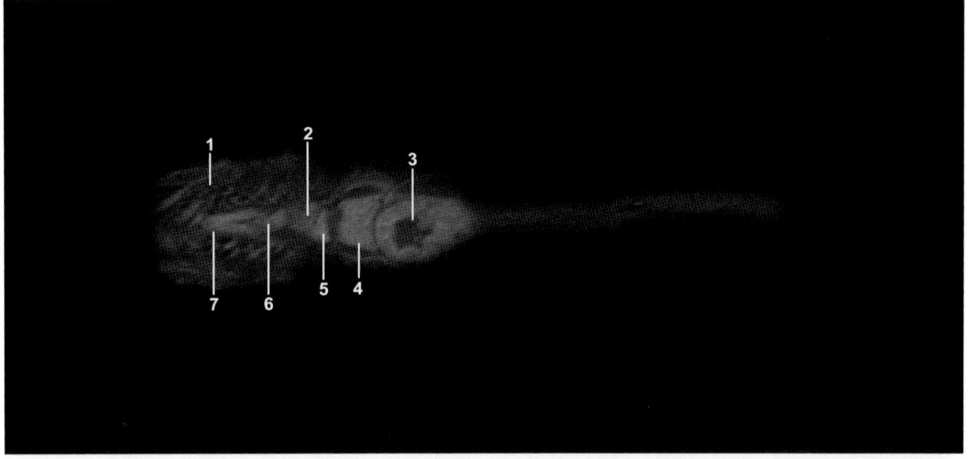

Figure 8.50

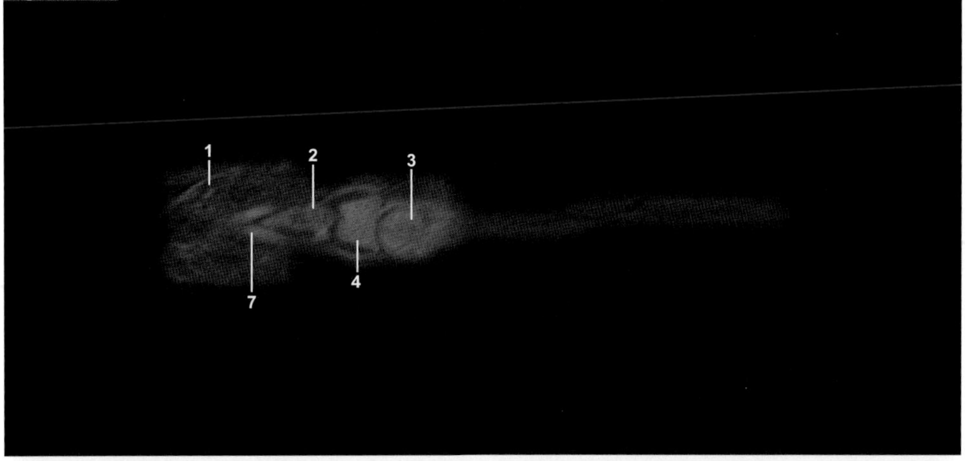

Figure 8.51

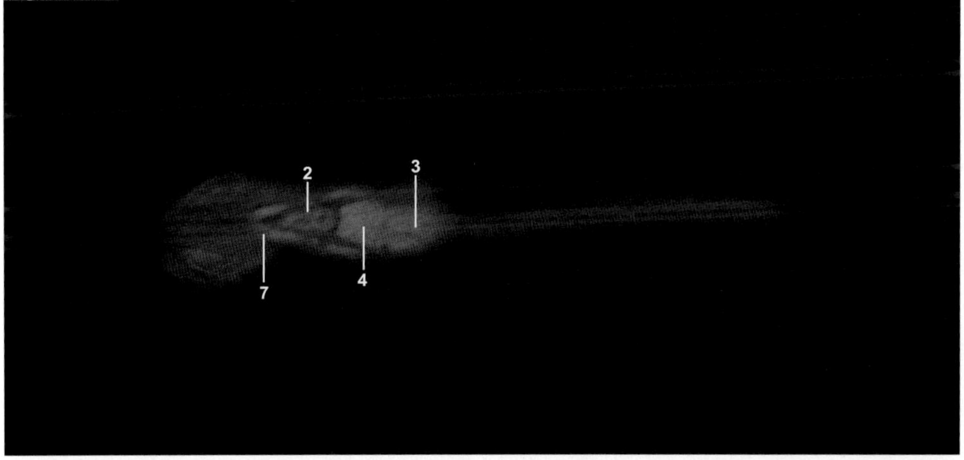

1. ceratohyal	**2.** cardiac ventricle	**3.** intestinal bulb	**4.** liver	**5.** atrium	**6.** bulbus arteriosus	**7.** sternohyoid

Transverse Sections

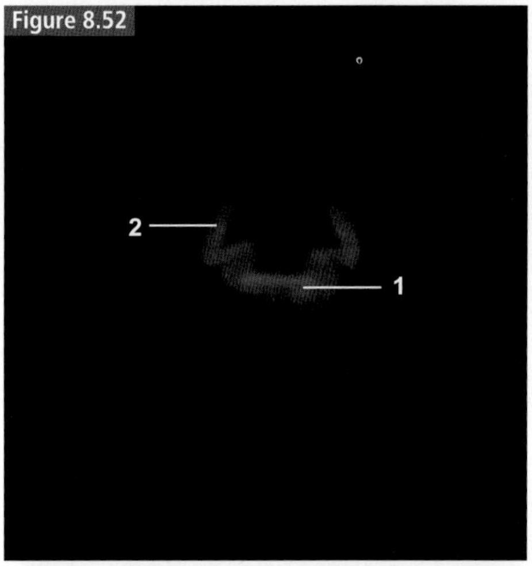

Figure 8.52

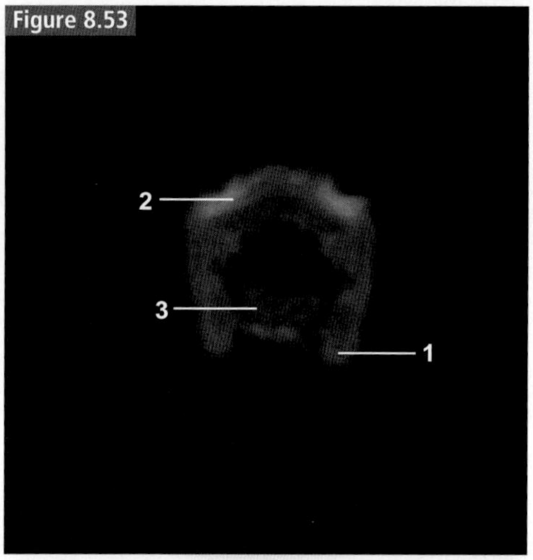

Figure 8.53

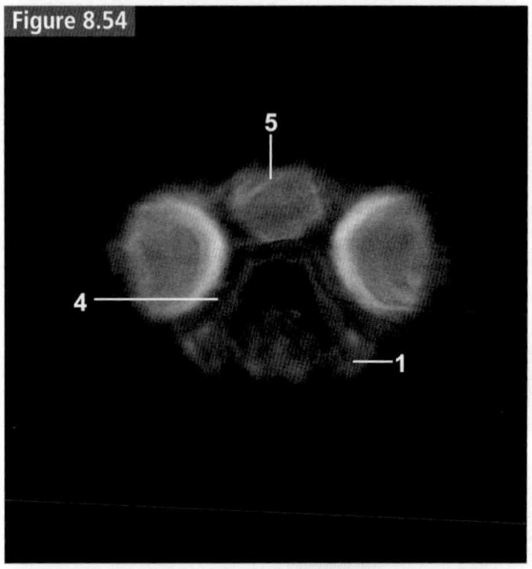

Figure 8.54

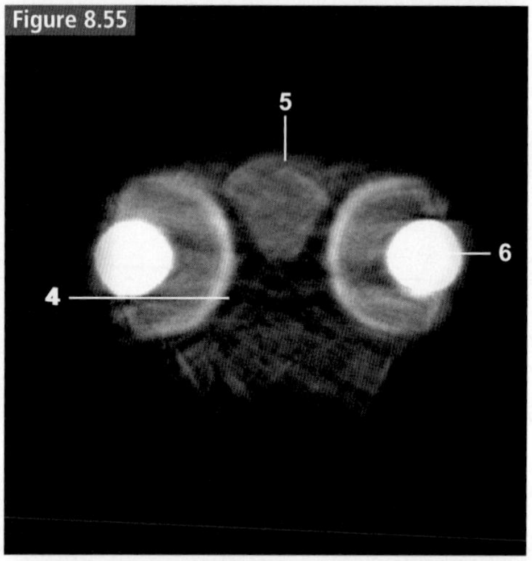

Figure 8.55

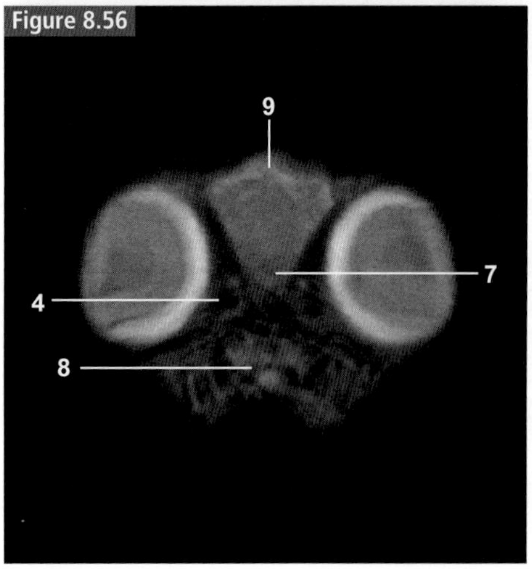

Figure 8.56

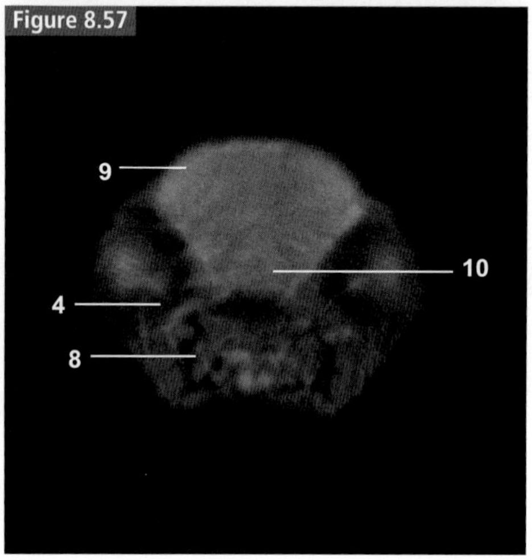

Figure 8.57

1. ventral mandibular arch	**3.** intermandibularis	**5.** telencephalon	**7.** hypothalamus	**9.** optic tectum
2. dorsal mandibular arch	**4.** optic chiasm	**6.** lens	**8.** gill filaments	**10.** inferior lobe

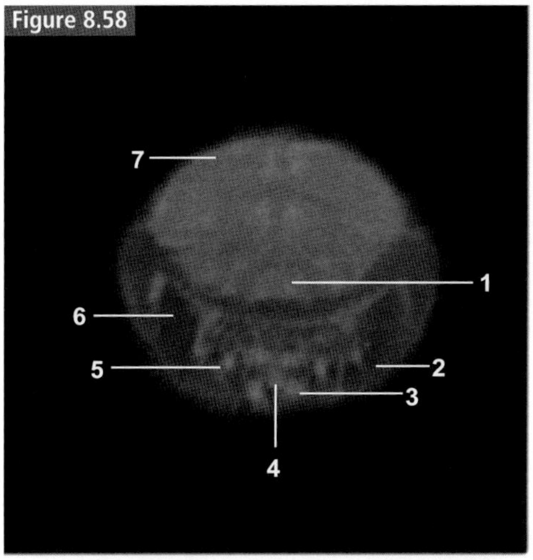

Figure 8.58

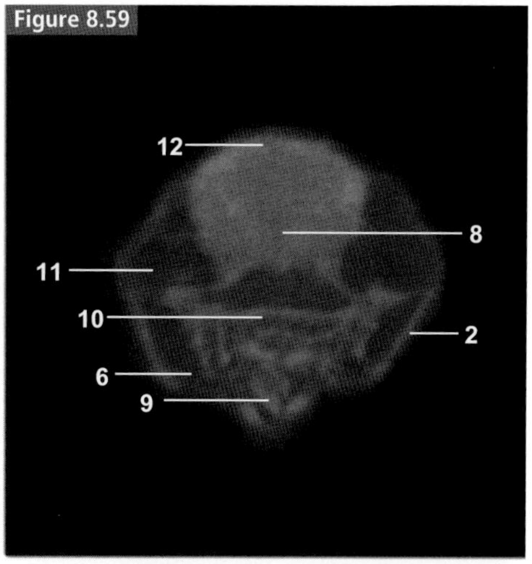

Figure 8.59

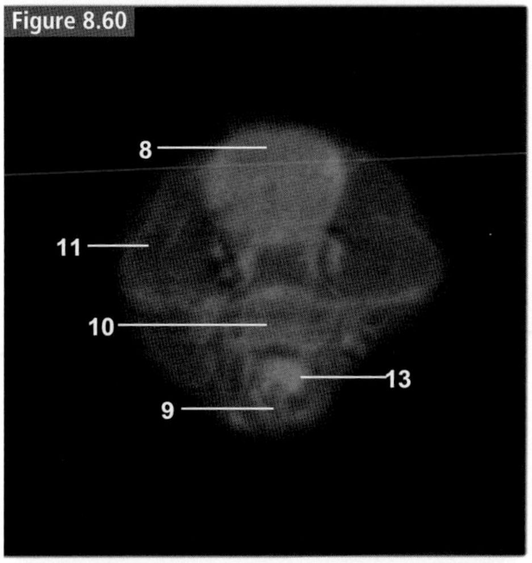

Figure 8.60

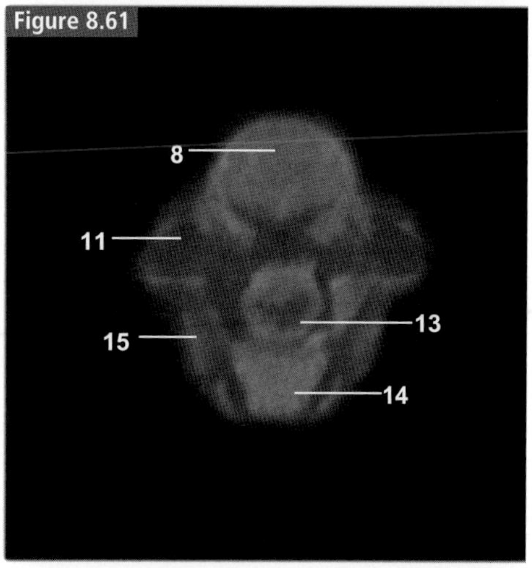

Figure 8.61

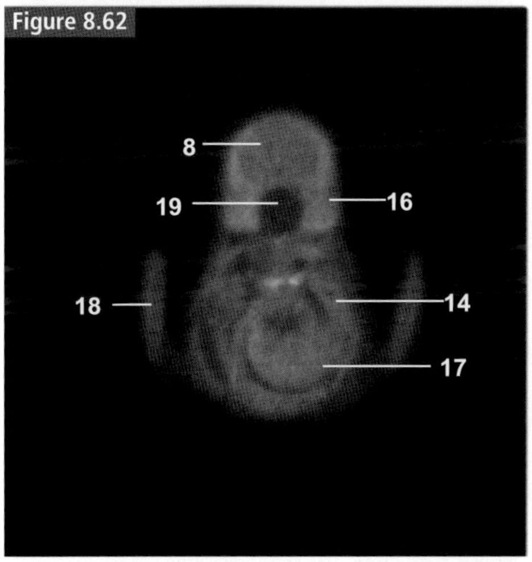

Figure 8.62

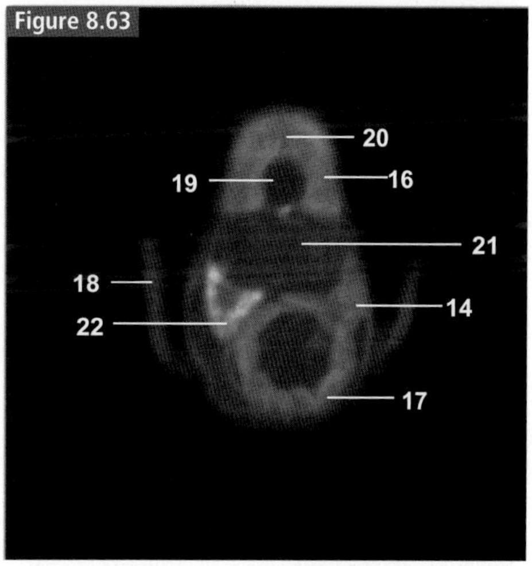

Figure 8.63

1. inferior lobe	**6.** opercular cavity	**11.** semicircular canal	**16.** myotome	**21.** swim bladder
2. opercle	**7.** optic tectum	**12.** cerebellum	**17.** intestinal bulb	**22.** pancreas
3. sternohyoid	**8.** medulla oblongata	**13.** atrium	**18.** pectoral fin	
4. bulbus arteriosus	**9.** cardiac ventricle	**14.** liver	**19.** notochord	
5. gill filaments	**10.** pharynx	**15.** pectoral fin musculature	**20.** spinal cord	

Figure 8.64

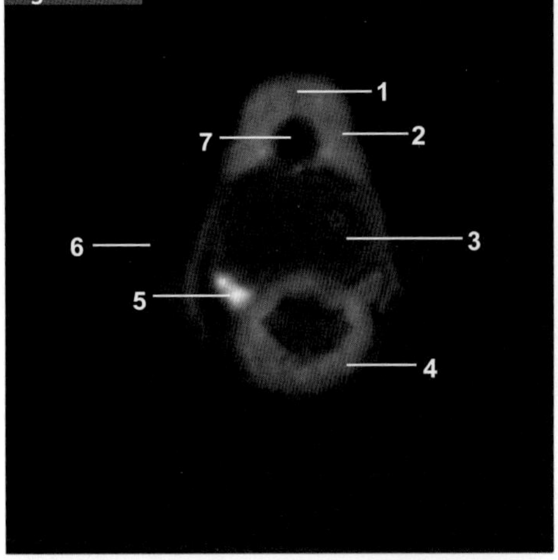

Figure 8.65

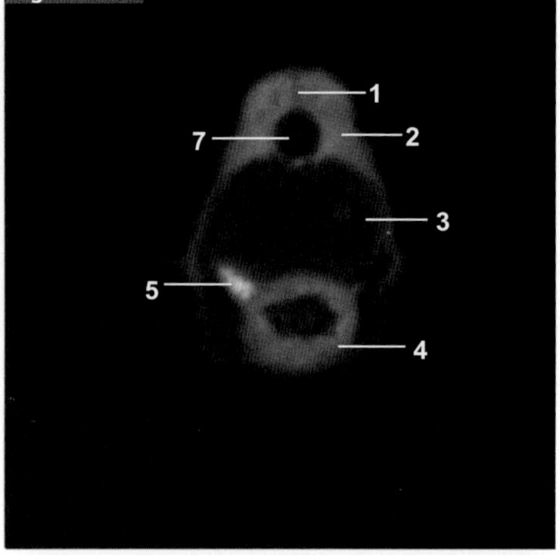

Figure 8.66

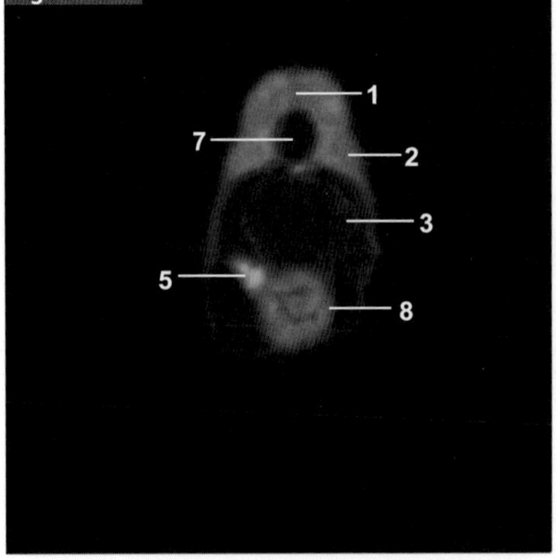

Figure 8.67

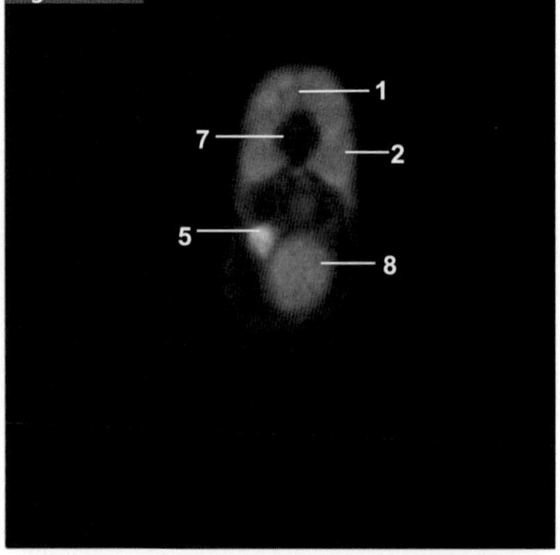

Figure 8.68

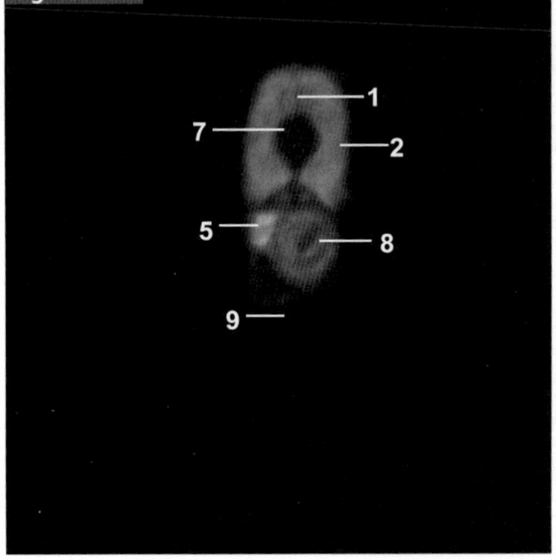

Figure 8.69

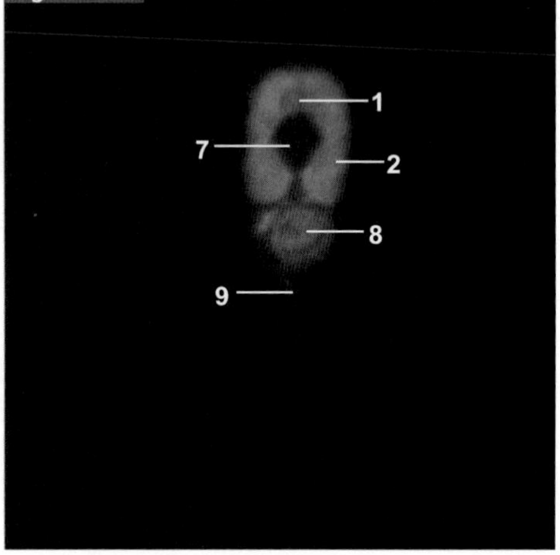

1. spinal cord **2.** myotome **3.** swim bladder **4.** intestinal bulb **5.** pancreas **6.** pectoral fin **7.** notochord **8.** mid intestine **9.** median fin fold

Figure 8.70

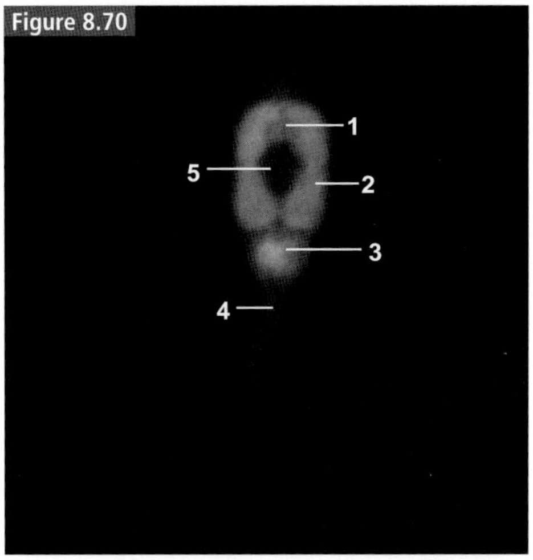

Figure 8.71

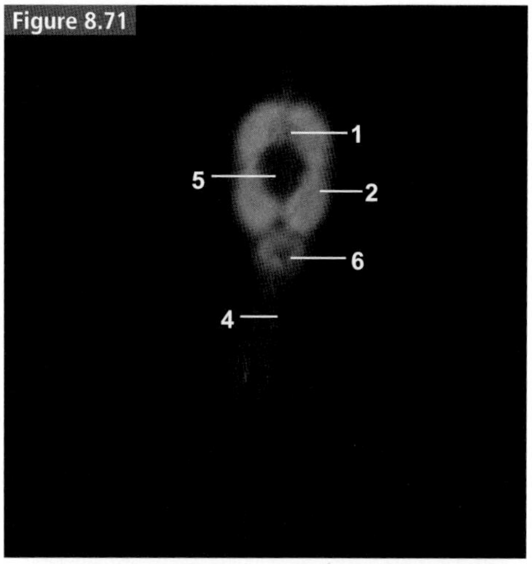

Figure 8.72

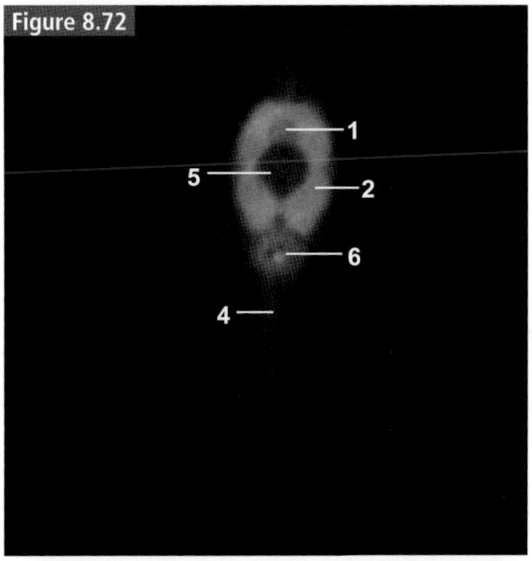

Figure 8.73

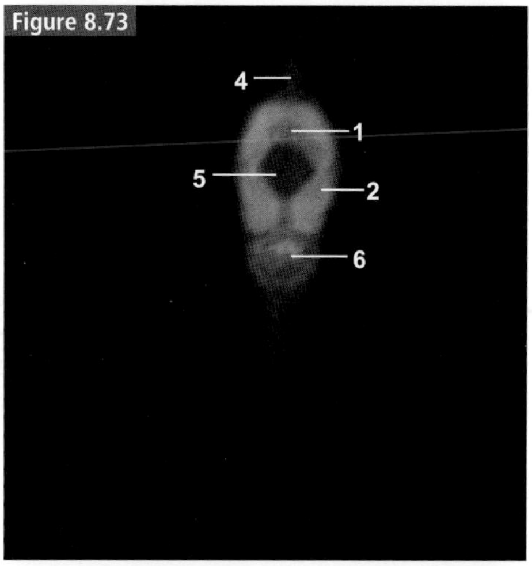

Figure 8.74

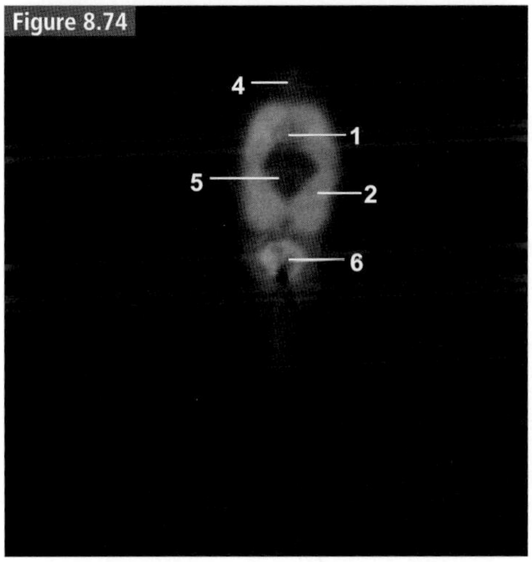

Figure 8.75

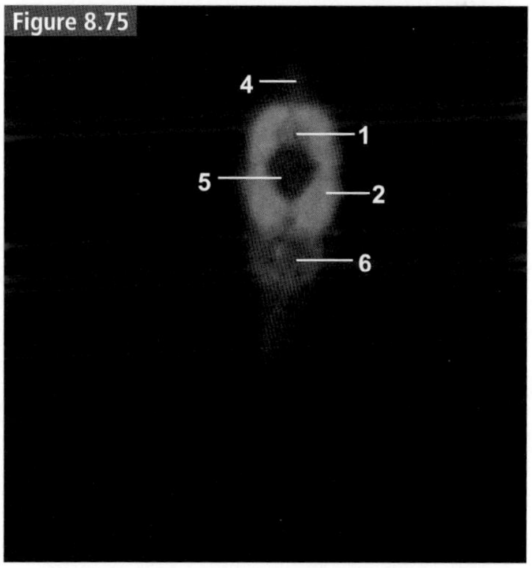

1. spinal cord	**2.** myotome	**3.** mid intestine	**4.** median fin fold	**5.** notochord	**6.** posterior intestine

PART II

THE DEVELOPING LARVAE

CHAPTER 9

6 MM

Sagittal Sections

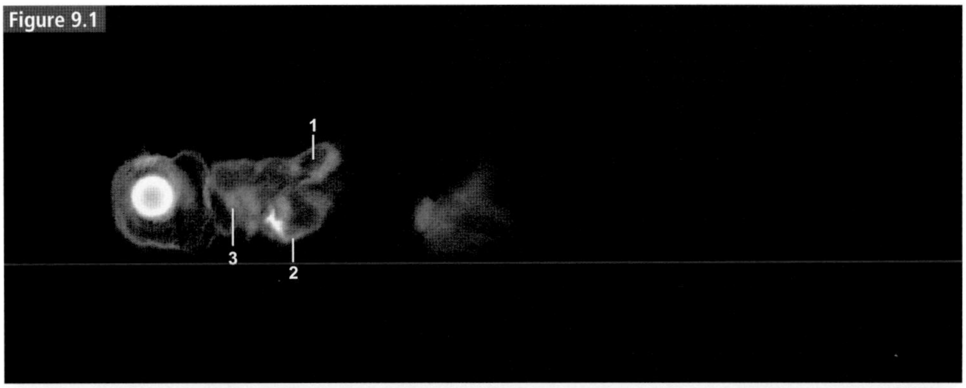

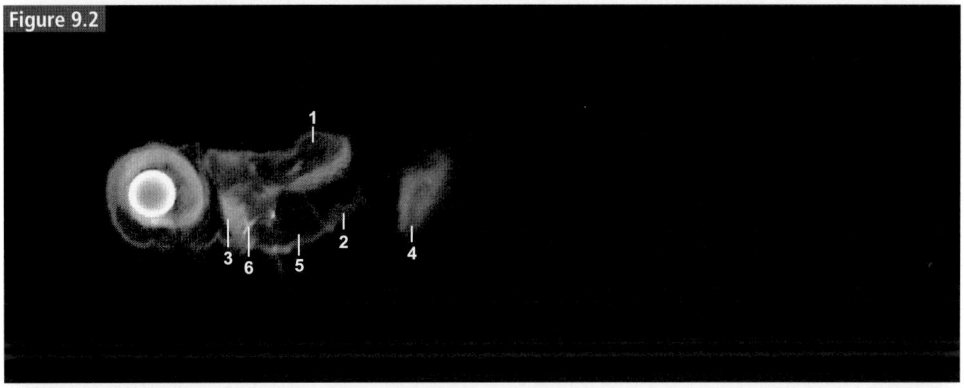

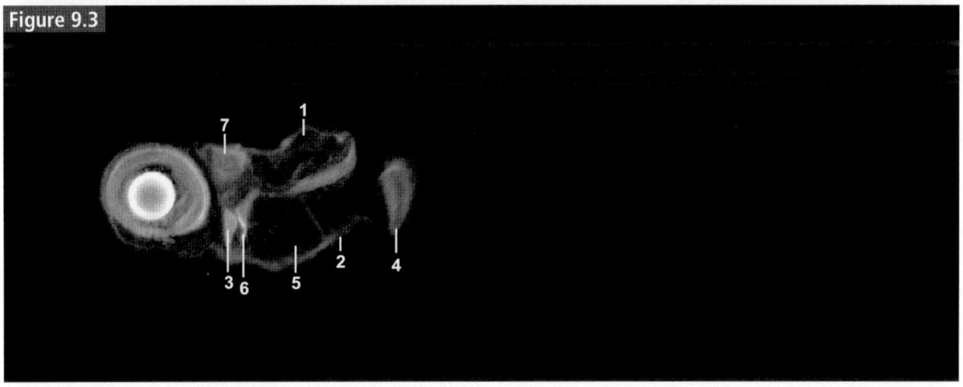

1. semicircular canal **2.** opercle **3.** dilator operculi **4.** pectoral fin musculature **5.** opercular cavity **6.** hyomandibula **7.** optic tectum

Atlas of Zebrafish Development. 10.1016/B978-0-12-374016-8.00009-1

Figure 9.4

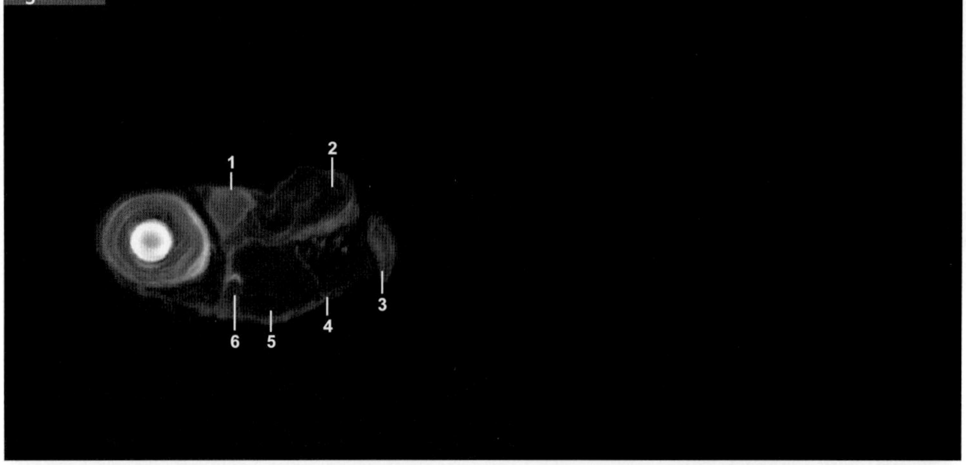

Figure 9.5

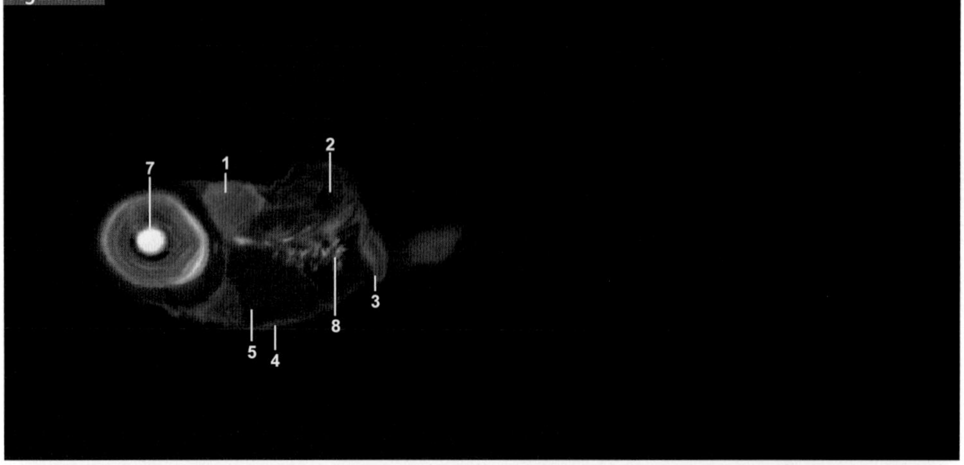

Figure 9.6

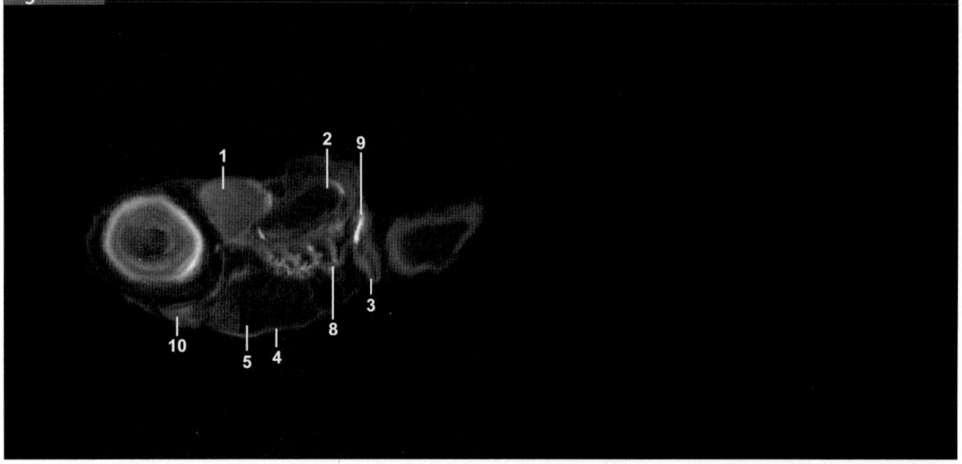

1. optic tectum	**3.** pectoral fin musculature	**5.** opercular cavity	**7.** lens	**9.** supracleithrum
2. semicircular canal	**4.** opercle	**6.** hyomandibula	**8.** gill filaments	**10.** adductor mandibulae

Figure 9.7

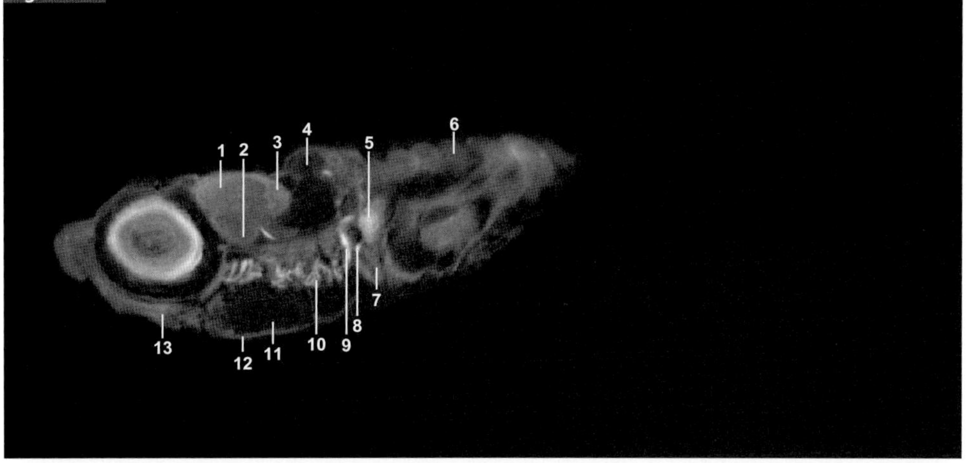

Figure 9.8

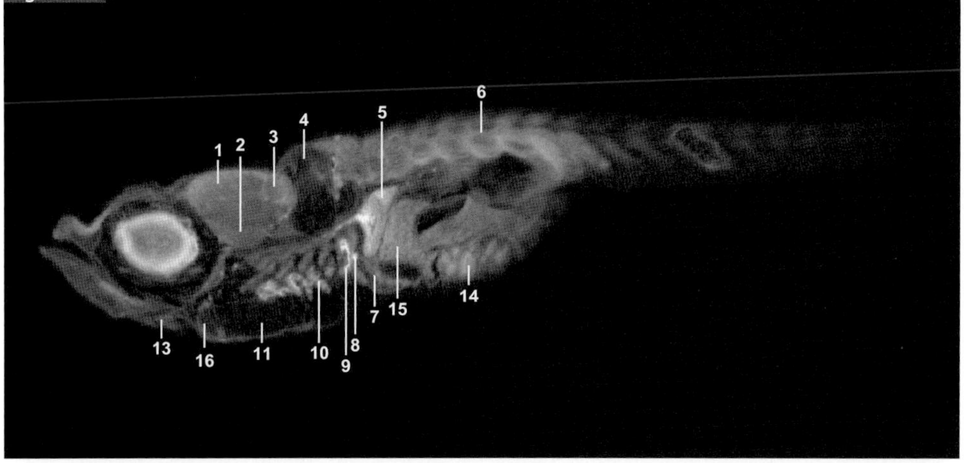

Figure 9.9

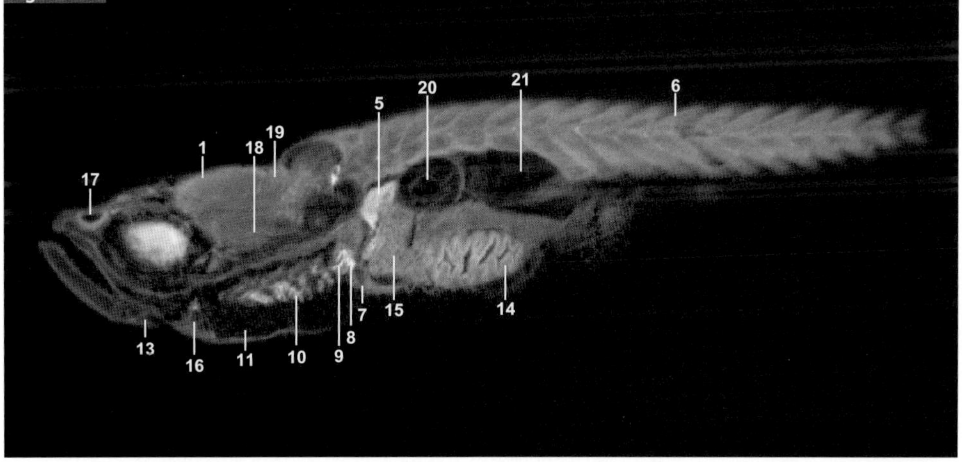

1. optic tectum
2. torus lateralis
3. eminentia granularis
4. semicircular canal
5. pronephros
6. myotome
7. pectoral fin musculature
8. cleithrum
9. ceratobranchial 5
10. gill filaments
11. opercular cavity
12. opercle
13. adductor mandibulae
14. intestinal bulb
15. liver
16. ceratohyal
17. olfactory pit
18. inferior lobe
19. cerebellum
20. anterior chamber swim bladder
21. posterior chamber swim bladder

Figure 9.10

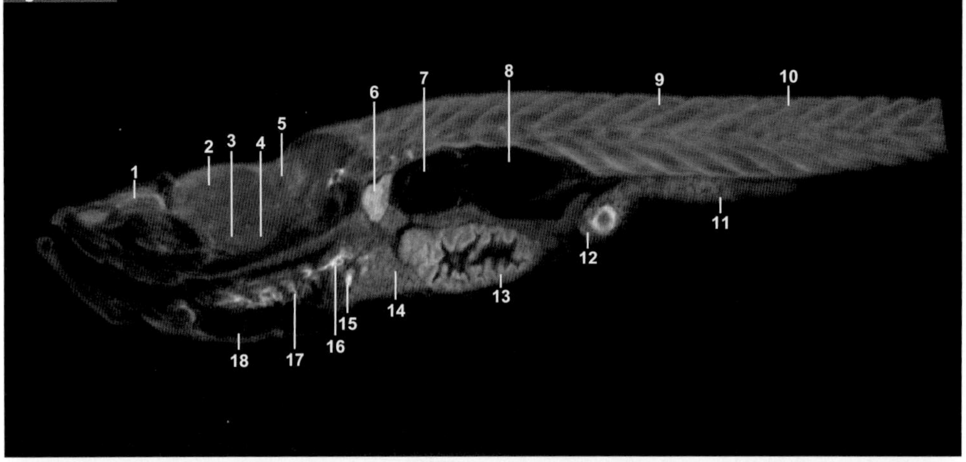

Figure 9.11

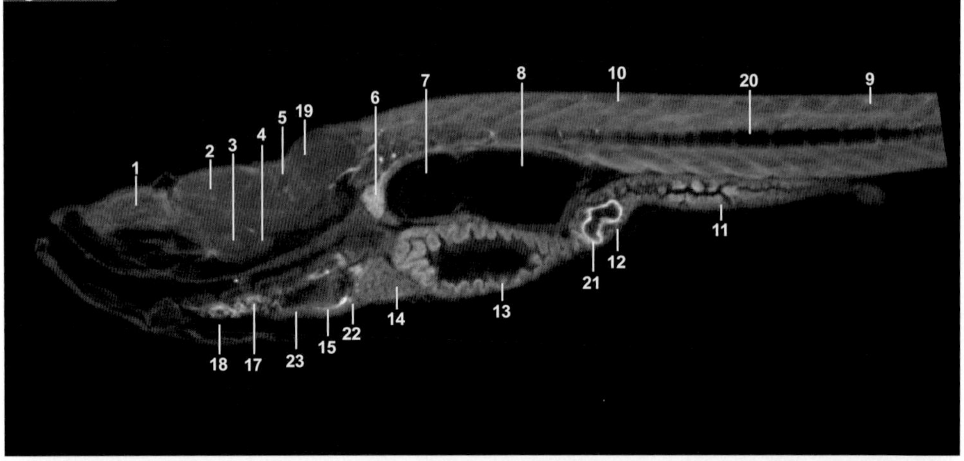

Figure 9.12

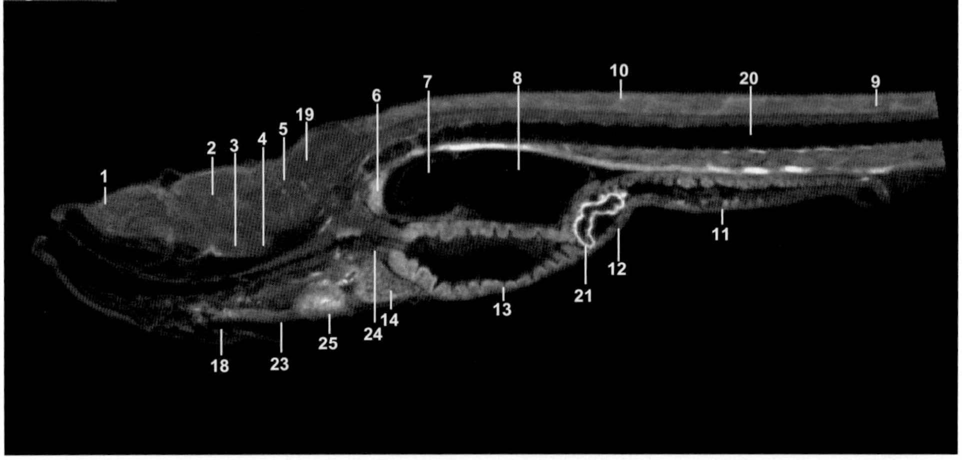

1. telencephalon	**6.** pronephros	**11.** posterior intestine	**16.** ceratobranchial 5	**21.** artemia cyst
2. optic tectum	**7.** anterior chamber swim bladder	**12.** mid intestine	**17.** gill filaments	**22.** pectoral fin musculature
3. hypothalamus	**8.** posterior chamber swim bladder	**13.** intestinal bulb	**18.** opercular cavity	**23.** sternohyoid
4. inferior lobe	**9.** myotome	**14.** liver	**19.** crista cerebellaris	**24.** esophagus
5. cerebellum	**10.** vertical myoseptum	**15.** cleithrum	**20.** vertebral column	**25.** cardiac ventricle

Figure 9.13

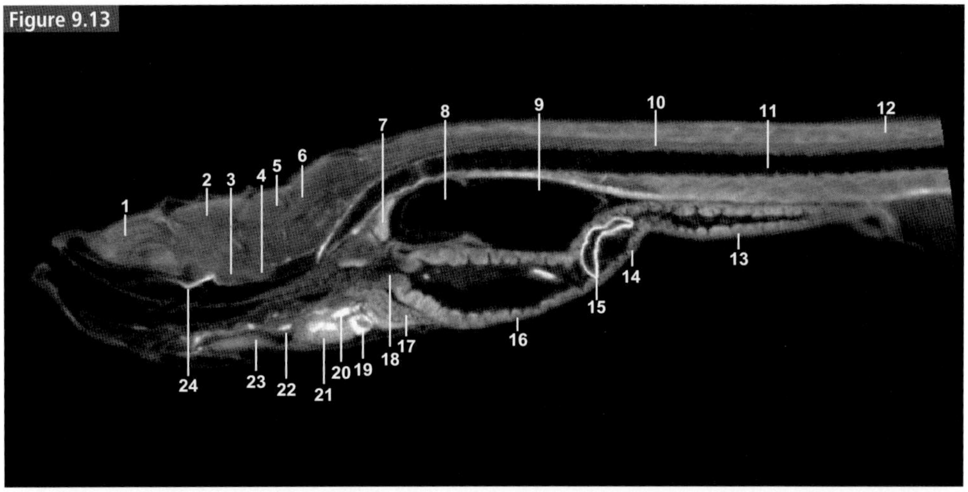

Figure 9.14

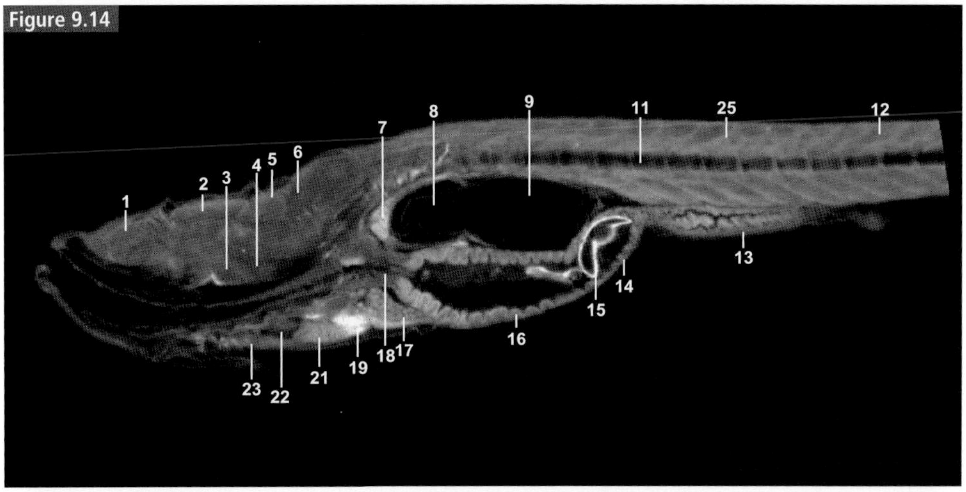

Figure 9.15

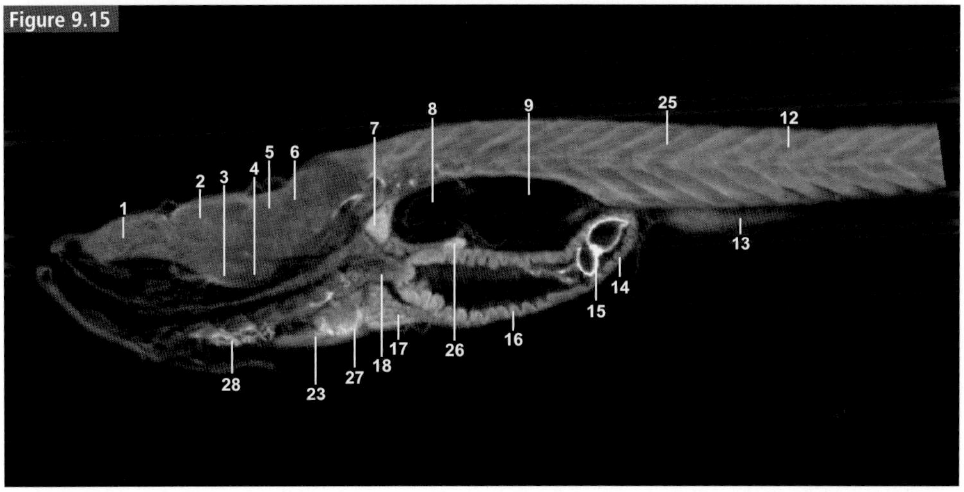

1. telencephalon	**7.** pronephros	**13.** posterior intestine	**19.** atrium	**25.** vertical myoseptum
2. optic tectum	**8.** anterior chamber swim bladder	**14.** mid intestine	**20.** sinus venosus	**26.** spleen
3. hypothalamus	**9.** posterior chamber swim bladder	**15.** artemia cyst	**21.** cardiac ventricle	**27.** cleithrum
4. inferior lobe	**10.** spinal cord	**16.** intestinal bulb	**22.** bulbus arteriosus	**28.** gill filaments
5. cerebellum	**11.** vertebral column	**17.** liver	**23.** sternohyoid	
6. crista cerebellaris	**12.** myotome	**18.** esophagus	**24.** parasphenoid	

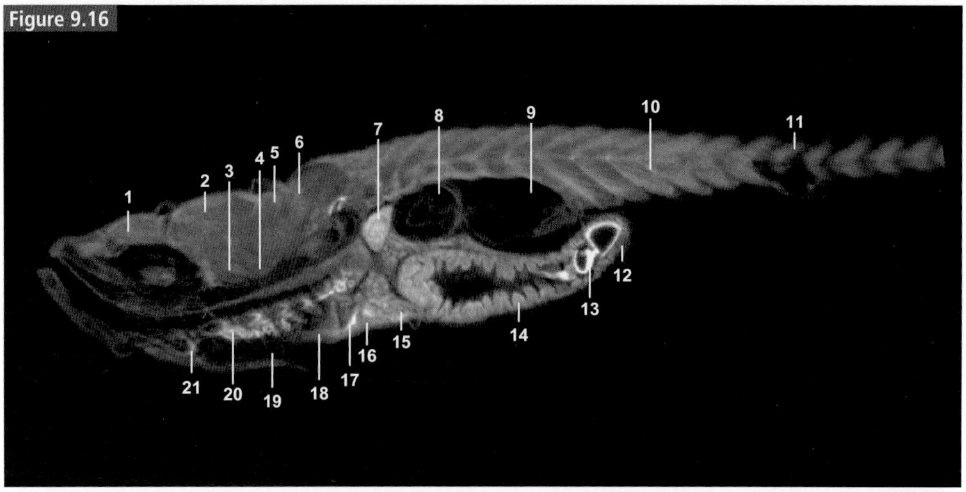

Figure 9.16

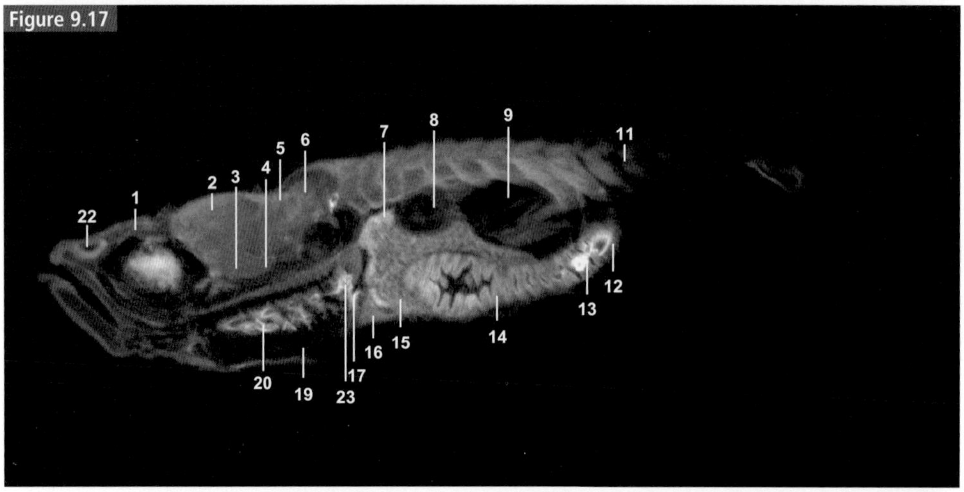

Figure 9.17

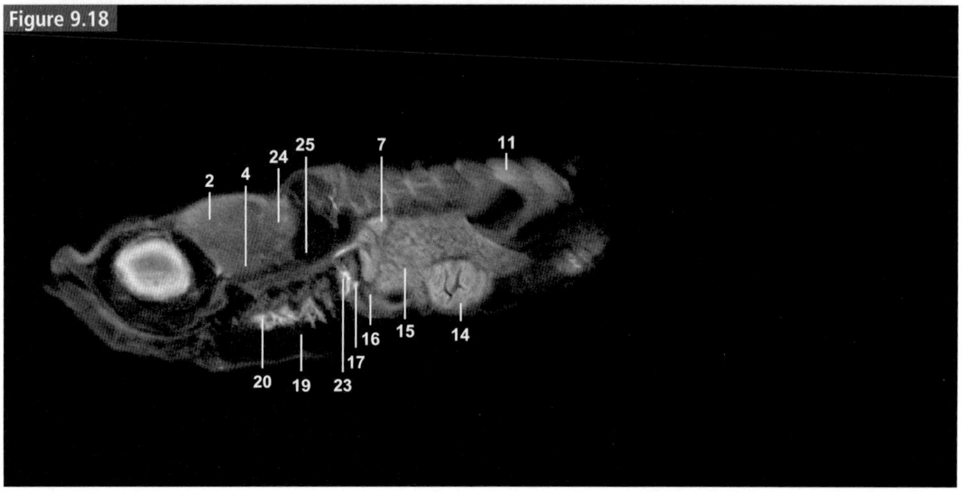

Figure 9.18

1. telencephalon	**6.** crista cerebellaris	**11.** myotome	**16.** pectoral fin musculature	**21.** retroarticular
2. optic tectum	**7.** pronephros	**12.** mid intestine	**17.** cleithrum	**22.** olfactory pit
3. hypothalamus	**8.** anterior chamber swim bladder	**13.** artemia cyst	**18.** sternohyoid	**23.** ceratobranchial 5
4. inferior lobe	**9.** posterior chamber swim bladder	**14.** intestinal bulb	**19.** opercular cavity	**24.** granular eminence
5. cerebellum	**10.** vertical myoseptum	**15.** liver	**20.** gill filaments	**25.** semicircular canal

Figure 9.19

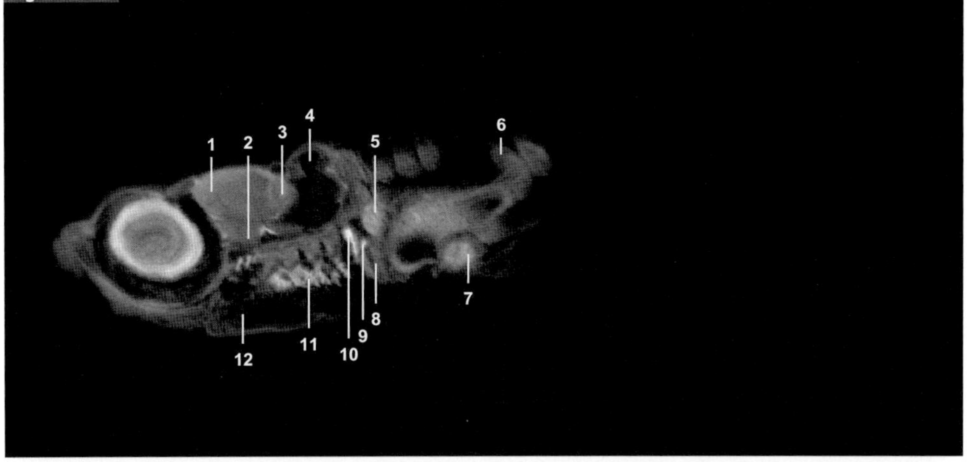

Figure 9.20

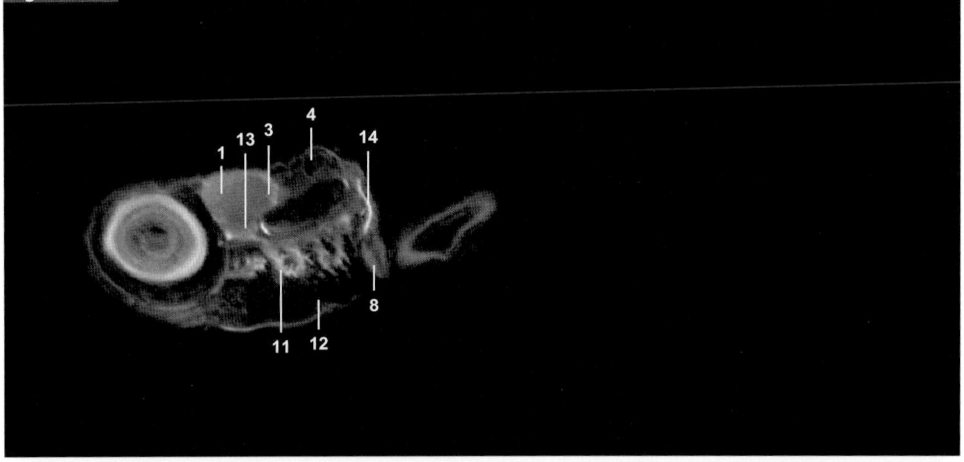

Figure 9.21

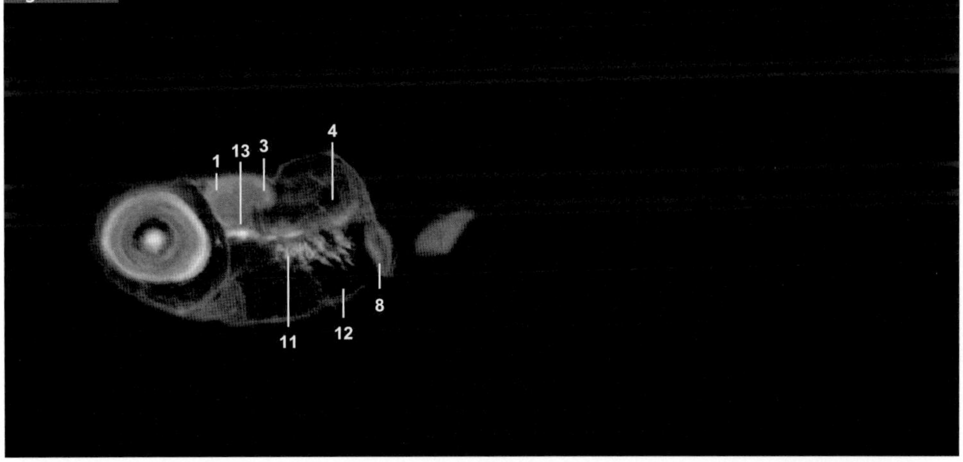

1. optic tectum
2. inferior lobe
3. granular eminence
4. semicircular canal
5. pronephros
6. myotome
7. intestinal bulb
8. pectoral fin musculature
9. cleithrum
10. ceratobranchial 5
11. gill filaments
12. opercular cavity
13. torus lateralis
14. supracleithrum

Figure 9.22

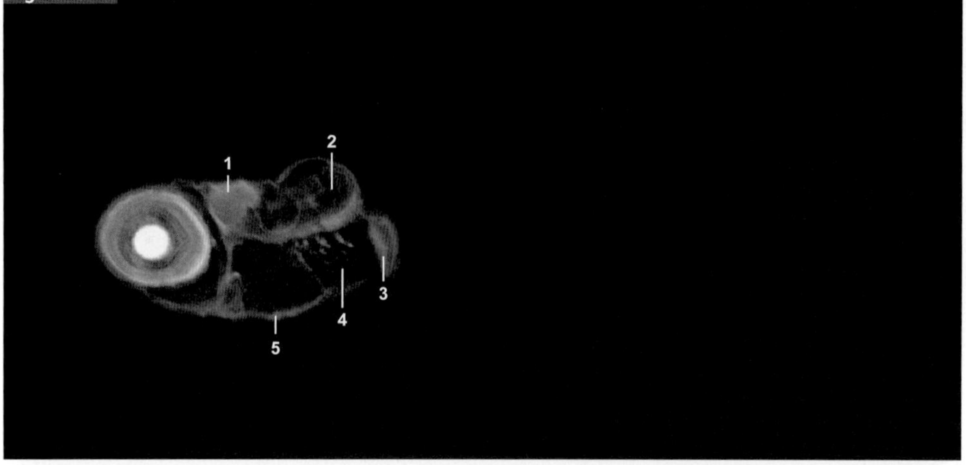

Figure 9.23

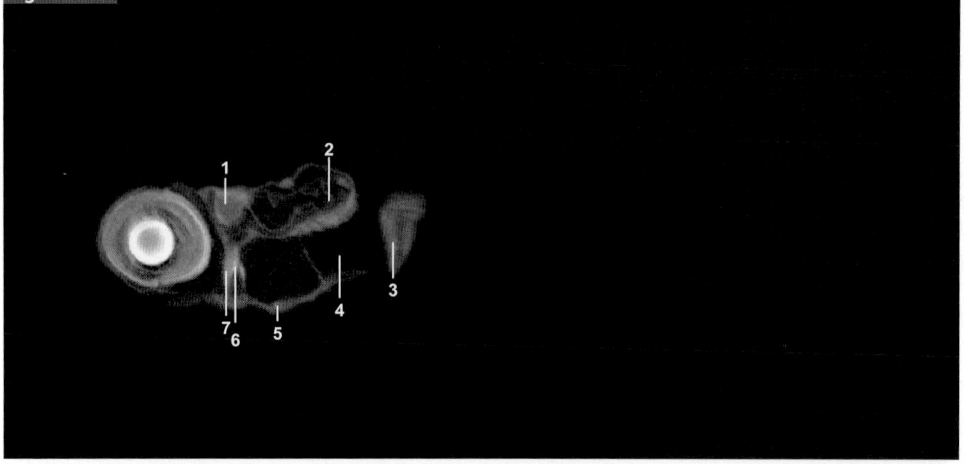

Figure 9.24

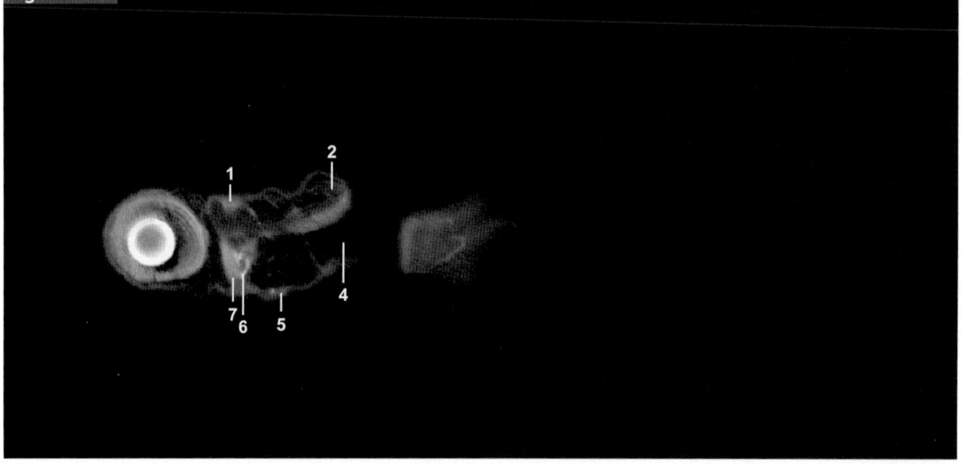

1. optic tectum **2.** semicircular canal **3.** pectoral fin musculature **4.** opercular cavity **5.** opercle **6.** hyomandibula **7.** dilator operculi

Coronal Sections

Figure 9.25

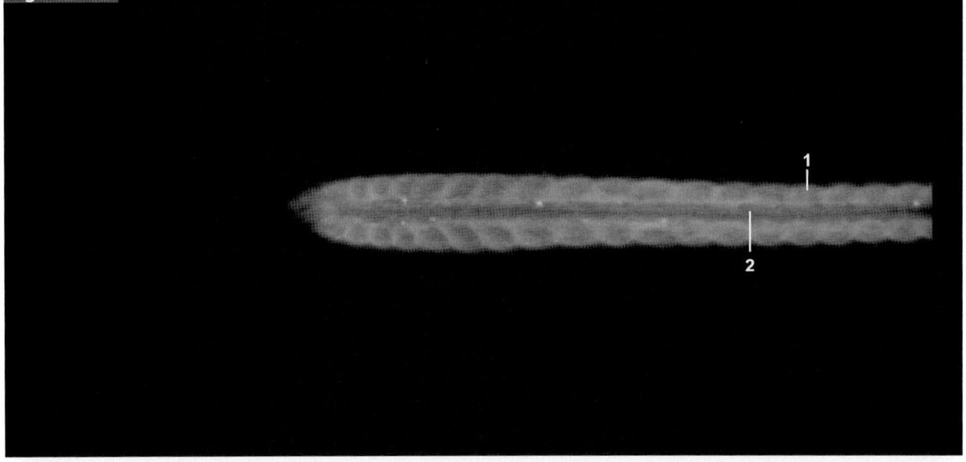

Figure 9.26

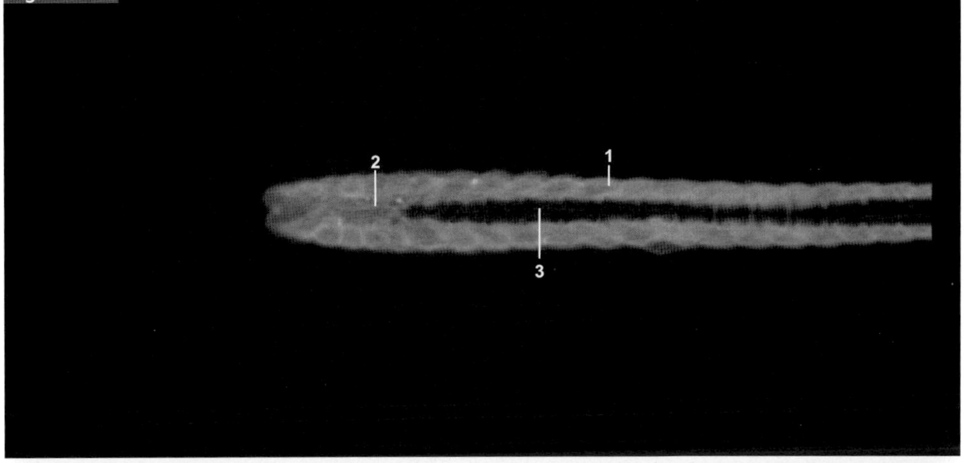

Figure 9.27

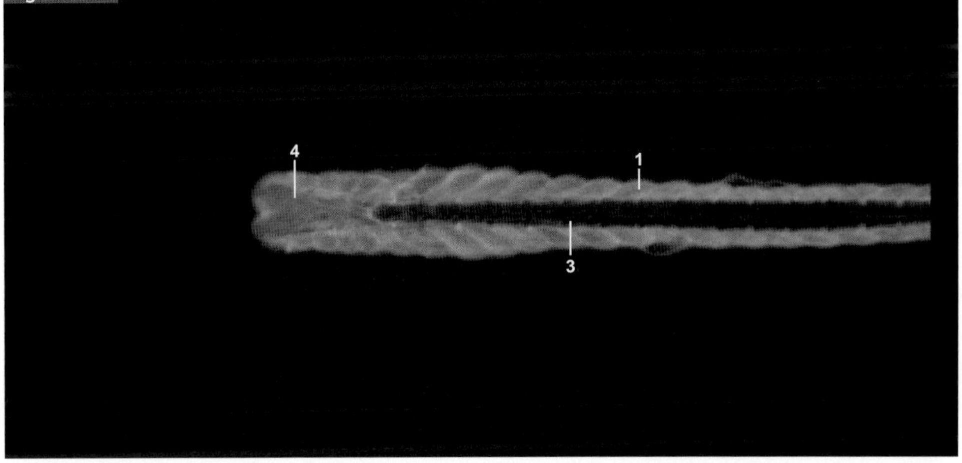

1. myotome	**2.** spinal cord	**3.** precaudal vertebra	**4.** vagal lobe

Figure 9.28

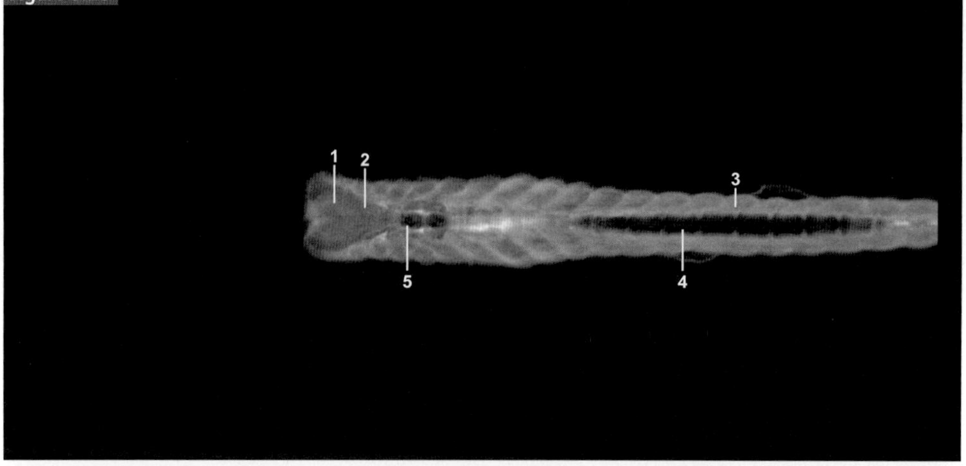

Figure 9.29

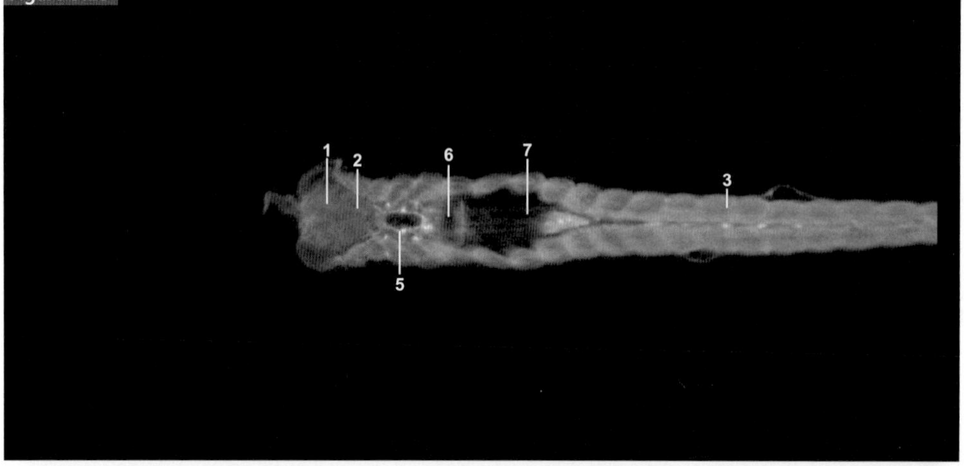

Figure 9.30

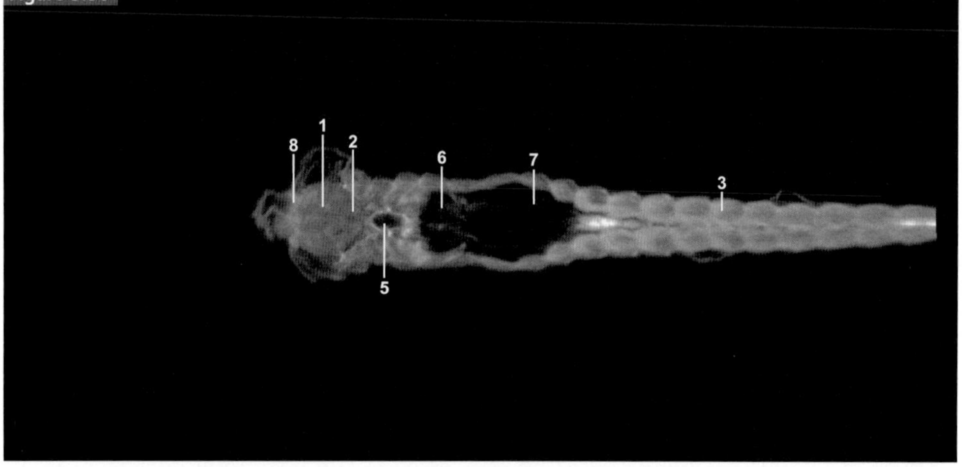

1. crista cerebellaris	**3.** myotome	**5.** Weberian apparatus	**7.** posterior chamber swim bladder
2. vagal lobe	**4.** precaudal vertebra	**6.** anterior chamber swim bladder	**8.** cerebellum

Figure 9.31

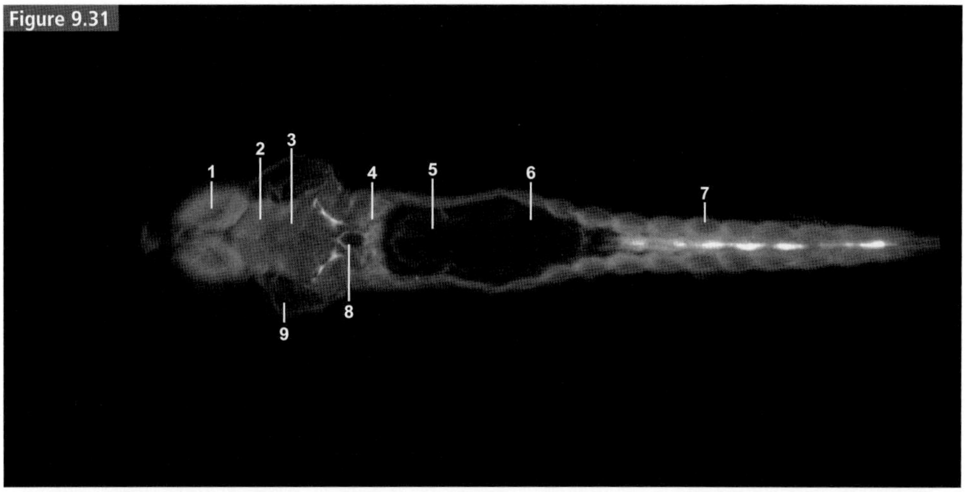

Figure 9.32

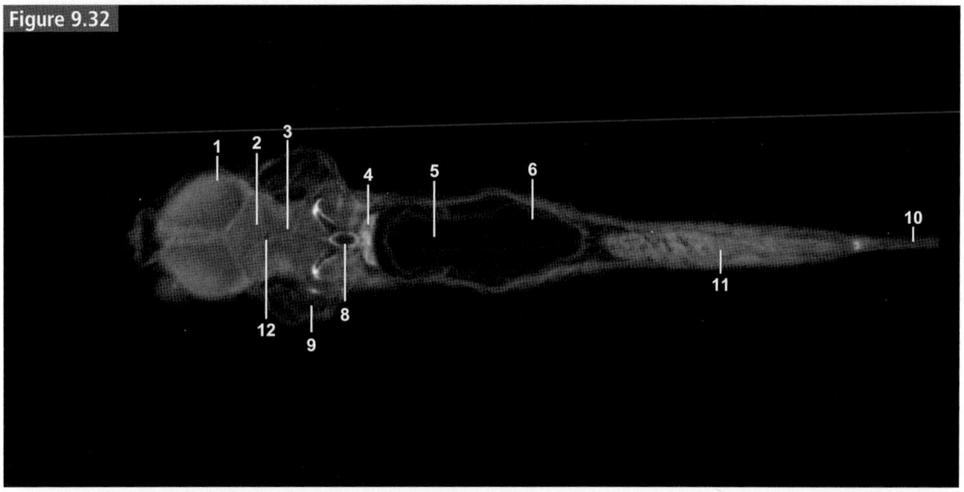

Figure 9.33

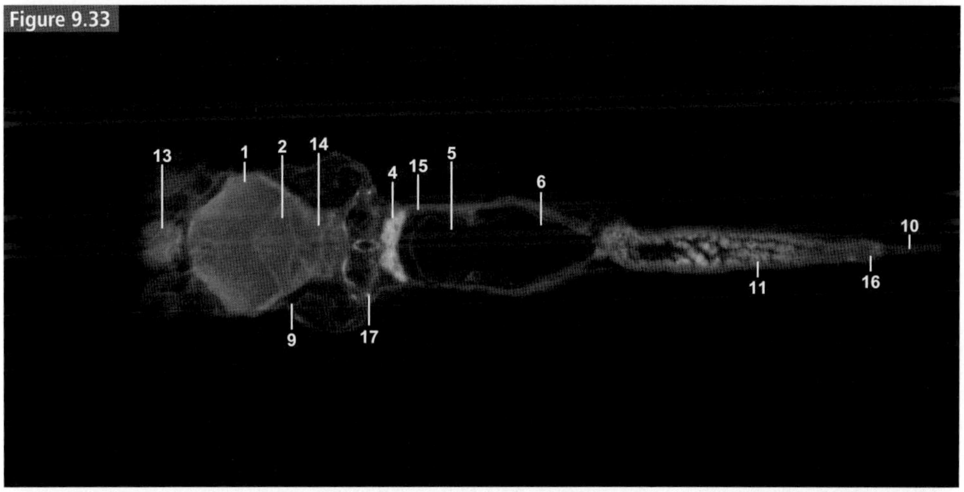

1. optic tectum	**5.** anterior chamber swim bladder	**9.** semicircular canal	**13.** telencephalon	**17.** supracleithrum
2. cerebellum	**6.** posterior chamber swim bladder	**10.** anal fin	**14.** medulla oblongata	
3. crista cerebellaris	**7.** myotome	**11.** posterior intestine	**15.** liver	
4. pronephros	**8.** Weberian apparatus	**12.** fourth ventricle	**16.** cloacal chamber	

Figure 9.34

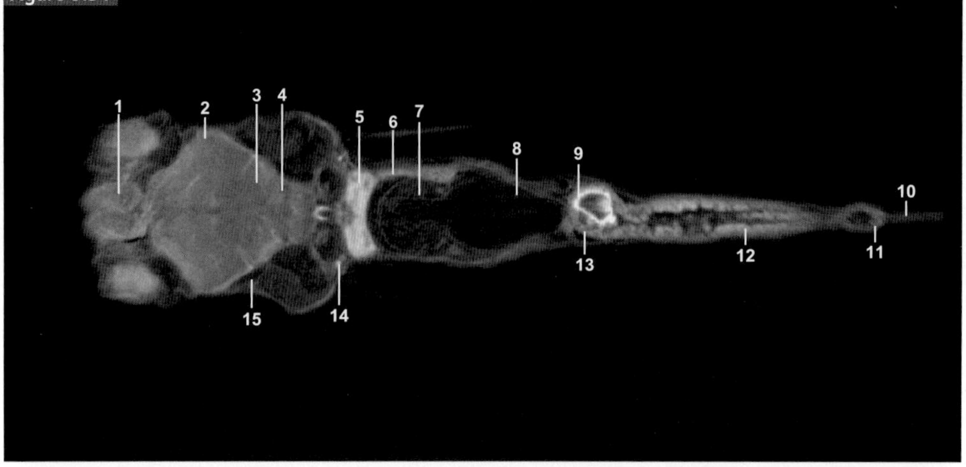

Figure 9.35

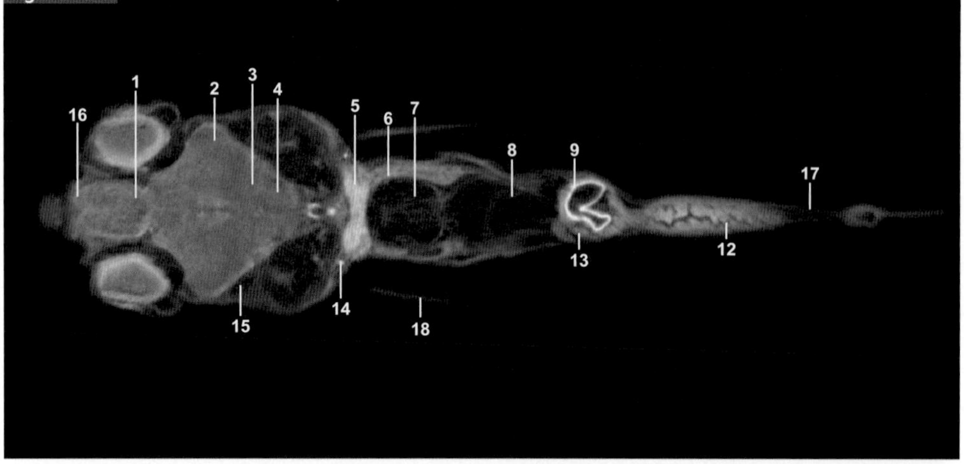

Figure 9.36

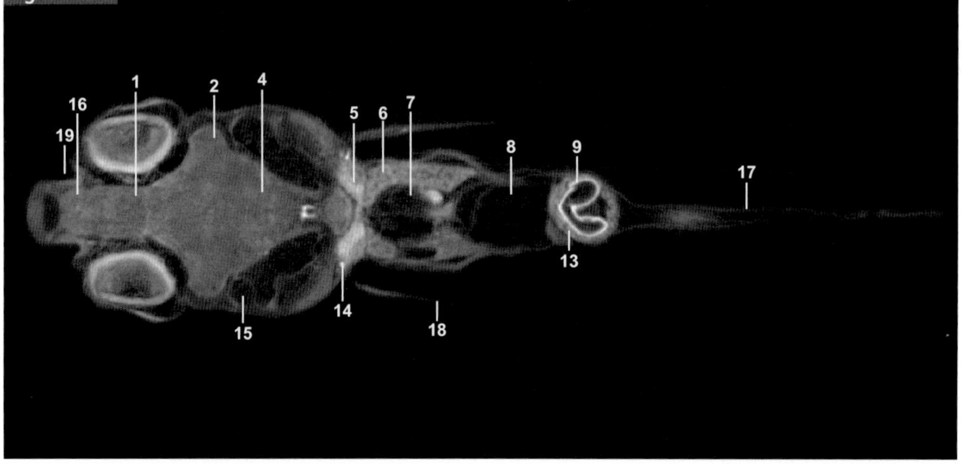

1. telencephalon	**5.** pronephros	**9.** artemia cyst	**13.** mid intestine	**17.** median fin
2. optic tectum	**6.** liver	**10.** anal fin	**14.** supracleithrum	**18.** pectoral fin
3. cerebellum	**7.** anterior chamber swim bladder	**11.** cloacal chamber	**15.** semicircular canal	**19.** olfactory pit
4. medulla oblongata	**8.** posterior chamber swim bladder	**12.** posterior intestine	**16.** olfactory bulb	

Figure 9.37

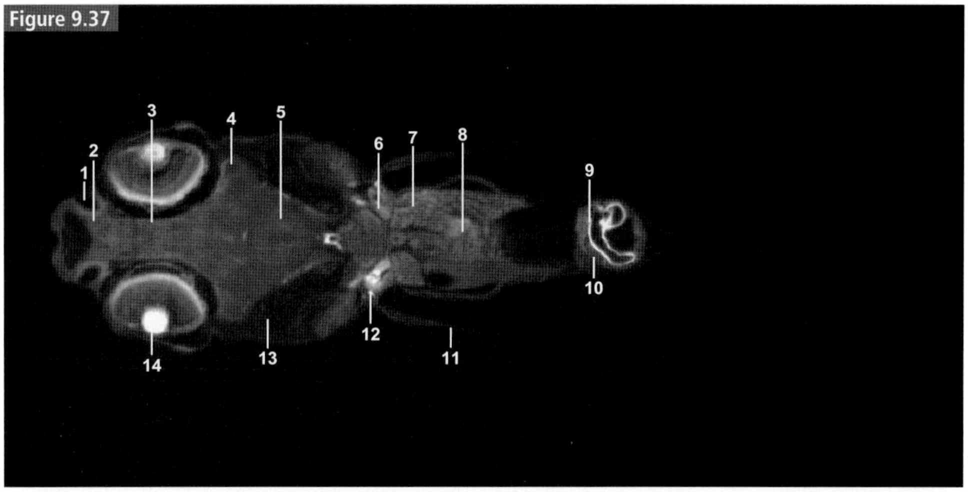

Figure 9.38

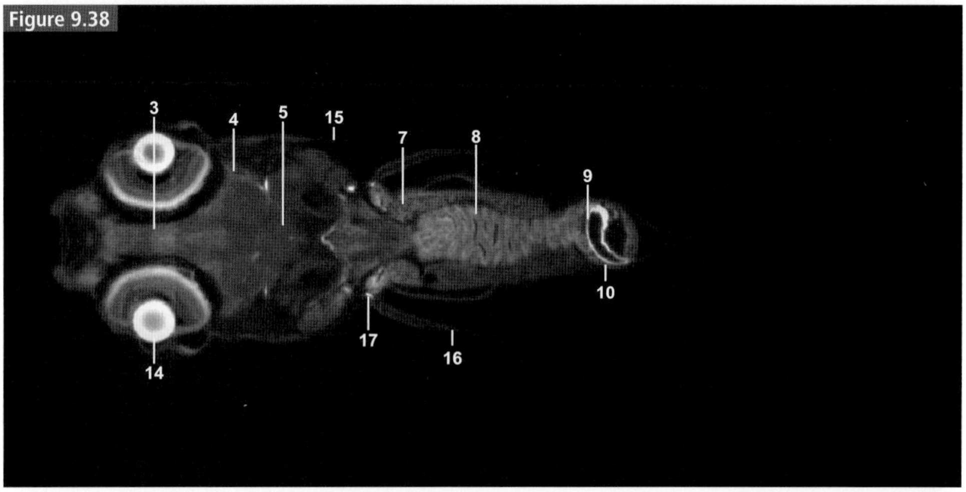

Figure 9.39

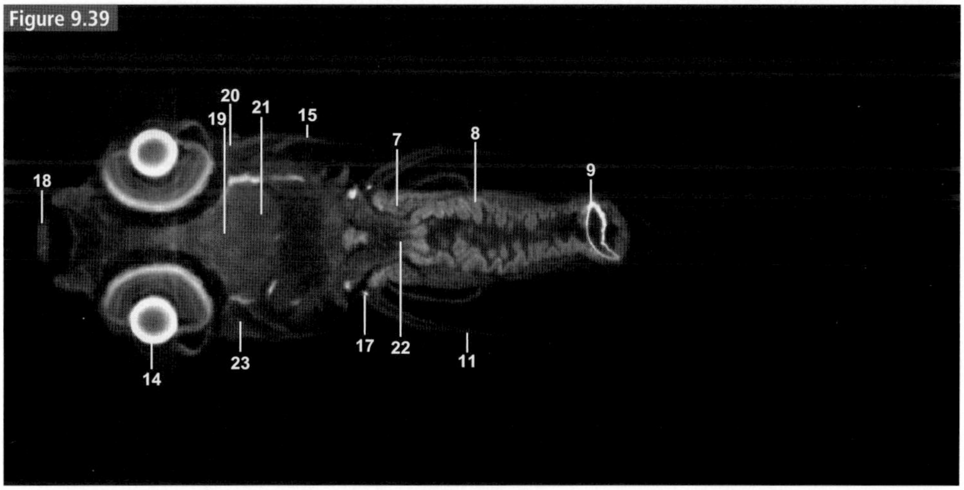

1. olfactory pit	**6.** pronephros	**11.** pectoral fin	**16.** pectoral fin	**21.** inferior lobe
2. olfactory bulb	**7.** liver	**12.** supracleithrum	**17.** cleithrum	**22.** esophagus
3. telencephalon	**8.** intestinal bulb	**13.** semicircular canal	**18.** dorsal mandibular arch	**23.** hyomandibula
4. optic tectum	**9.** artemia cyst	**14.** lens	**19.** hypothalamus	
5. medulla oblongata	**10.** mid intestine	**15.** opercle	**20.** dilator operculi	

Figure 9.40

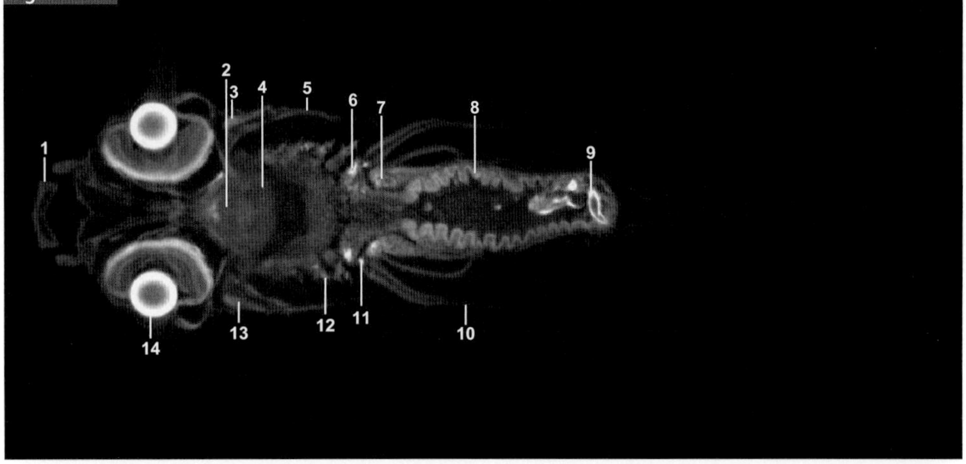

Figure 9.41

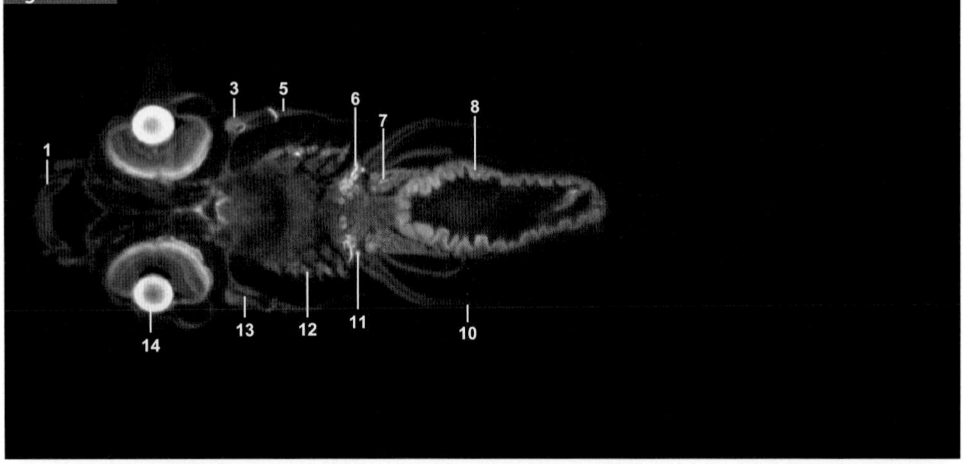

Figure 9.42

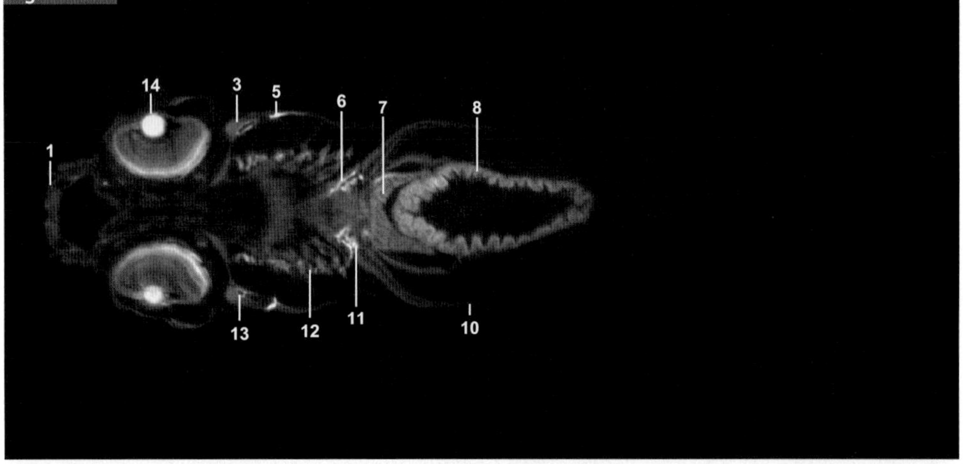

1. dorsal mandibular arch	**4.** inferior lobe	**7.** liver	**10.** pectoral fin	**13.** hyomandibula
2. hypothalamus	**5.** opercle	**8.** intestinal bulb	**11.** cleithrum	**14.** lens
3. dilator operculi	**6.** ceratobranchial 5	**9.** artemia cyst	**12.** gill filaments	

Figure 9.43

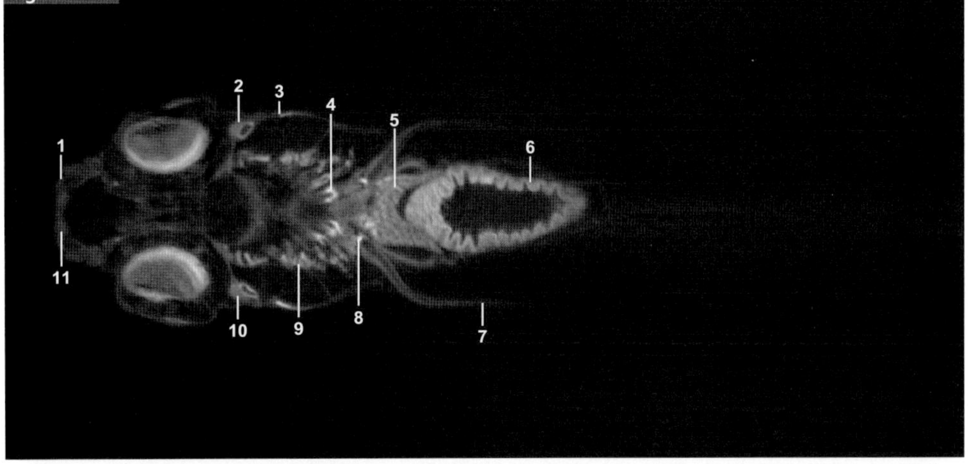

Figure 9.44

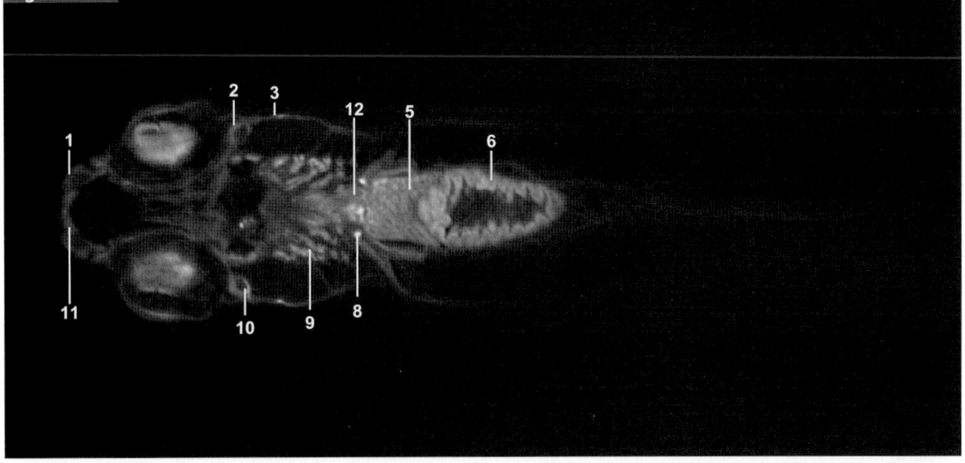

Figure 9.45

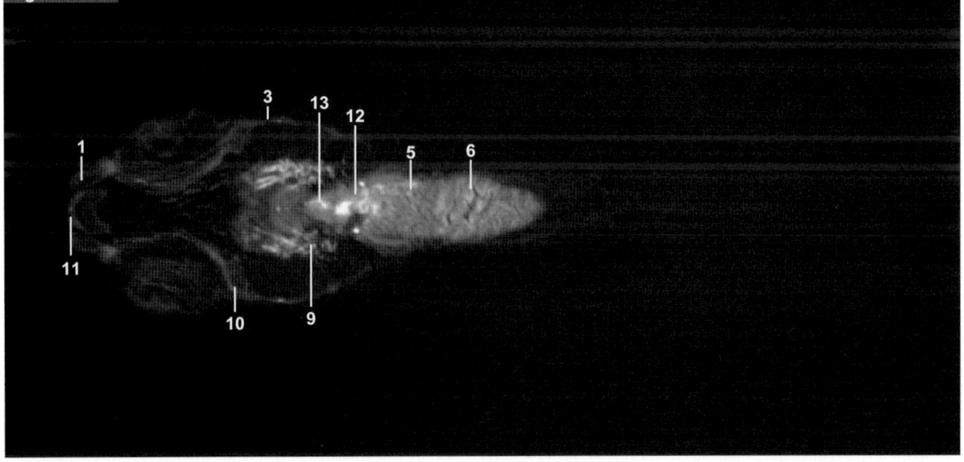

1. dorsal mandibular arch	**4.** ceratobranchial 5	**7.** pectoral fin	**10.** hyomandibula	**13.** cardiac ventricle
2. dilator operculi	**5.** liver	**8.** cleithrum	**11.** intermandibularis	
3. opercle	**6.** intestinal bulb	**9.** gill filaments	**12.** atrium	

Figure 9.46

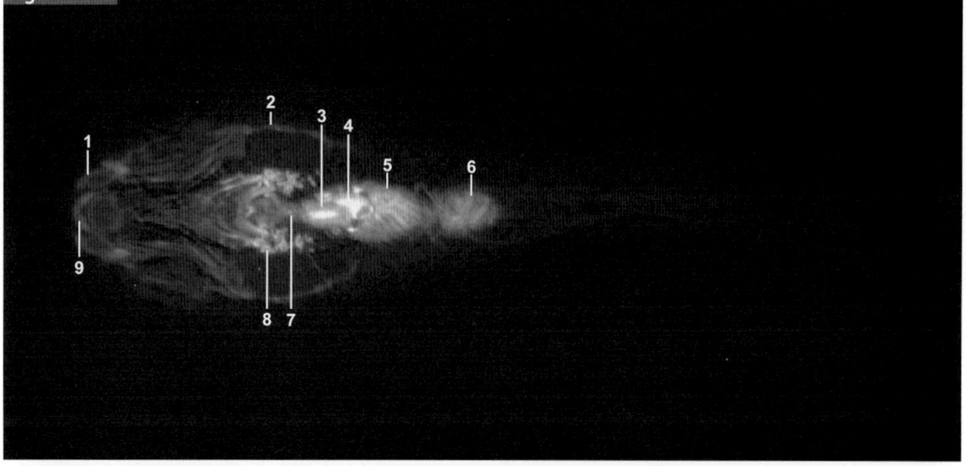

Figure 9.47

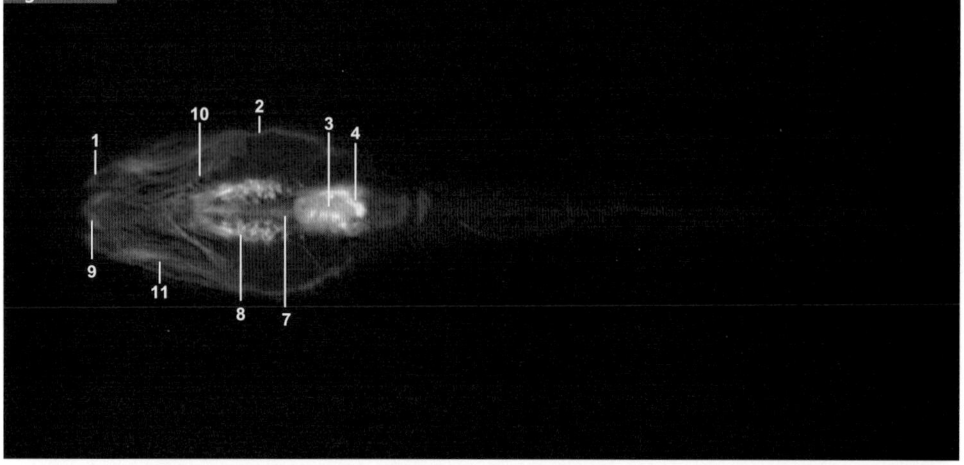

Figure 9.48

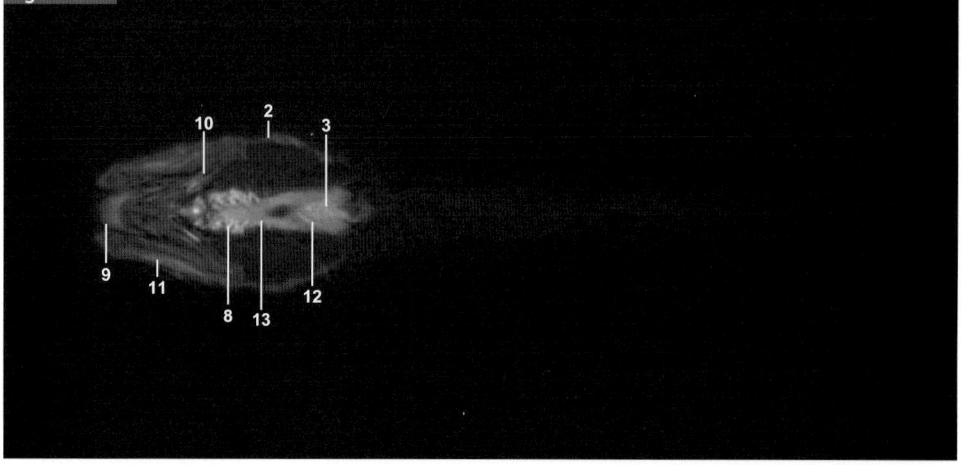

1. dorsal mandibular arch	**4.** atrium	**7.** bulbus arteriosus	**10.** ceratohyal	**13.** sternohyoid
2. opercle	**5.** liver	**8.** gill filaments	**11.** adductor mandibulae complex	
3. cardiac ventricle	**6.** intestinal bulb	**9.** interhyal	**12.** cleithrum	

Transverse Sections

Figure 9.49

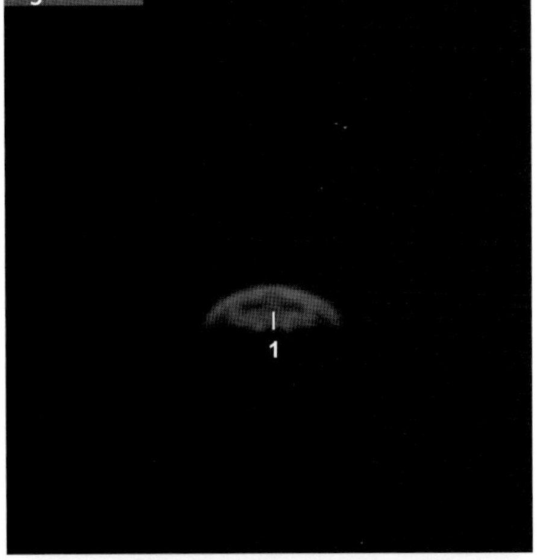

Figure 9.50

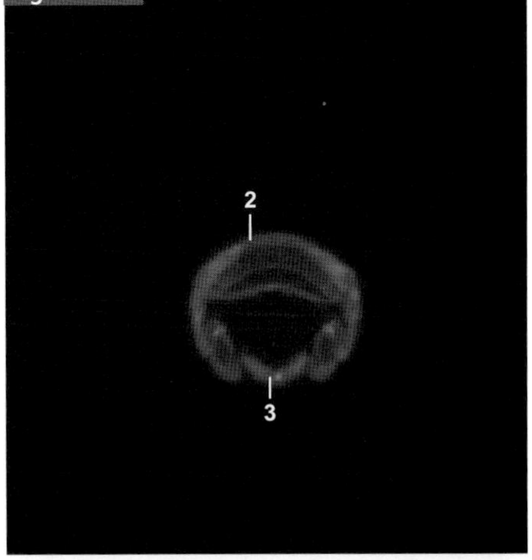

Figure 9.51

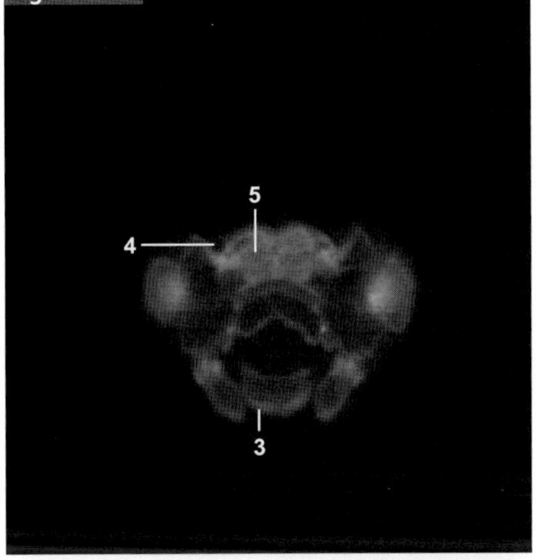

Figure 9.52

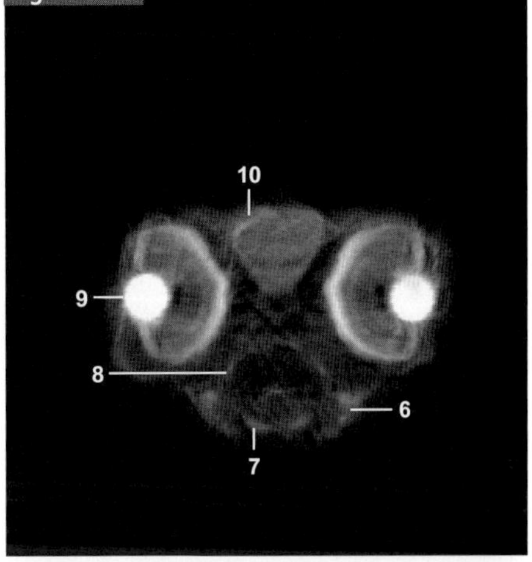

Figure 9.53

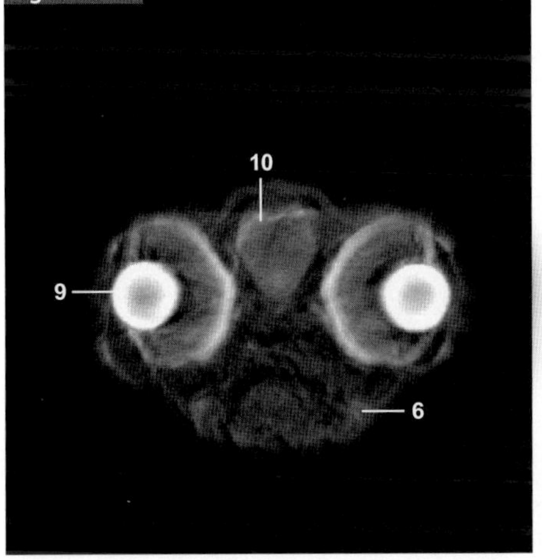

Figure 9.54

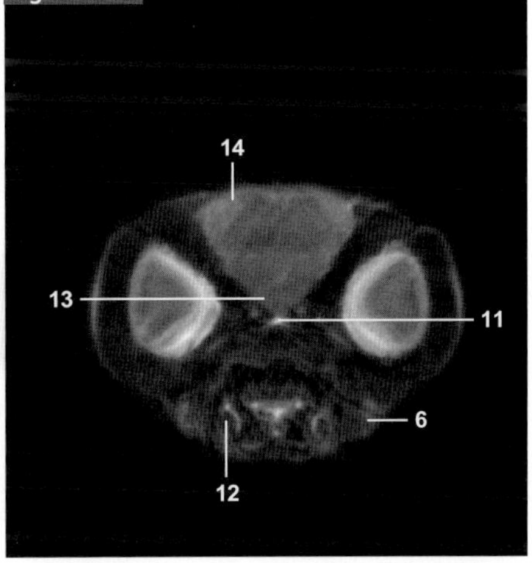

1. ventral mandibular arch	**4.** olfactory pit	**7.** interhyal	**10.** telencephalon	**13.** hypothalamus
2. dorsal mandibular arch	**5.** olfactory bulb	**8.** palatoquadrate	**11.** parasphenoid	**14.** optic tectum
3. intermandibularis	**6.** adductor mandibulae complex	**9.** lens	**12.** ceratohyal	

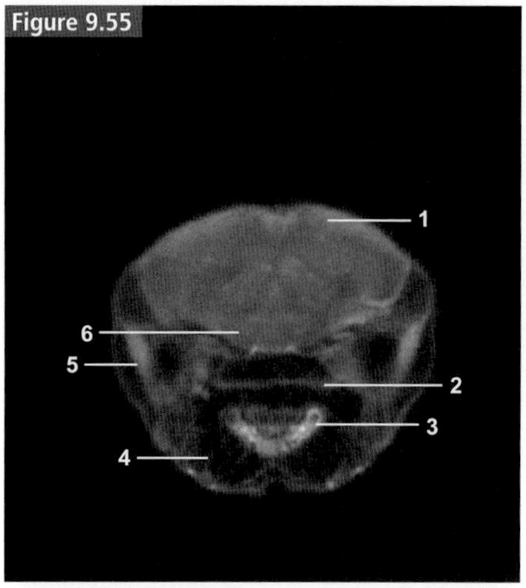

Figure 9.55

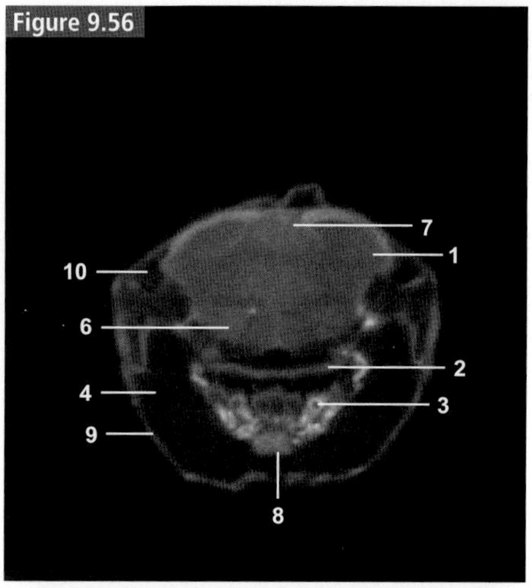

Figure 9.56

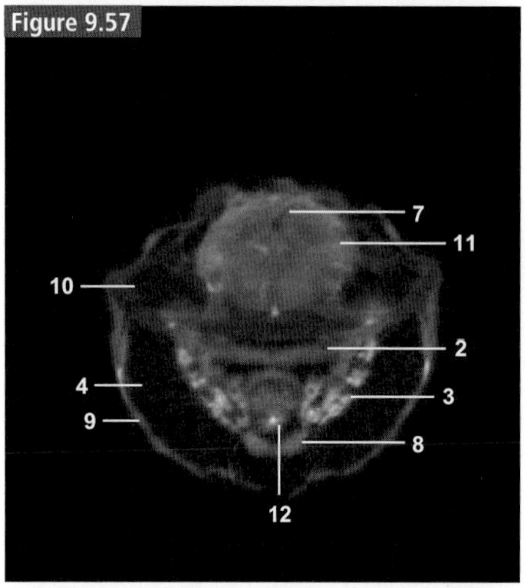

Figure 9.57

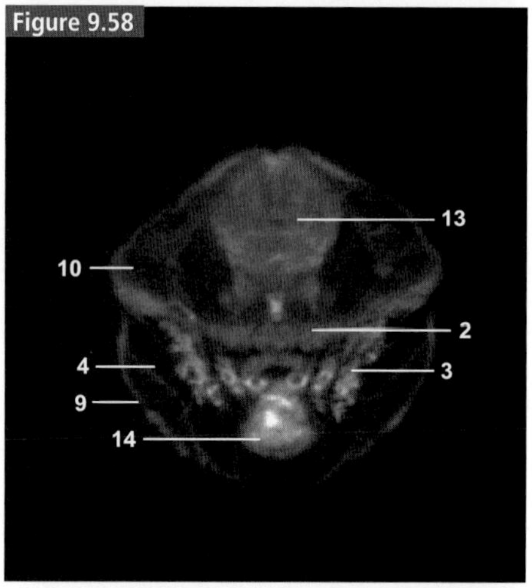

Figure 9.58

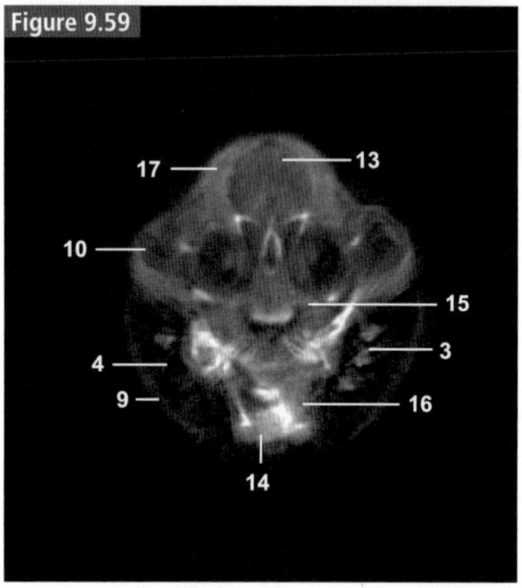

Figure 9.59

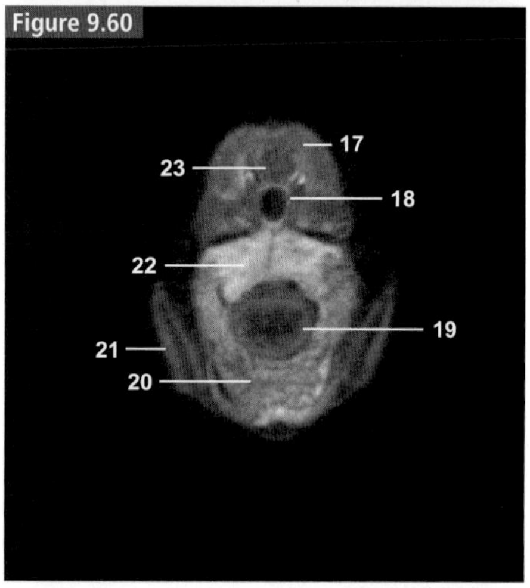

Figure 9.60

1. optic tectum
2. pharynx
3. gill filaments
4. opercular cavity
5. dilator operculi
6. inferior lobe
7. cerebellum
8. sternohyoid
9. opercle
10. semicircular canal
11. granular eminence
12. bulbus arteriosus
13. medulla oblongata
14. cardiac ventricle
15. esophagus
16. atrium
17. myotome
18. caudal vertebra
19. intestinal bulb
20. liver
21. pectoral fin
22. pronephros
23. spinal cord

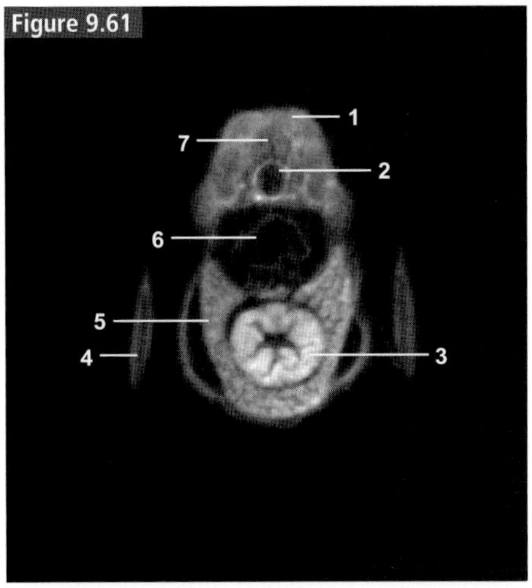

Figure 9.61

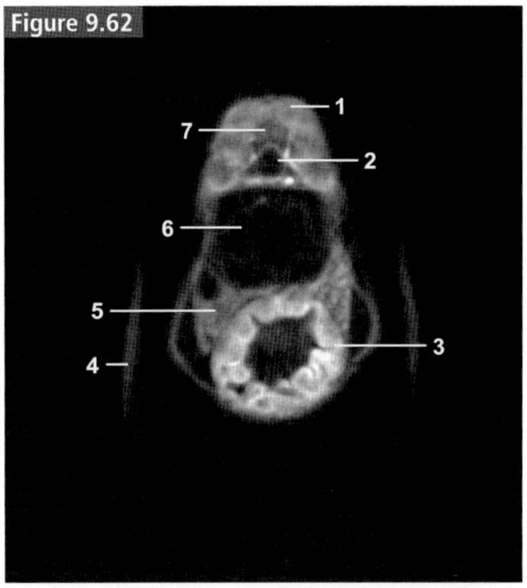

Figure 9.62

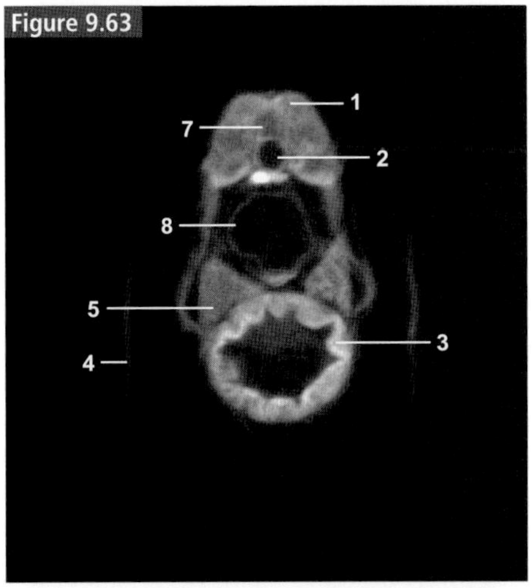

Figure 9.63

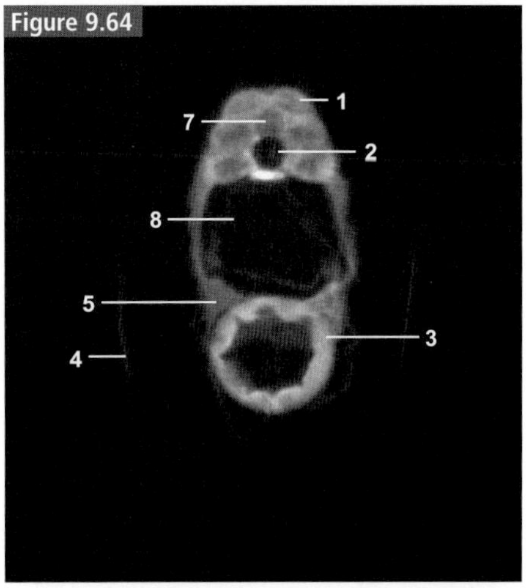

Figure 9.64

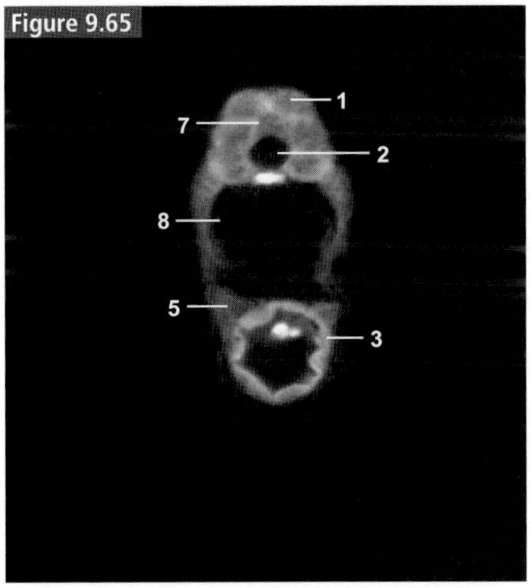

Figure 9.65

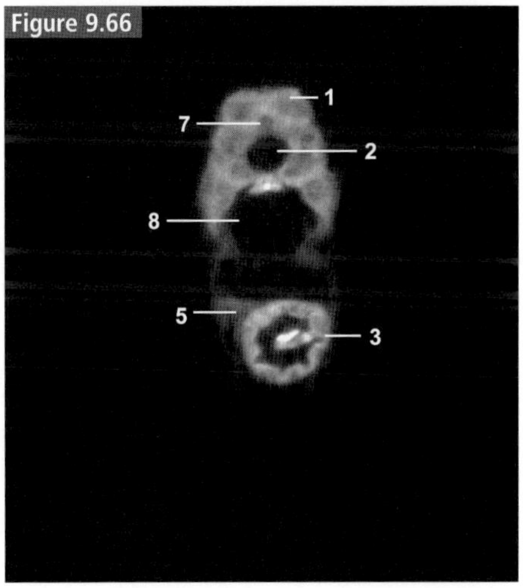

Figure 9.66

1. myotome	**3.** intestinal bulb	**5.** liver	**7.** spinal cord
2. caudal vertebra	**4.** pectoral fin	**6.** anterior chamber swim bladder	**8.** posterior chamber swim bladder

Figure 9.67

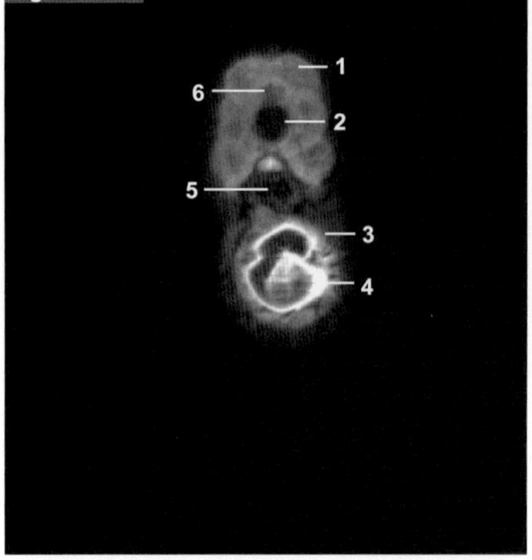

Figure 9.68

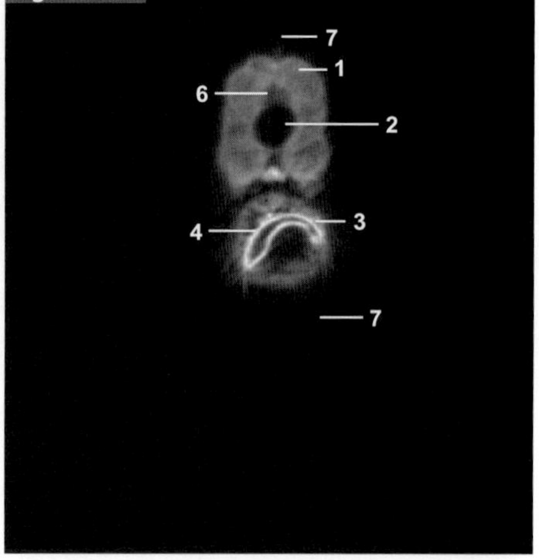

Figure 9.69

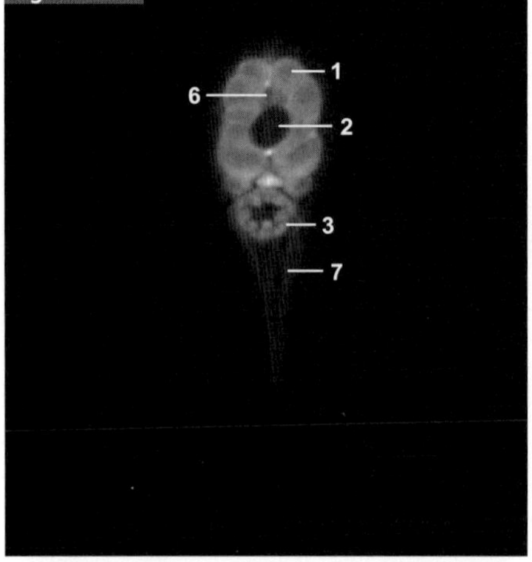

Figure 9.70

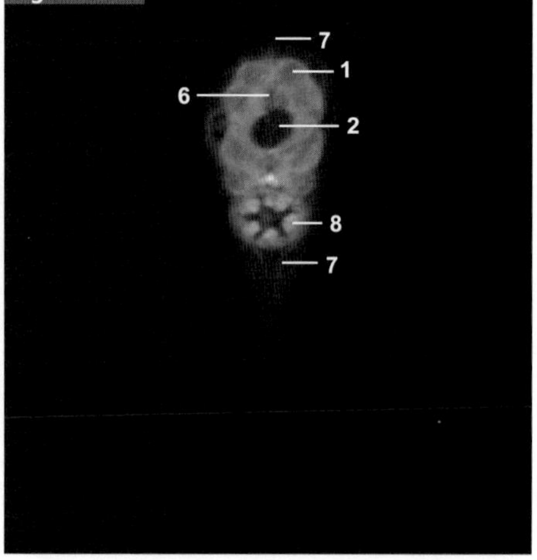

Figure 9.71

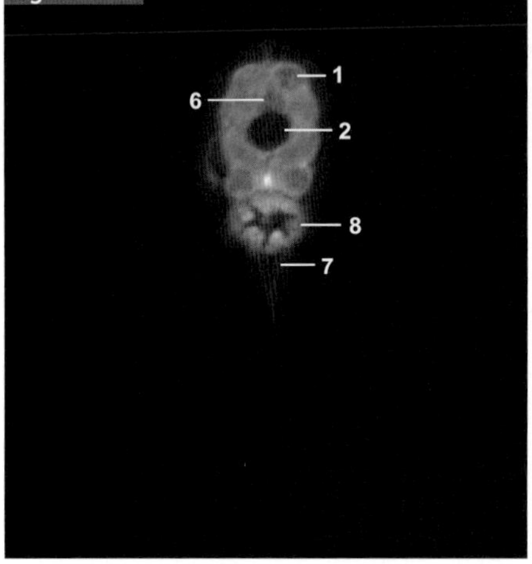

Figure 9.72

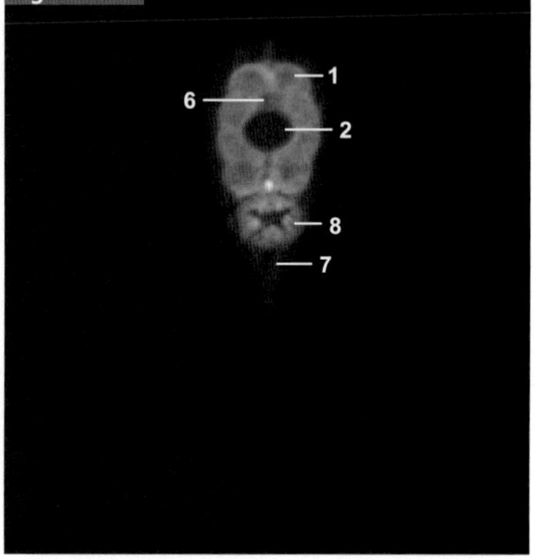

1. myotome	**3.** mid intestine	**5.** posterior chamber swim bladder	**7.** median fin
2. caudal vertebra	**4.** artemia cyst	**6.** spinal cord	**8.** posterior intestine

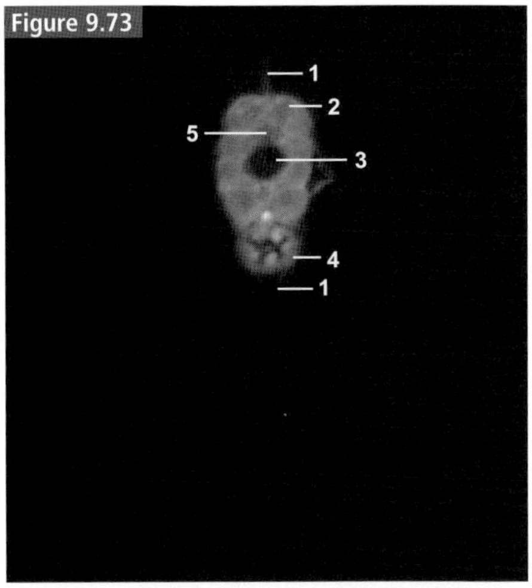

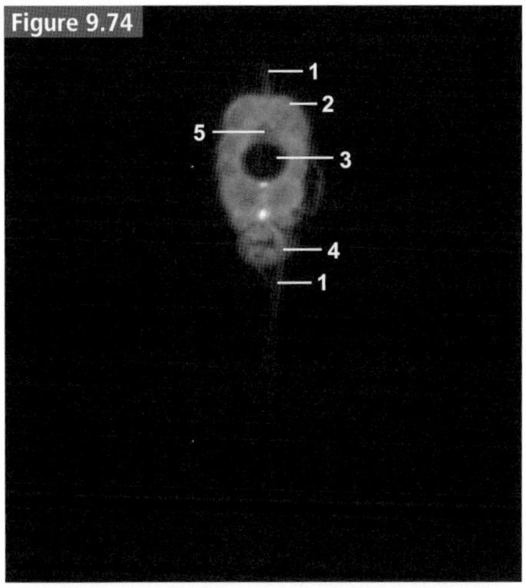

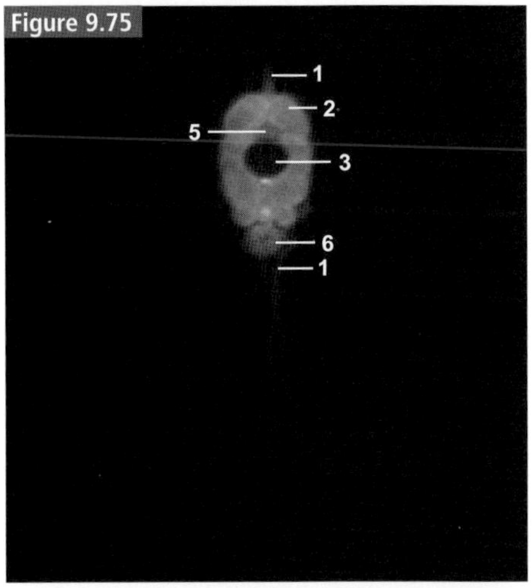

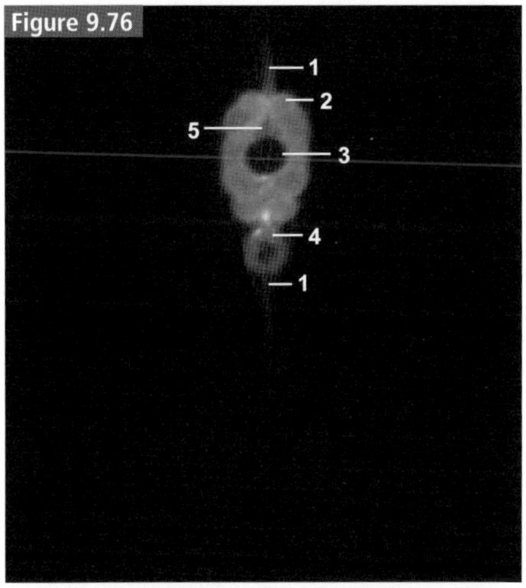

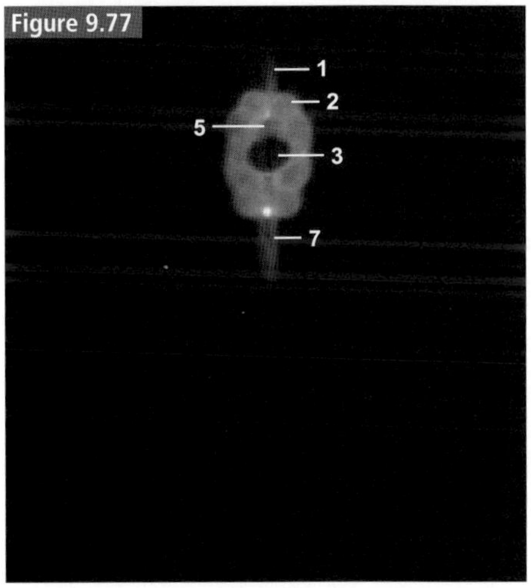

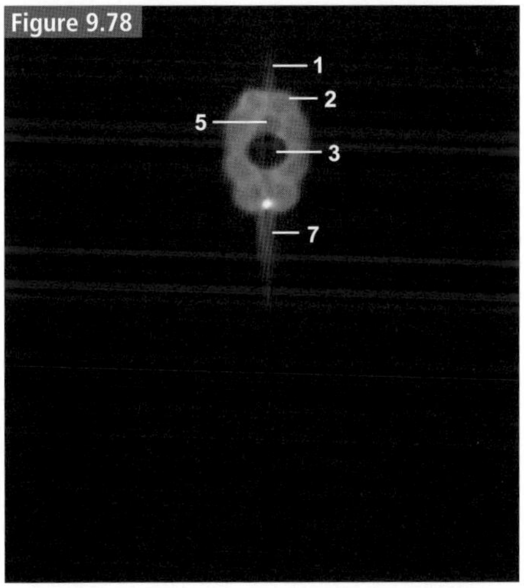

1. median fin	**3.** caudal vertebra	**5.** spinal cord	**6.** cloacal chamber	**7.** anal fin
2. myotome	**4.** posterior intestine			

CHAPTER 10

9 MM

Sagittal Sections

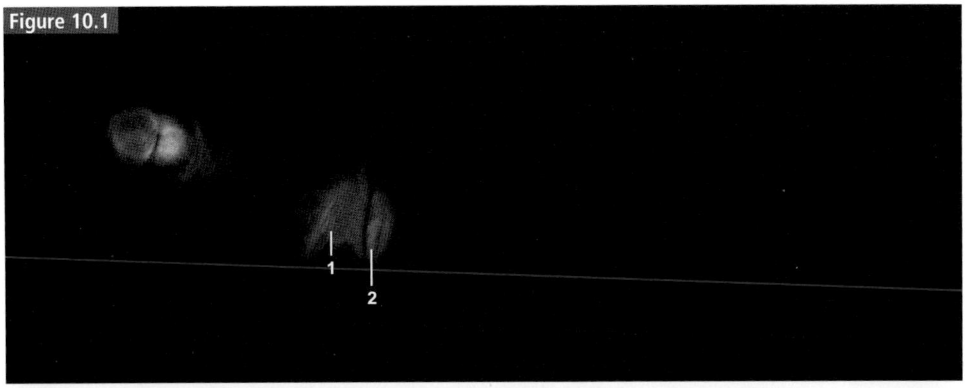

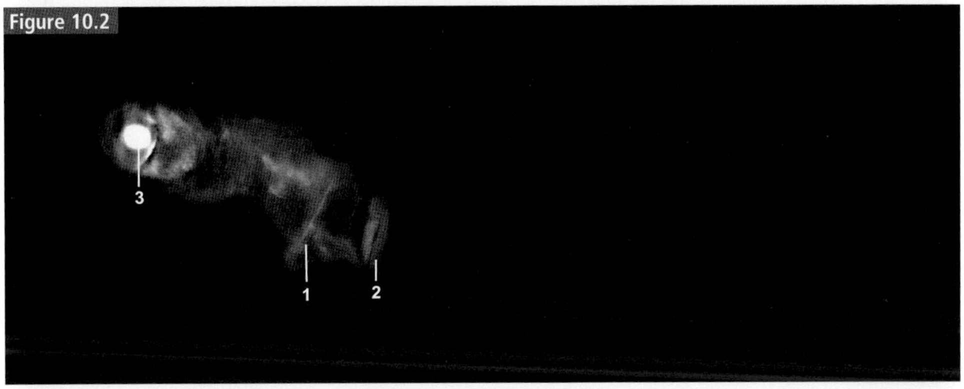

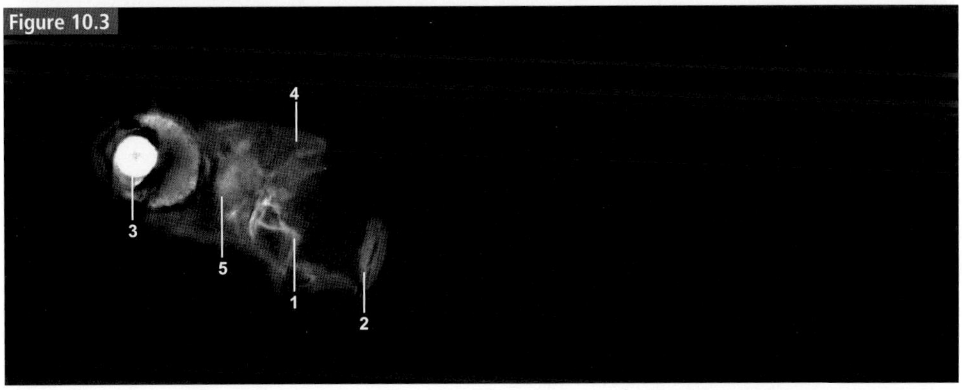

1. opercle	**2.** pectoral fin musculature	**3.** lens	**4.** semicircular canal	**5.** dilator operculi

Atlas of Zebrafish Development. 10.1016/B978-0-12-374016-8.00010-8

Figure 10.4

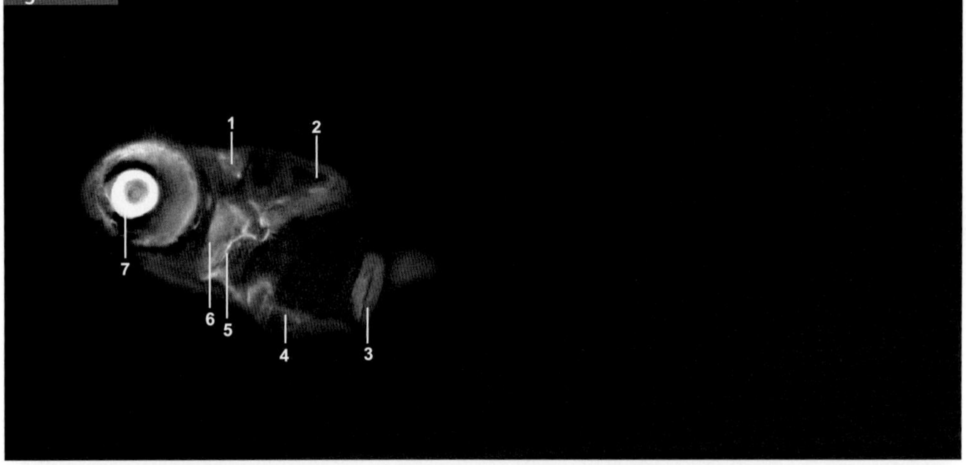

Figure 10.5

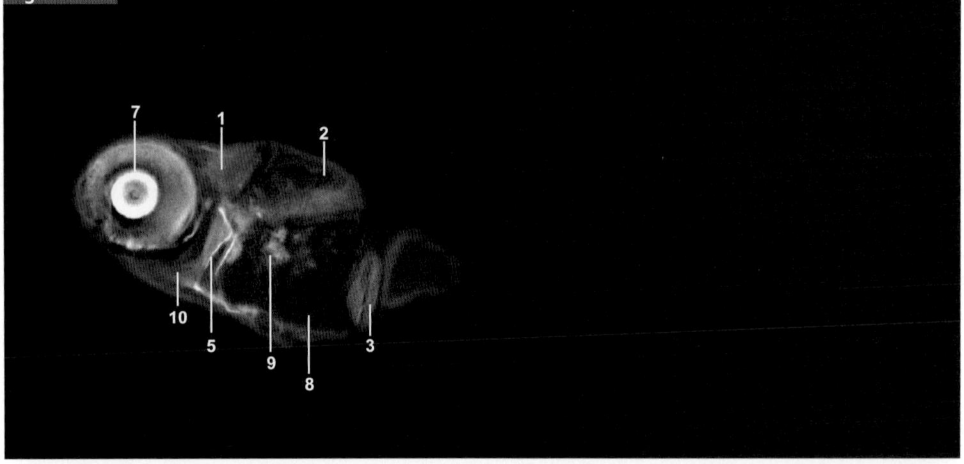

Figure 10.6

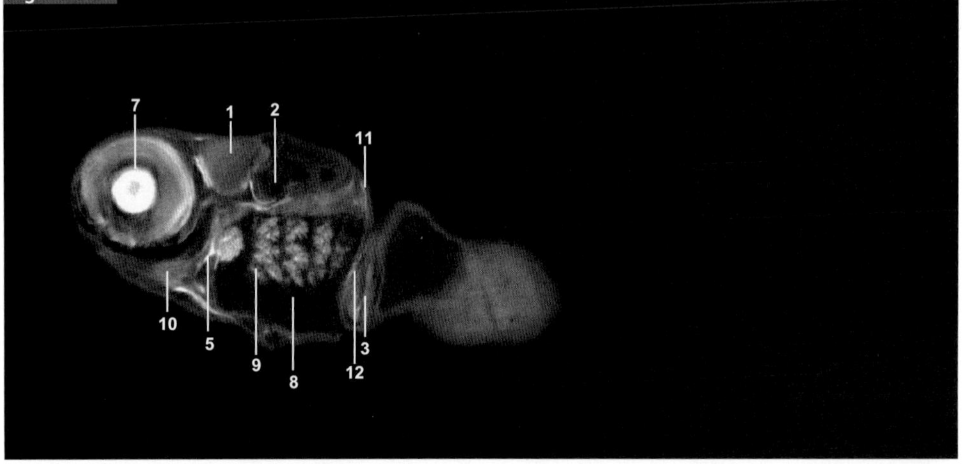

1. optic tectum	**4.** opercle	**7.** lens	**10.** adductor mandibulae
2. semicircular canal	**5.** hyomandibula	**8.** opercular cavity	**11.** supracleithrum
3. pectoral fin musculature	**6.** dilator operculi	**9.** gill filaments	**12.** cleithrum

Figure 10.7

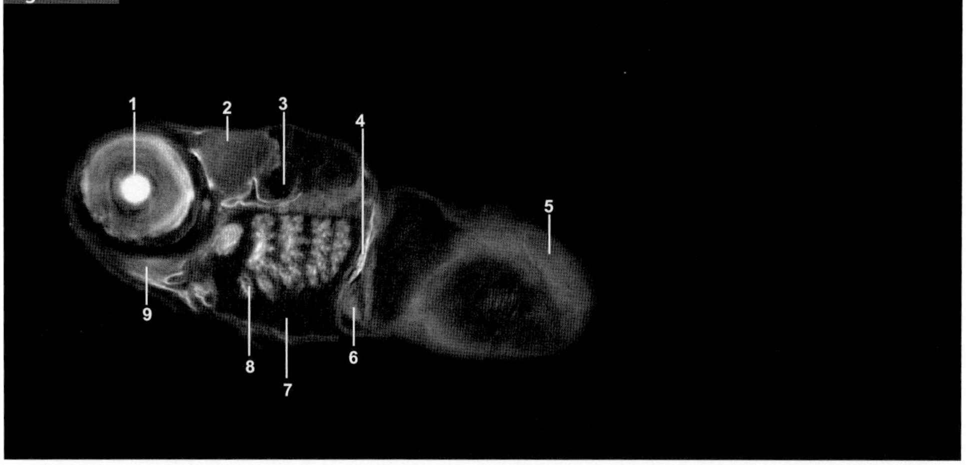

Figure 10.8

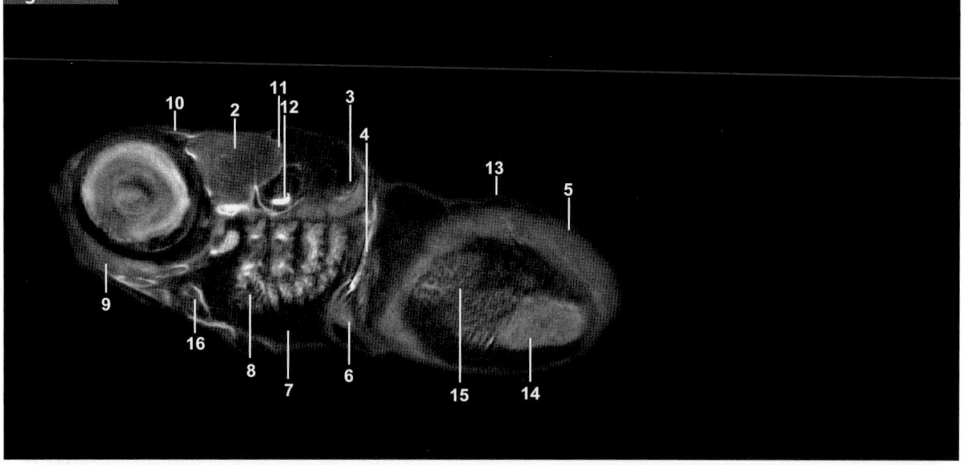

Figure 10.9

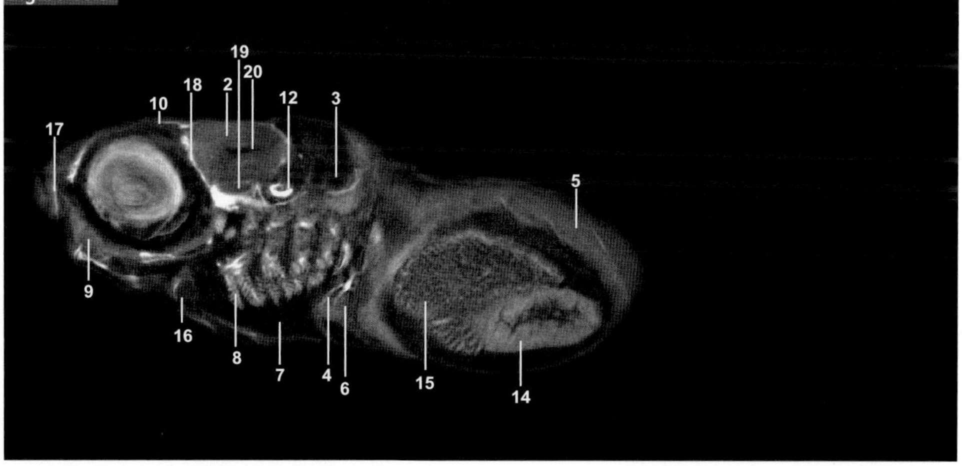

1. lens	**6.** pectoral fin musculature	**11.** eminentia granularis	**16.** ceratohyal
2. optic tectum	**7.** opercular cavity	**12.** otolith	**17.** maxilla
3. semicircular canal	**8.** gill filaments	**13.** scales	**18.** infraorbital
4. cleithrum	**9.** adductor mandibulae	**14.** intestinal bulb	**19.** torus lateralis
5. myotome	**10.** frontal bone	**15.** liver	**20.** tectal ventricle

Figure 10.10

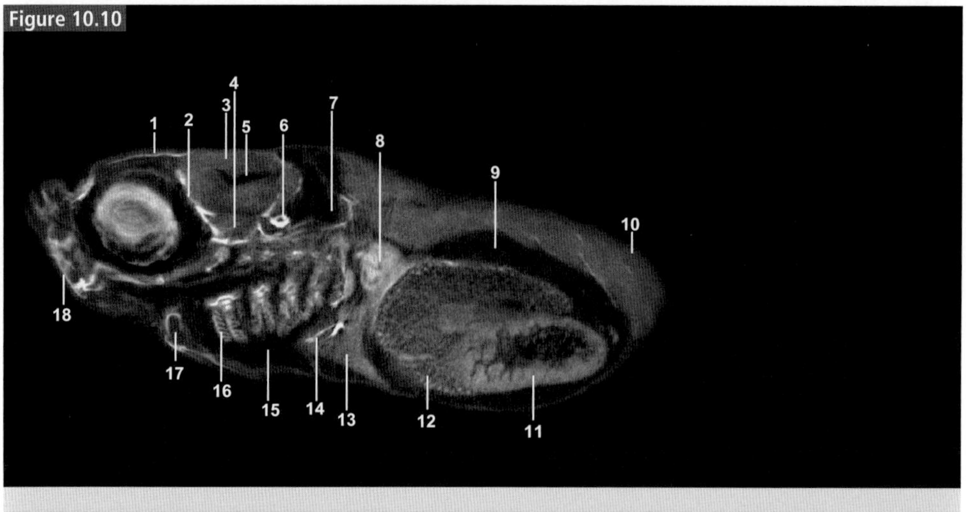

Figure 10.11

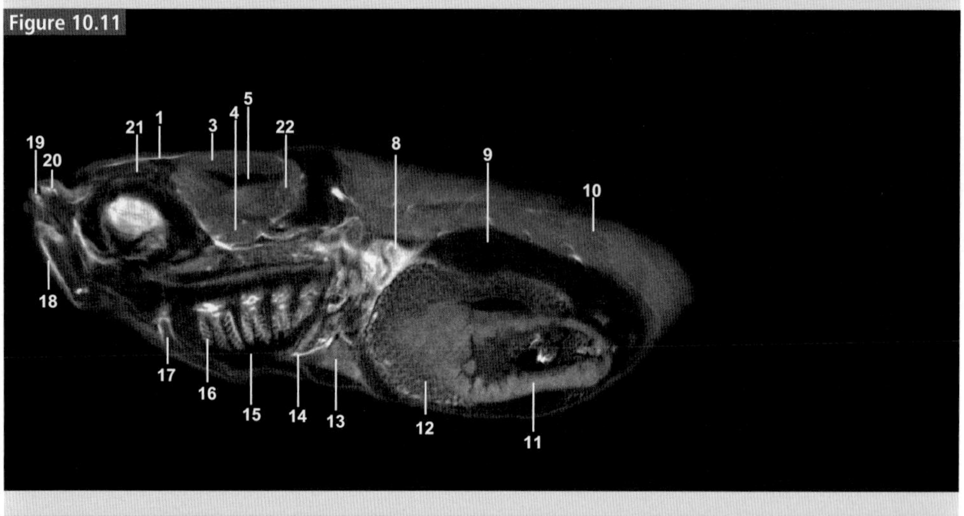

Figure 10.12

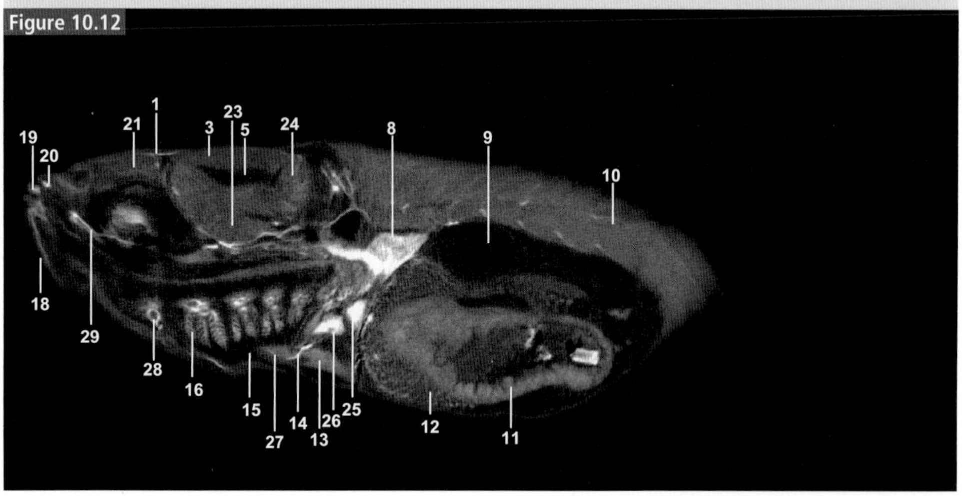

1. frontal bone
2. infraorbital
3. optic tectum
4. torus lateralis
5. tectal ventricle
6. otolith
7. semicircular canal
8. pronephros
9. anterior chamber swim bladder
10. myotome
11. intestinal bulb
12. liver
13. pectoral fin musculature
14. cleithrum
15. opercular cavity
16. gill filaments
17. ceratohyal
18. dentary
19. premaxilla
20. maxilla
21. telencephalon
22. eminentia granularis
23. inferior lobe
24. cerebellum
25. sinus venosus
26. atrium
27. sternohyoid
28. retroarticular
29. infraorbital

Figure 10.13

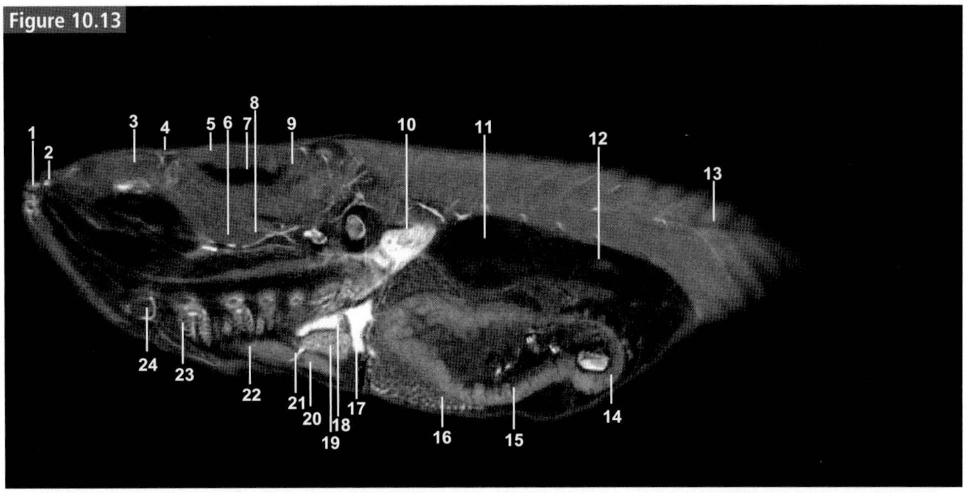

Figure 10.14

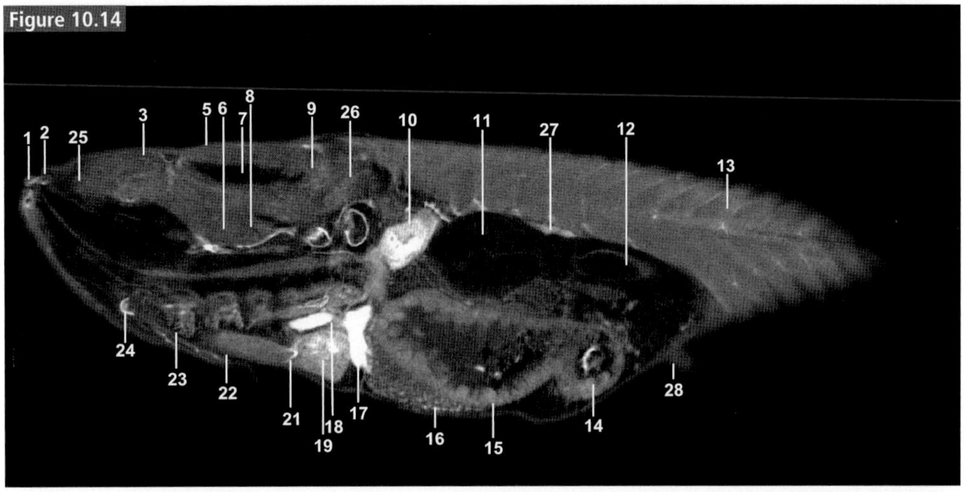

Figure 10.15

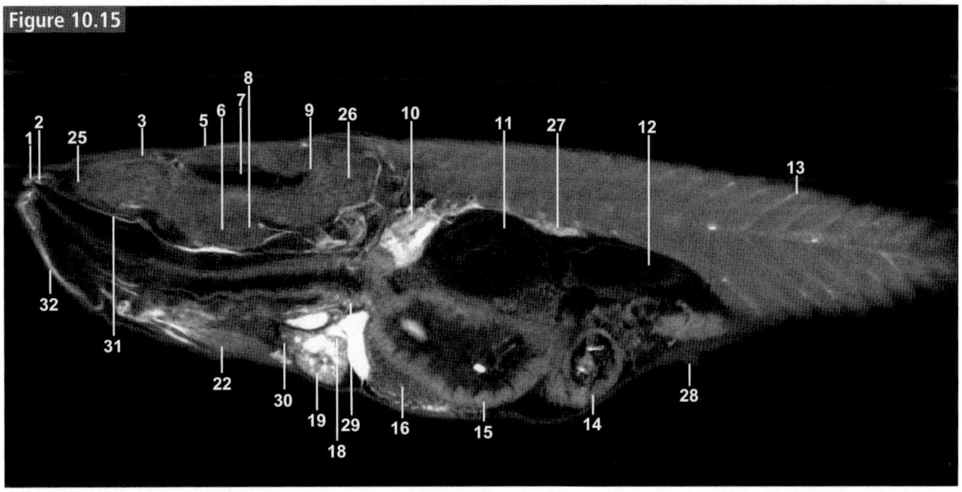

1. premaxilla
2. maxilla
3. telencephalon
4. frontal bone
5. optic tectum
6. hypothalamus
7. tectal ventricle
8. inferior lobe
9. cerebellum
10. pronephros
11. anterior chamber swim bladder
12. posterior chamber swim bladder
13. myotome
14. mid intestine
15. intestinal bulb
16. liver
17. sinus venosus
18. atrium
19. cardiac ventricle
20. pectoral fin musculature
21. cleithrum
22. sternohyoid
23. gill filaments
24. retroarticular
25. olfactory bulb
26. crista cerebellaris
27. mesonephros
28. pelvic fin
29. esophagus
30. bulbus arteriosus
31. parasphenoid
32. dentary

Figure 10.16

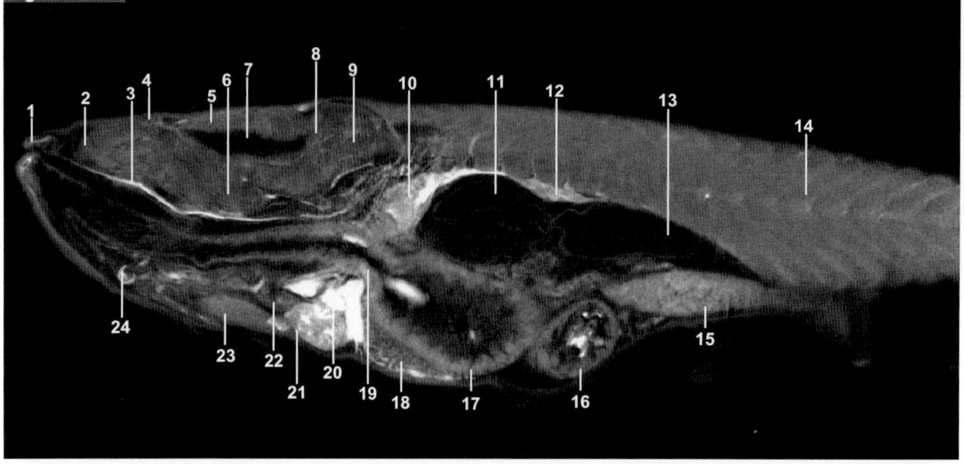

Figure 10.17

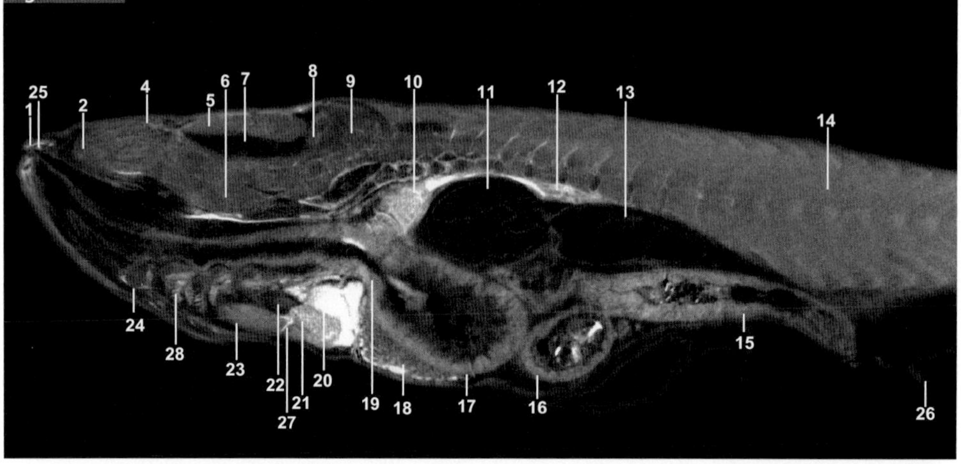

Figure 10.18

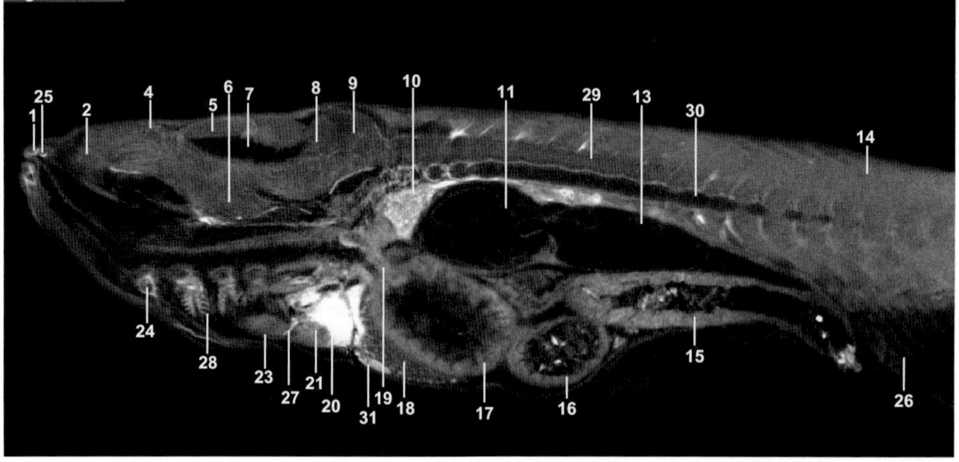

1. premaxilla
2. olfactory bulb
3. parasphenoid
4. telencephalon
5. optic tectum
6. hypothalamus
7. tectal ventricle
8. cerebellum
9. crista cerebellaris
10. pronephros
11. anterior chamber swim bladder
12. mesonephros
13. posterior chamber swim bladder
14. myotome
15. posterior intestine
16. mid intestine
17. intestinal bulb
18. liver
19. esophagus
20. atrium
21. cardiac ventricle
22. bulbus arteriosus
23. sternohyoid
24. retroarticular
25. maxilla
26. anal fin
27. cleithrum
28. gill filaments
29. spinal cord
30. vertebral column
31. hepatic portal vein

Figure 10.19

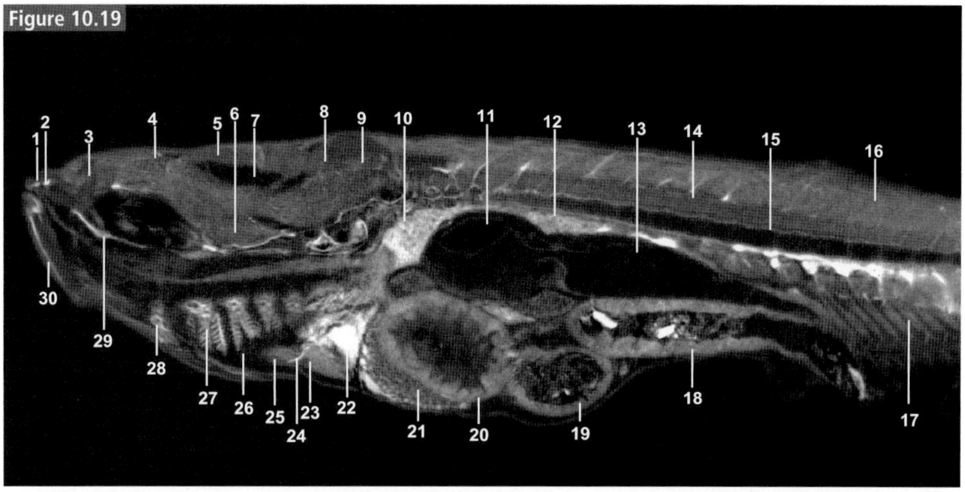

Figure 10.20

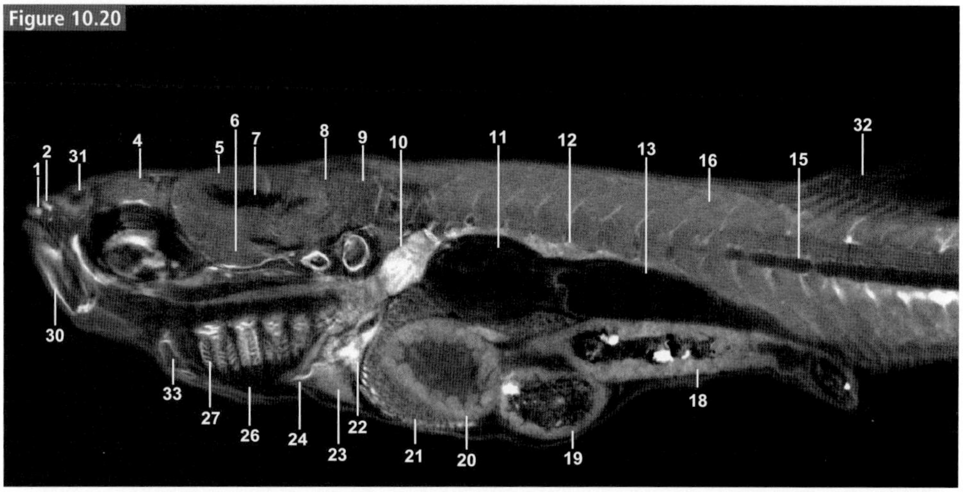

Figure 10.21

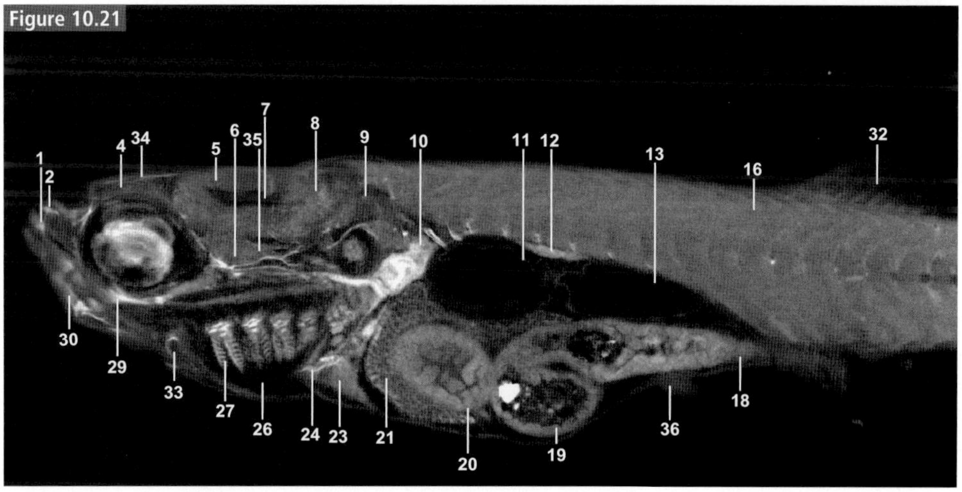

1. premaxilla	**9.** crista cerebellaris	**17.** anal fin	**25.** sternohyoid	**33.** ceratohyal
2. maxilla	**10.** pronephros	**18.** posterior intestine	**26.** opercular cavity	**34.** frontal bone
3. olfactory bulb	**11.** anterior chamber swim bladder	**19.** mid intestine	**27.** gill filaments	**35.** inferior lobe
4. telencephalon	**12.** mesonephros	**20.** intestinal bulb	**28.** retroarticular	**36.** pelvic fin
5. optic tectum	**13.** posterior chamber swim bladder	**21.** liver	**29.** infraorbital	
6. hypothalamus	**14.** spinal cord	**22.** atrium	**30.** dentary	
7. tectal ventricle	**15.** vertebral column	**23.** pectoral fin musculature	**31.** olfactory pit	
8. cerebellum	**16.** myotome	**24.** cleithrum	**32.** dorsal fin	

Figure 10.22

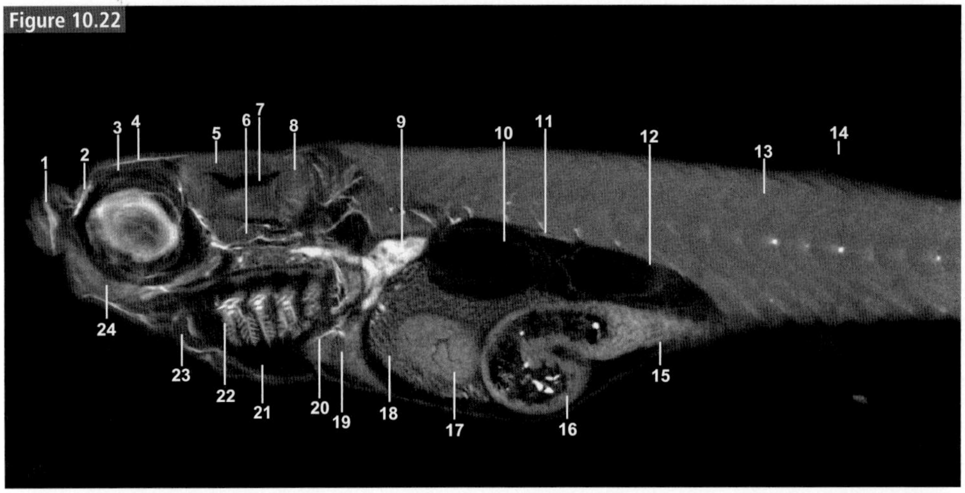

Figure 10.23

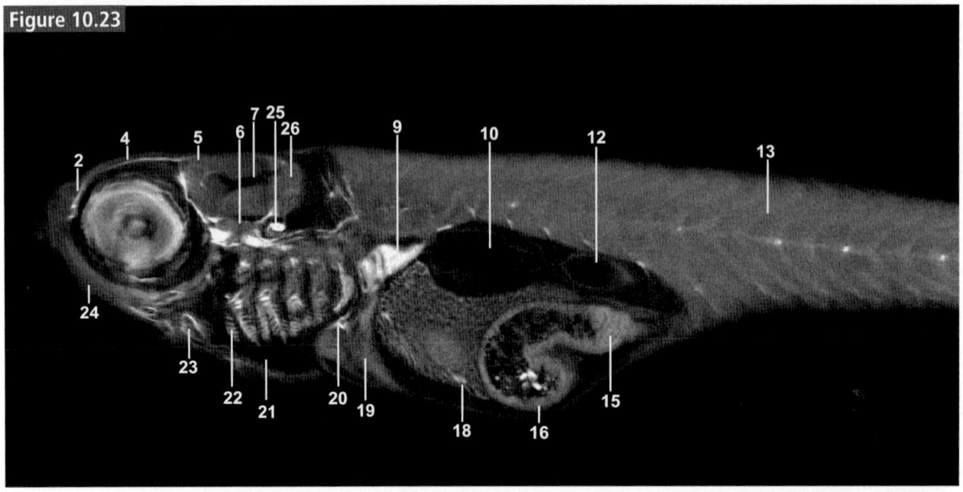

Figure 10.24

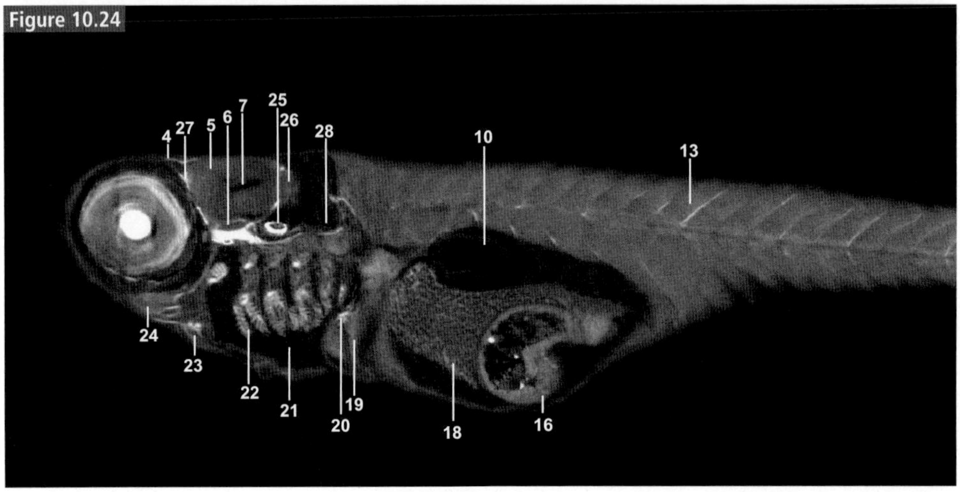

1. maxilla
2. lateral ethmoid
3. telencephalon
4. frontal bone
5. optic tectum
6. inferior lobe
7. tectal ventricle
8. cerebellum
9. pronephros
10. anterior chamber swim bladder
11. mesonephros
12. posterior chamber swim bladder
13. myotome
14. dorsal fin
15. posterior intestine
16. mid intestine
17. intestinal bulb
18. liver
19. pectoral fin musculature
20. cleithrum
21. opercular cavity
22. gill filaments
23. ceratohyal
24. adductor mandibulae
25. otolith
26. eminentia granularis
27. infraorbital
28. semicircular canal

Figure 10.25

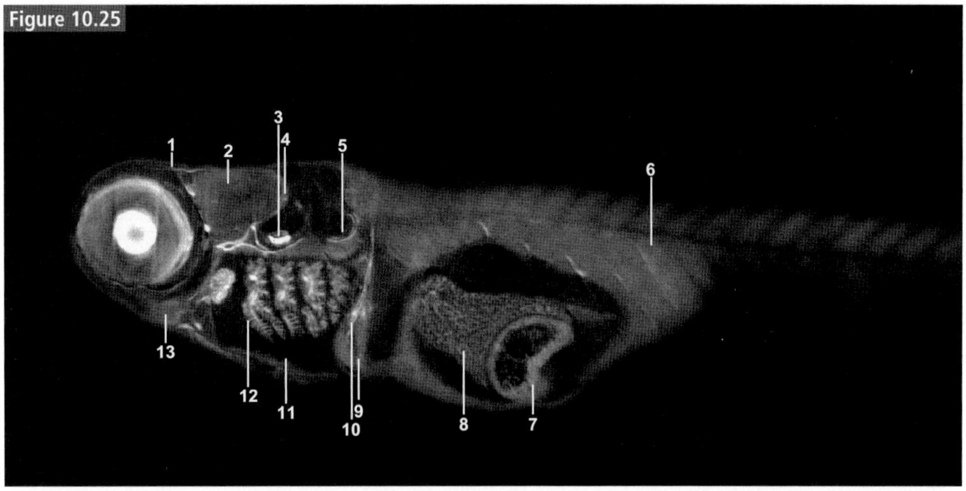

Figure 10.26

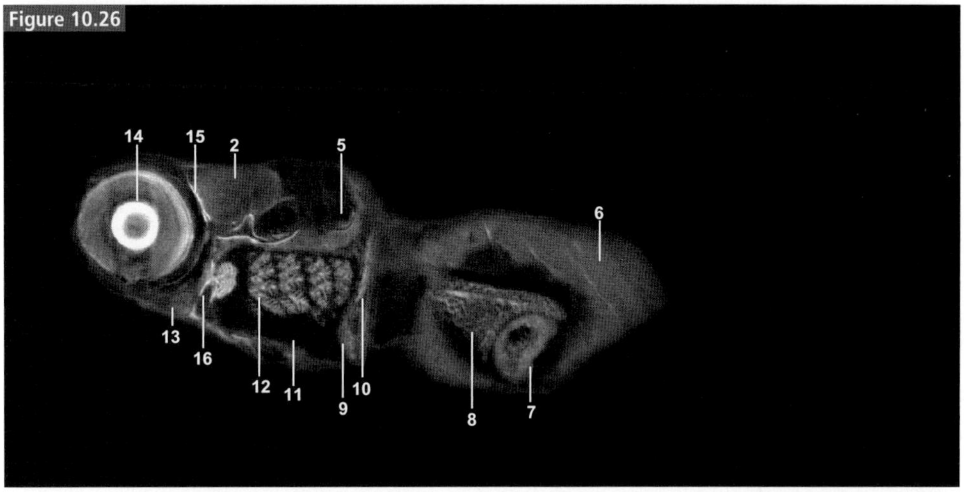

Figure 10.27

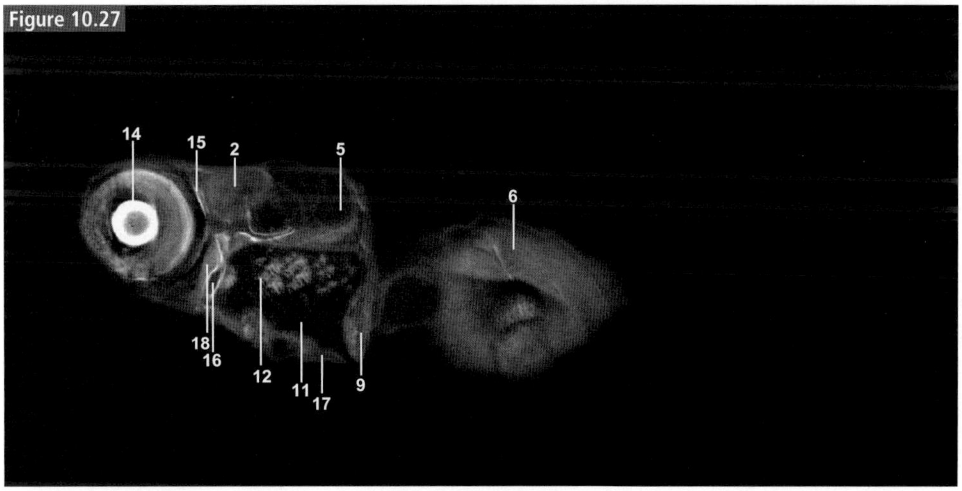

1. frontal bone
2. optic tectum
3. otolith
4. eminentia granularis
5. semicircular canal
6. myotome
7. mid intestine
8. liver
9. pectoral fin musculature
10. cleithrum
11. opercular cavity
12. gill filaments
13. adductor mandibulae
14. lens
15. infraorbital
16. hyomandibula
17. opercle
18. dilator operculi

Figure 10.28

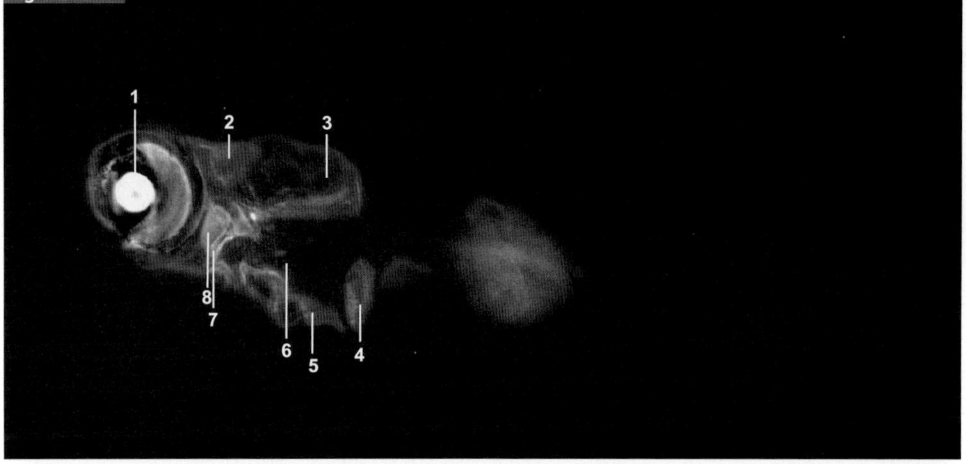

Figure 10.29

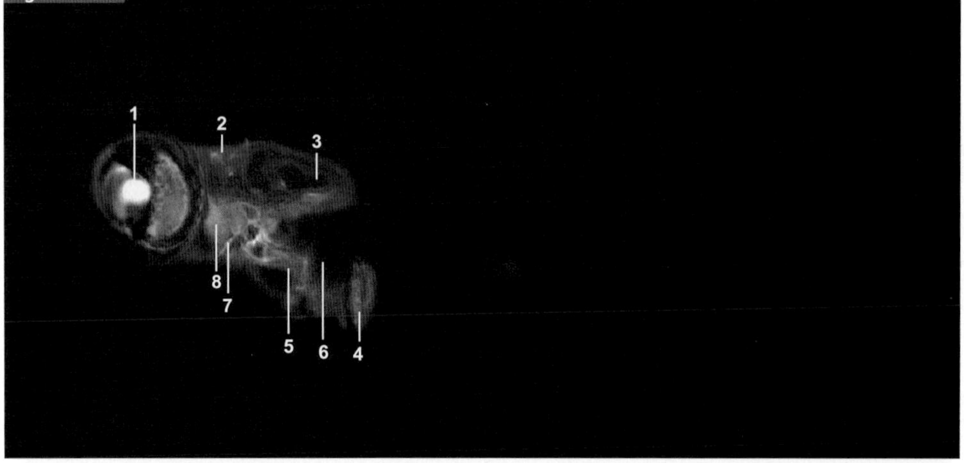

Figure 10.30

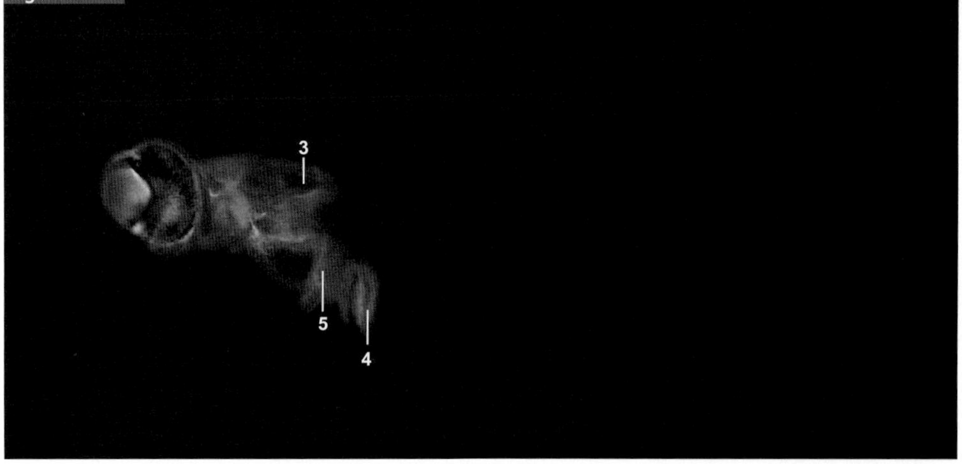

1. lens	**3.** semicircular canal	**5.** opercle	**7.** hyomandibula
2. optic tectum	**4.** pectoral fin musculature	**6.** opercular cavity	**8.** dilator operculi

Coronal Sections

Figure 10.31

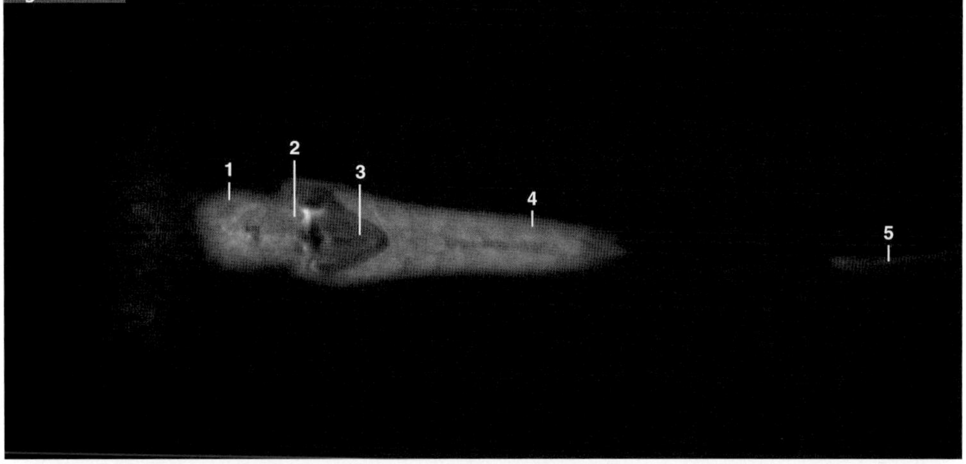

Figure 10.32

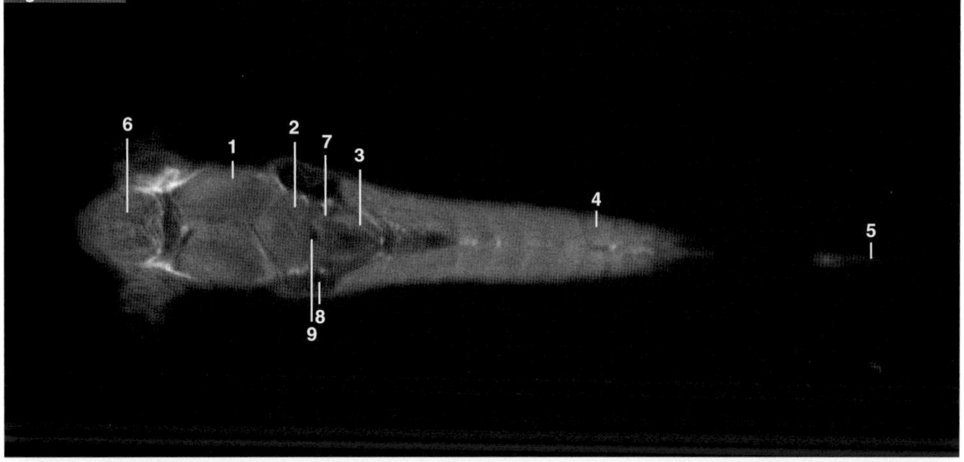

Figure 10.33

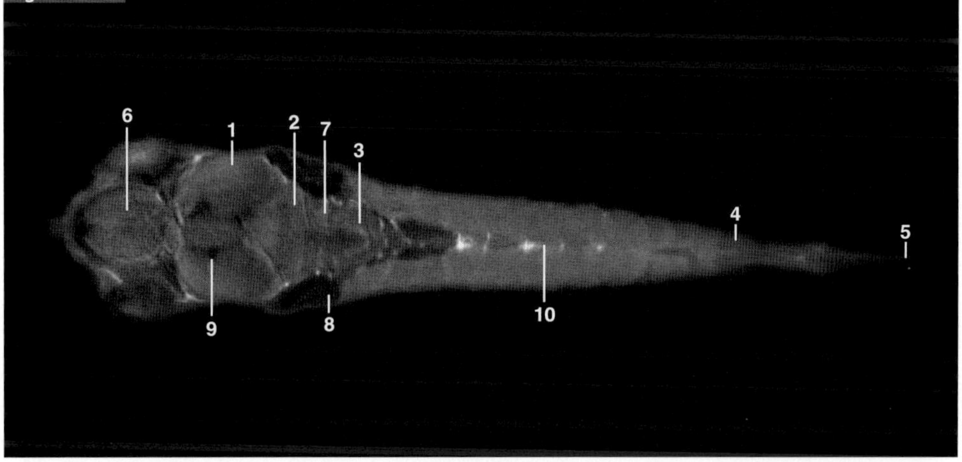

1. optic tectum	**3.** vagal lobe	**5.** dorsal fin	**7.** crista cerebellaris	**9.** tectal ventricle
2. cerebellum	**4.** myotome	**6.** telencephalon	**8.** semicircular canal	**10.** spinal cord

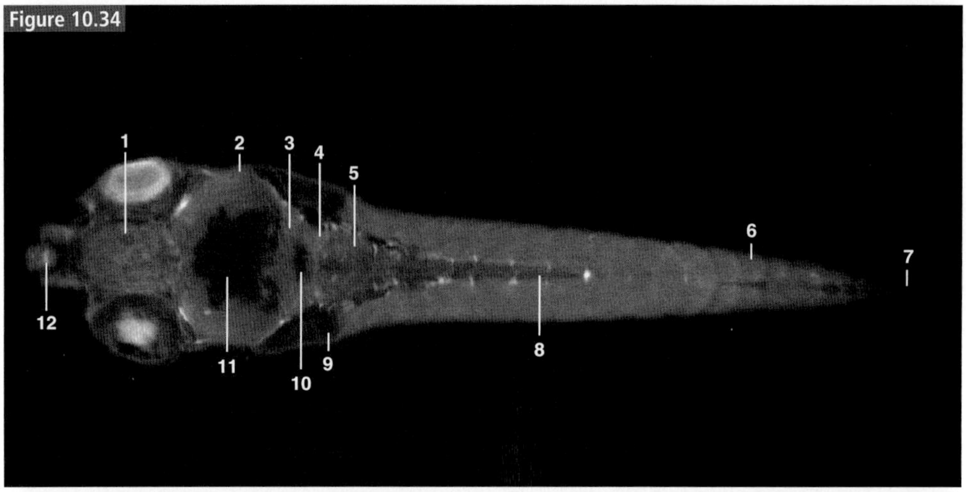

Figure 10.34

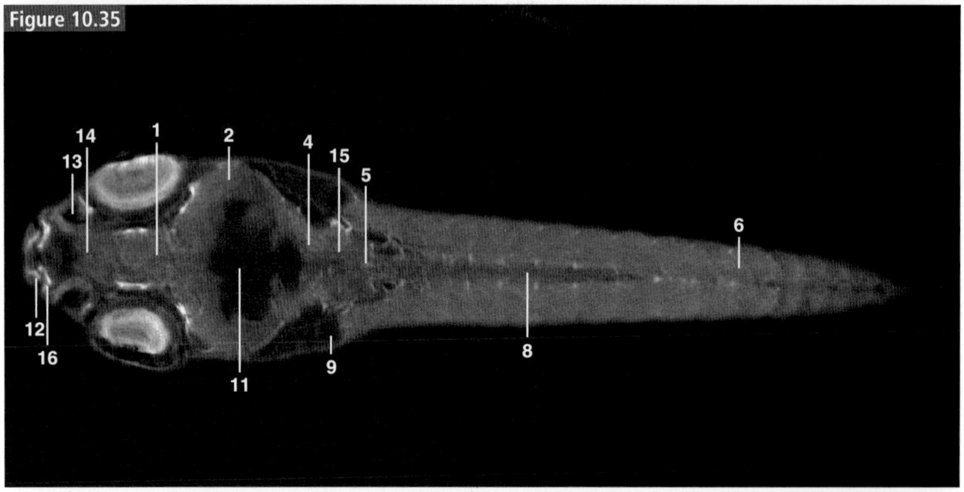

Figure 10.35

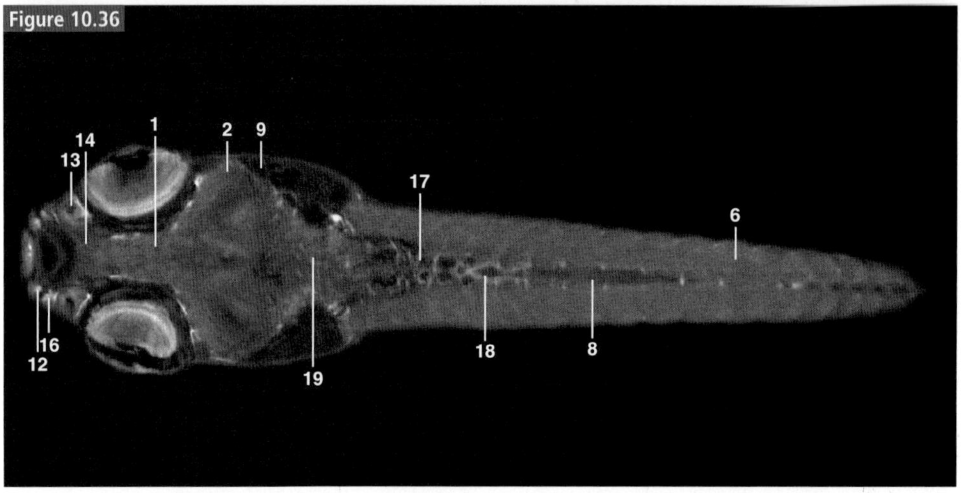

Figure 10.36

1. telencephalon
2. optic tectum
3. cerebellum
4. crista cerebellaris
5. vagal lobe
6. myotome
7. dorsal fin
8. spinal cord
9. semicircular canal
10. fourth ventricle
11. tectal ventricle
12. premaxilla
13. olfactory pit
14. olfactory bulb
15. facial lobe
16. maxilla
17. Weberian apparatus
18. precaudal vertebra
19. medulla oblongata

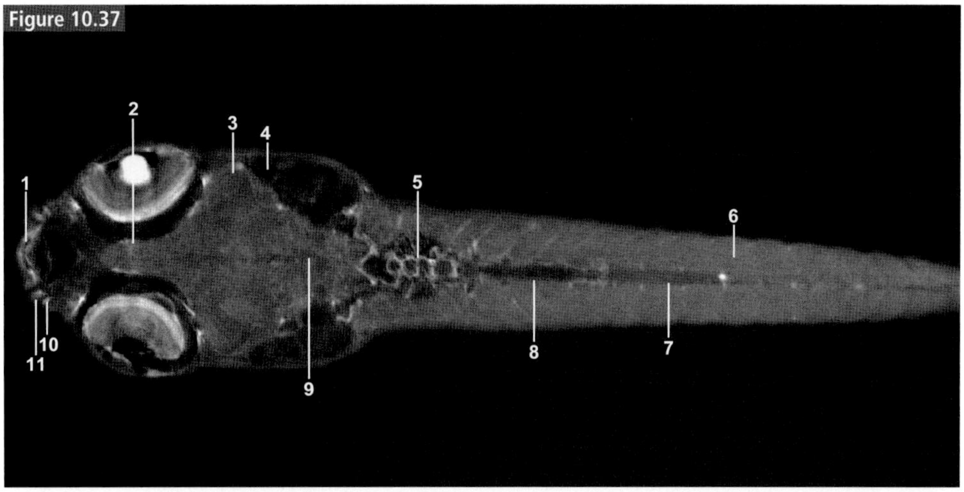

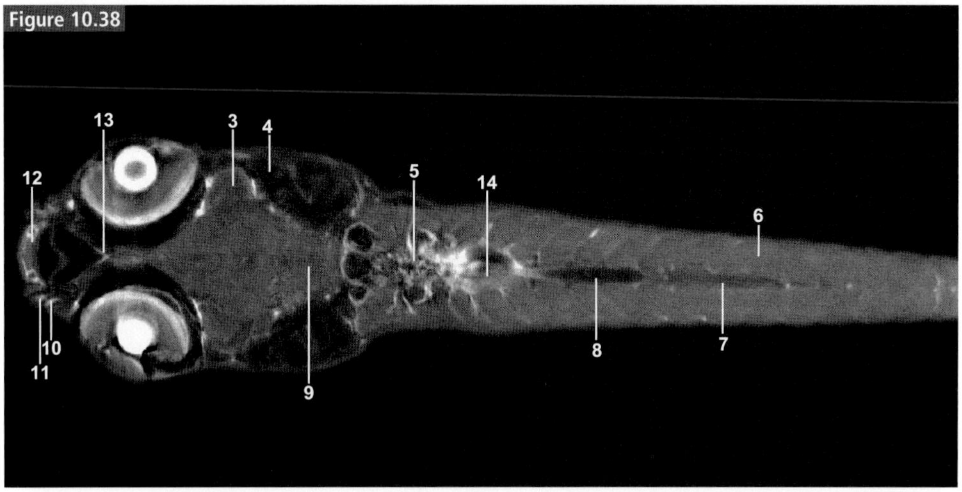

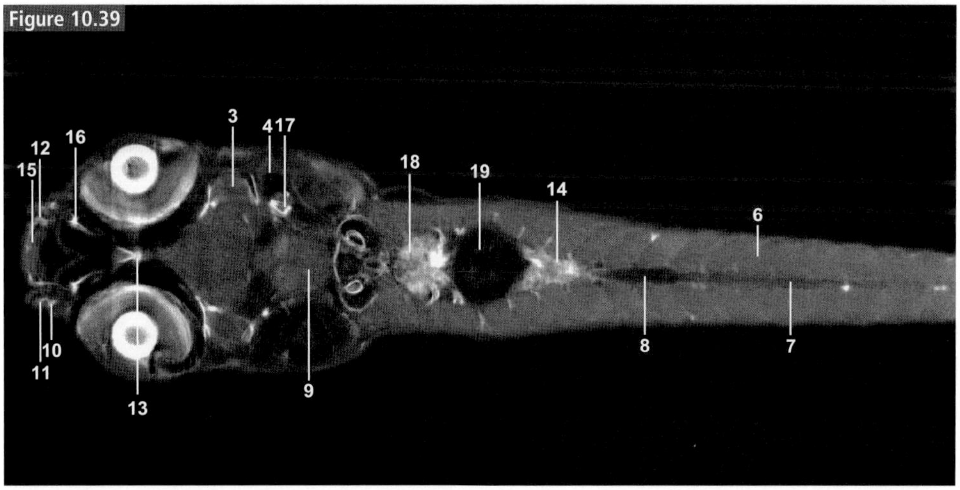

1. Meckel's cartilage	**5.** Weberian apparatus	**9.** medulla oblongata	**13.** parasphenoid	**17.** otolith
2. telencephalon	**6.** myotome	**10.** maxilla	**14.** mesonephros	**18.** pronephros
3. optic tectum	**7.** spinal cord	**11.** premaxilla	**15.** dentary	**19.** anterior chamber swim bladder
4. semicircular canal	**8.** precaudal vertebra	**12.** intermandibularis	**16.** palatoquadrate	**20.** parasphenoid

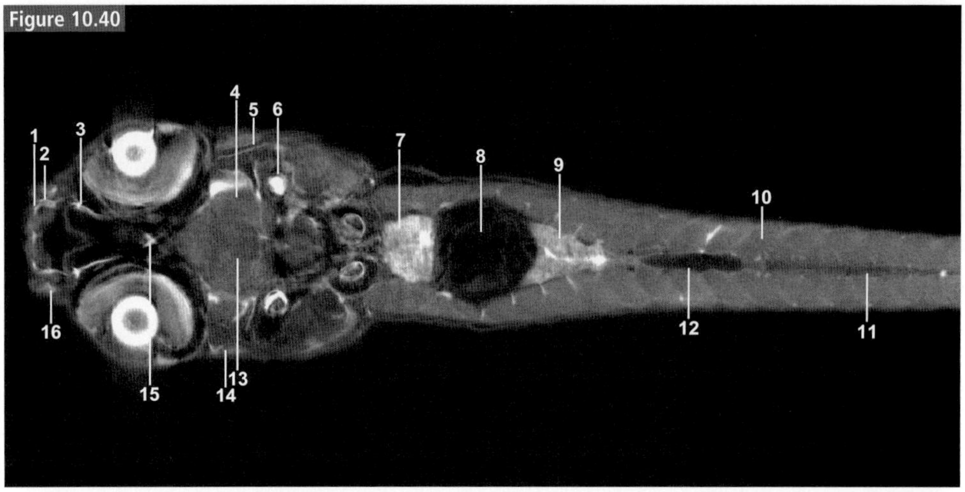

Figure 10.40

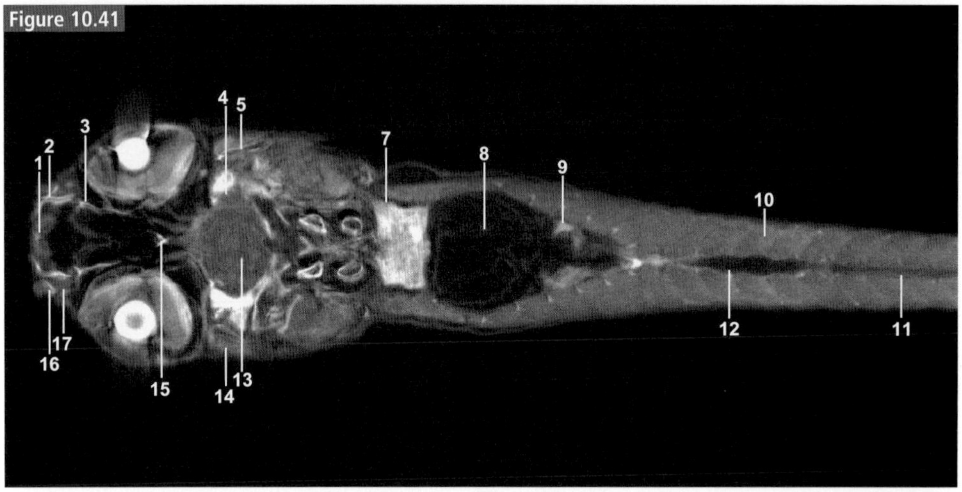

Figure 10.41

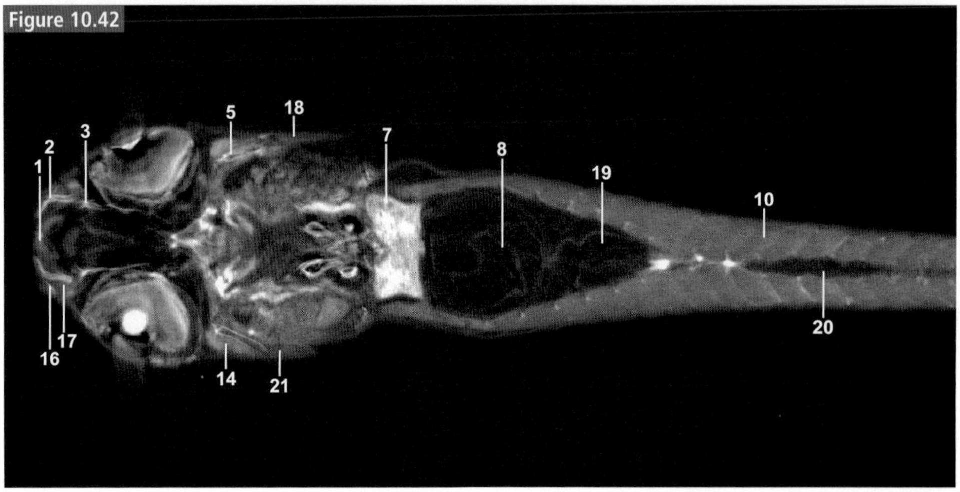

Figure 10.42

1. intermandibularis
2. dentary
3. palatoquadrate
4. inferior lobe
5. hyomandibula
6. otolith
7. pronephros
8. anterior chamber swim bladder
9. mesonephros
10. myotome
11. spinal cord
12. precaudal vertebra
13. hypothalamus
14. dilator operculi
15. parasphenoid
16. maxilla
17. adductor mandibulae complex
18. opercle
19. posterior chamber swim bladder
20. caudal vertebra
21. adductor operculi

Figure 10.43

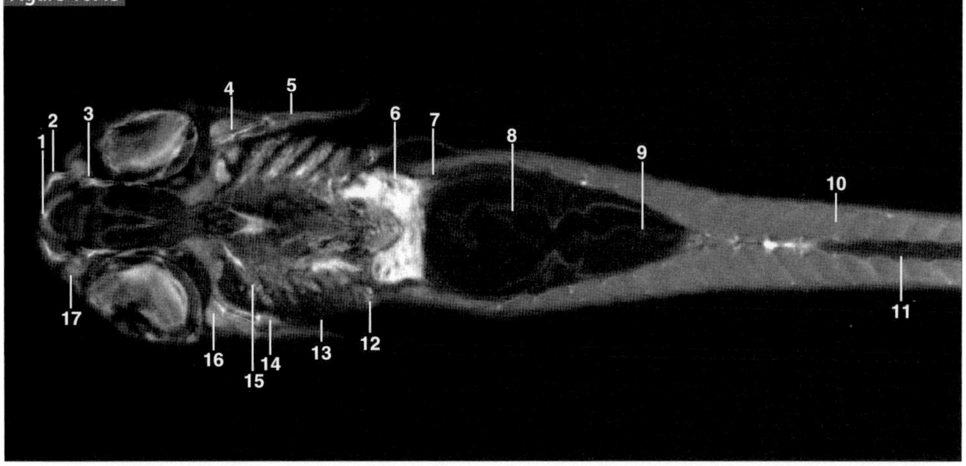

Figure 10.44

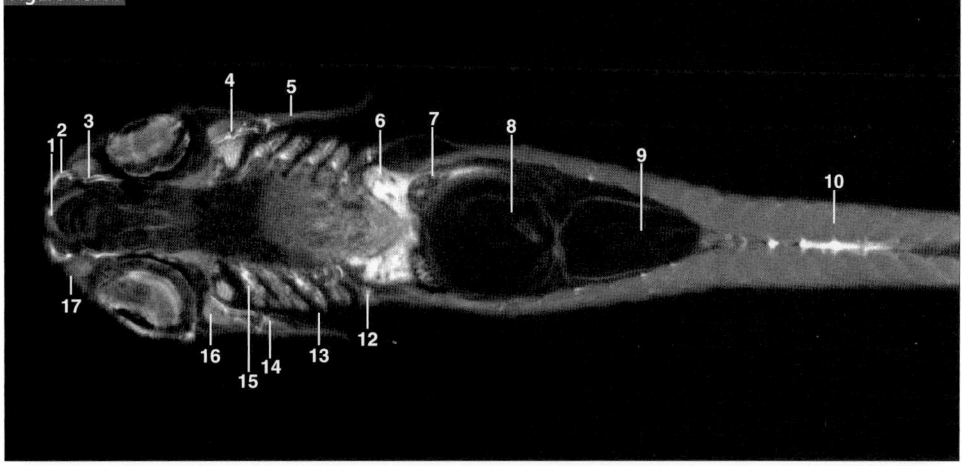

Figure 10.45

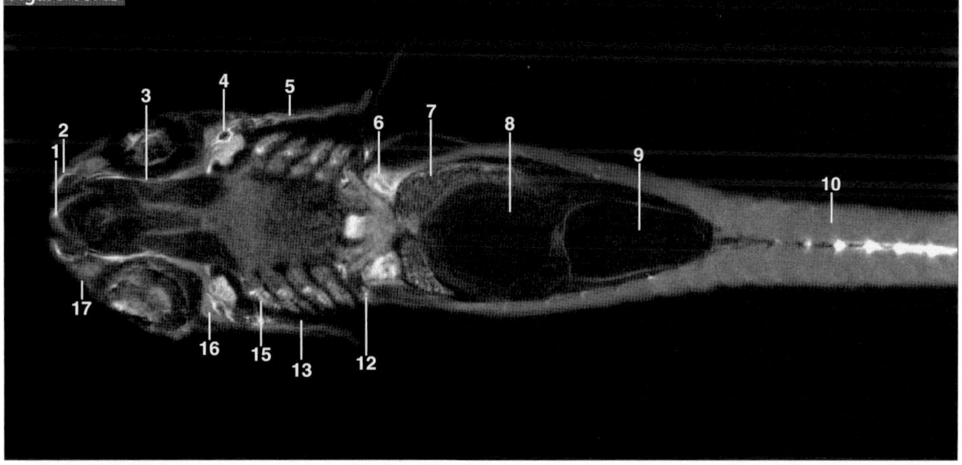

1. intermandibularis
2. dentary
3. palatoquadrate
4. hyomandibula
5. opercle
6. pronephros
7. liver
8. anterior chamber swim bladder
9. posterior chamber swim bladder
10. myotome
11. caudal vertebra
12. cleithrum
13. opercular cavity
14. adductor operculi
15. gill filaments
16. dilator operculi
17. adductor mandibulae complex

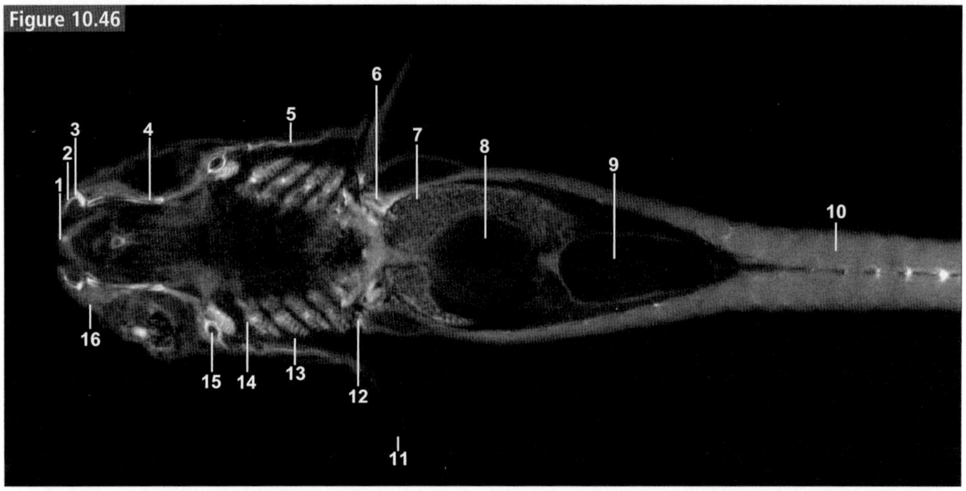

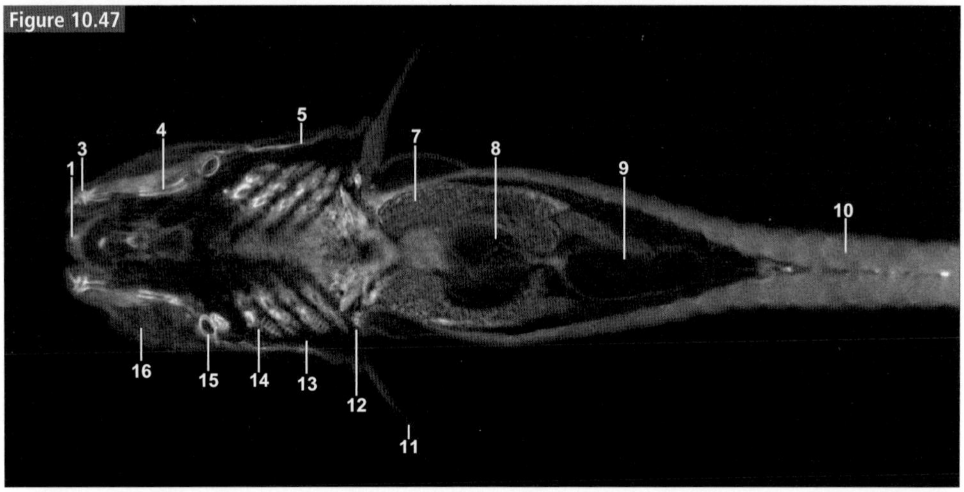

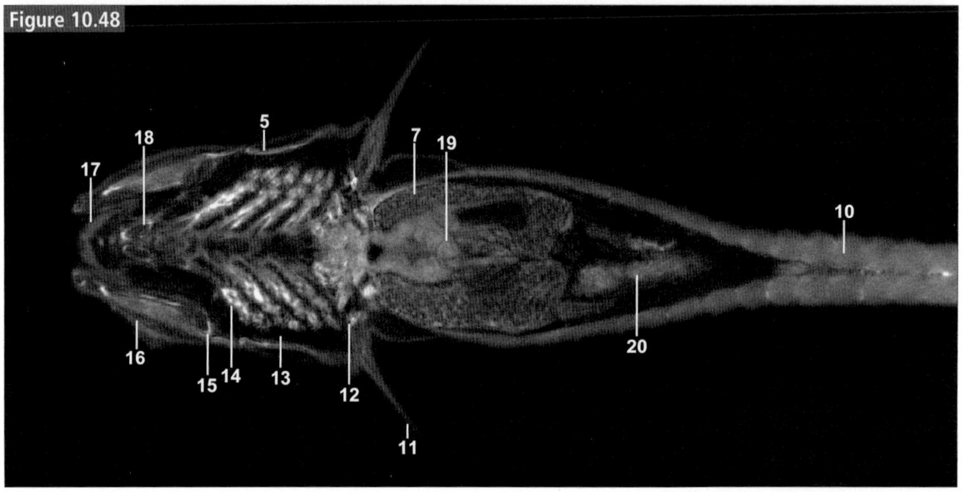

1. intermandibularis	**5.** opercle	**9.** posterior chamber swim bladder	**13.** opercular cavity	**17.** interhyal
2. dentary	**6.** pronephros	**10.** myotome	**14.** gill filaments	**18.** ceratohyal
3. retroarticular	**7.** liver	**11.** pectoral fin	**15.** hyomandibula	**19.** intestinal bulb
4. palatoquadrate	**8.** anterior chamber swim bladder	**12.** cleithrum	**16.** adductor mandibulae complex	**20.** posterior intestine

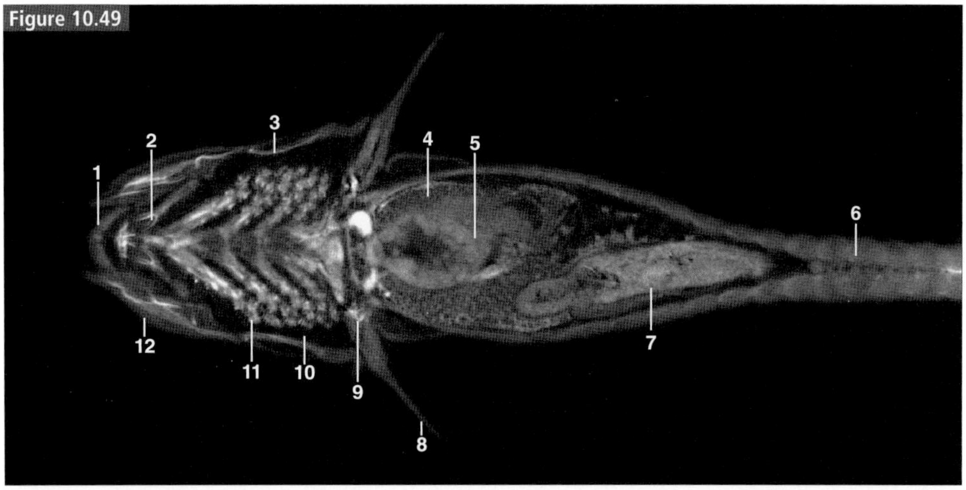

Figure 10.49

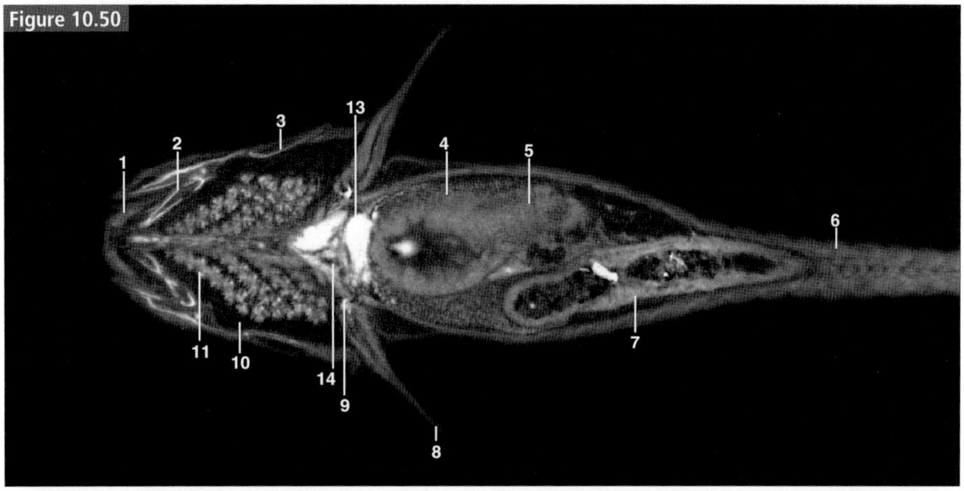

Figure 10.50

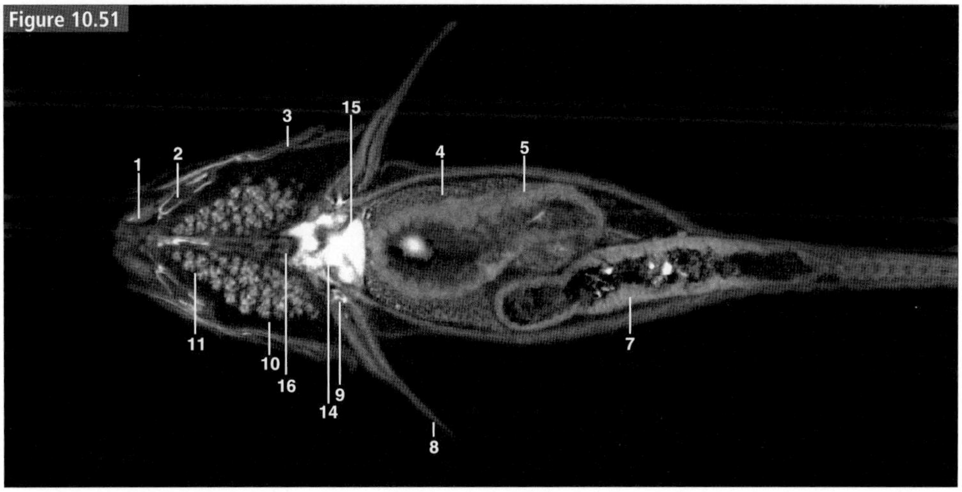

Figure 10.51

1. interhyal	**5.** intestinal bulb	**9.** cleithrum	**13.** common cardinal vein
2. ceratohyal	**6.** myotome	**10.** opercular cavity	**14.** atrium
3. opercle	**7.** posterior intestine	**11.** gill filaments	**15.** sinus venosus
4. liver	**8.** pectoral fin	**12.** adductor mandibulae complex	**16.** bulbus arteriosus

Figure 10.52

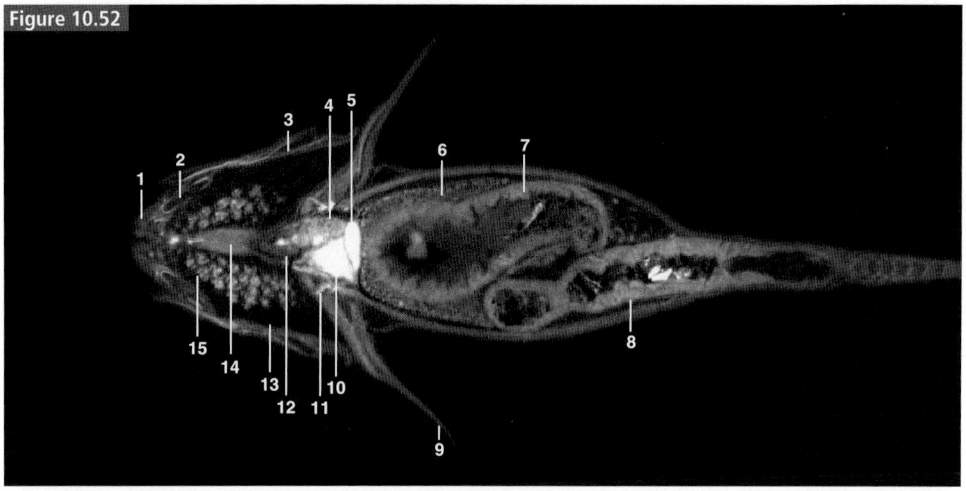

Figure 10.53

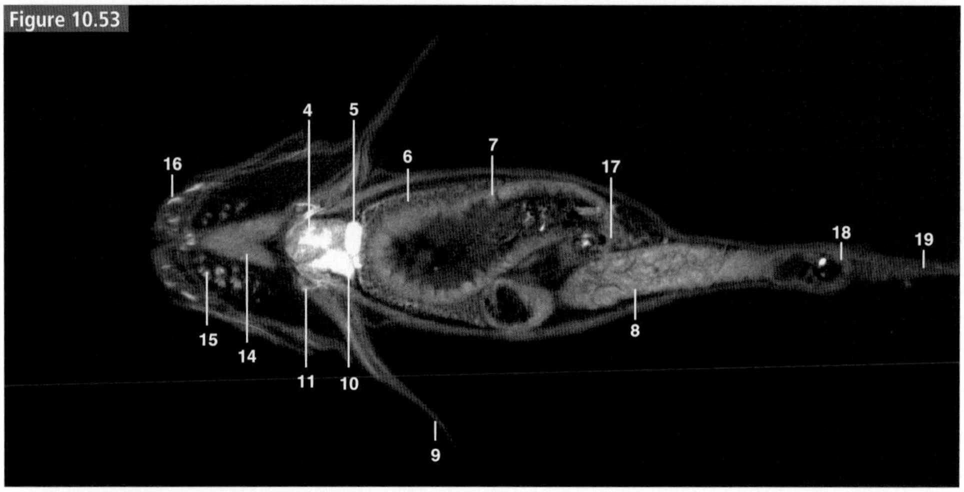

Figure 10.54

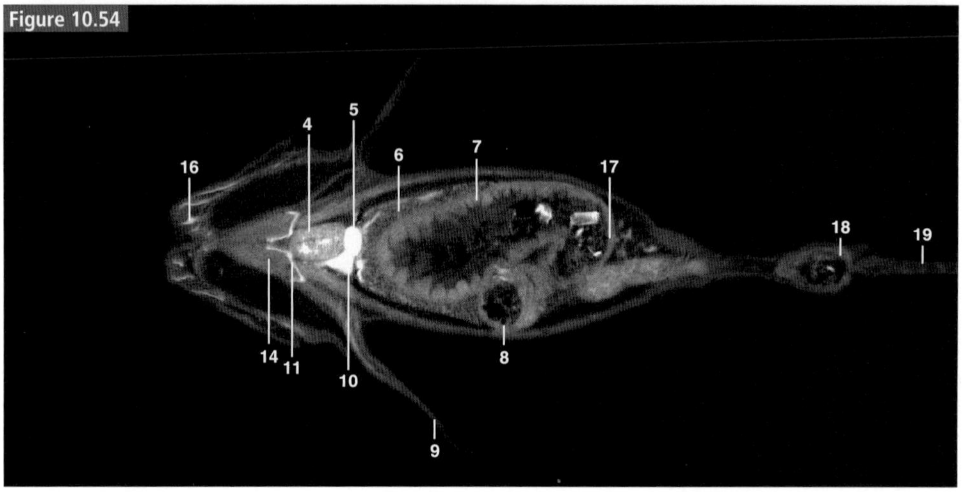

1. interhyal	**5.** sinus venosus	**9.** pectoral fin	**13.** opercular cavity	**17.** mid intestine
2. ceratohyal	**6.** liver	**10.** atrium	**14.** sternohyoid	**18.** cloacal chamber
3. opercle	**7.** intestinal bulb	**11.** cleithrum	**15.** gill filaments	**19.** anal fin
4. cardiac ventricle	**8.** posterior intestine	**12.** bulbus arteriosus	**16.** branchiostegal ray	

Figure 10.55

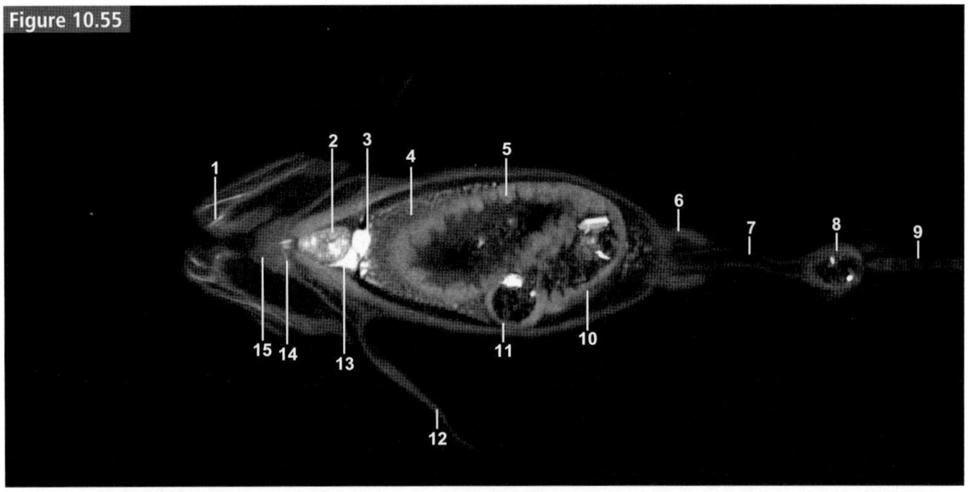

Figure 10.56

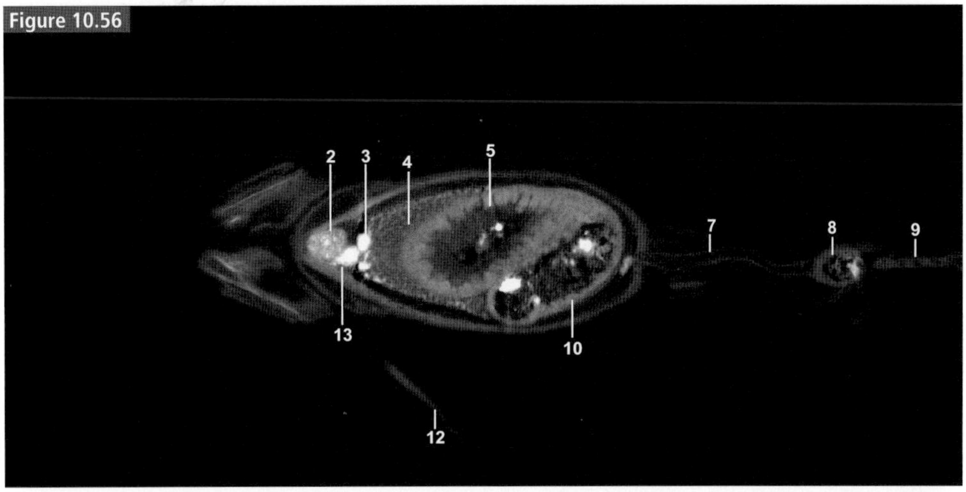

Figure 10.57

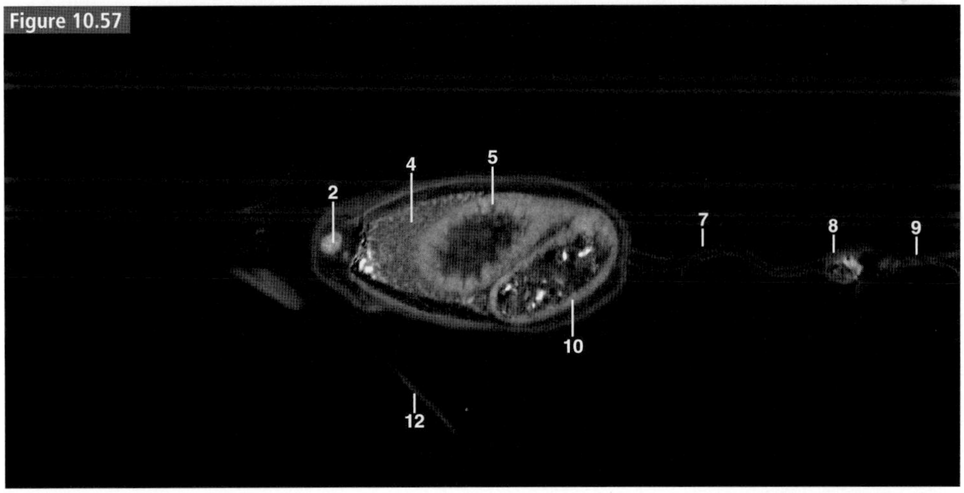

1. branchiostegal ray	**4.** liver	**7.** median fin fold	**10.** mid intestine	**13.** atrium
2. cardiac ventricle	**5.** intestinal bulb	**8.** cloacal chamber	**11.** posterior intestine	**14.** cleithrum
3. sinus venosus	**6.** pelvic fin musculature	**9.** anal fin	**12.** pectoral fin	**15.** sternohyoid

Figure 10.58

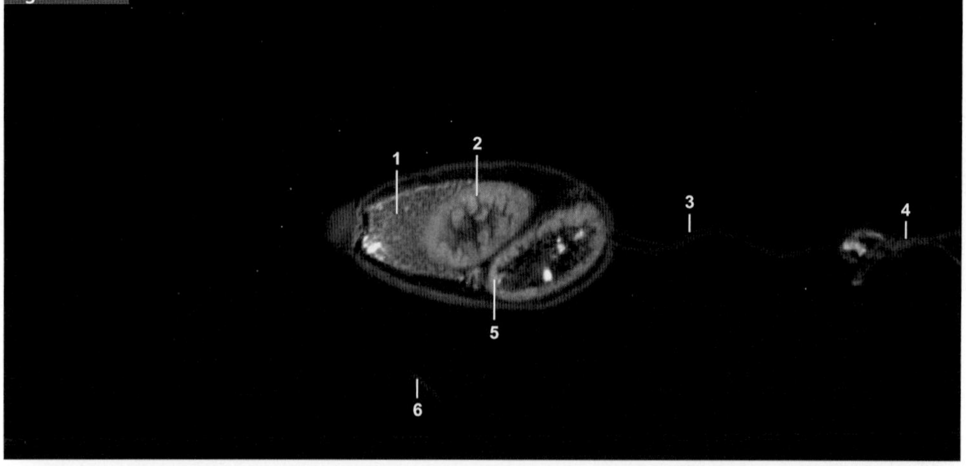

Figure 10.59

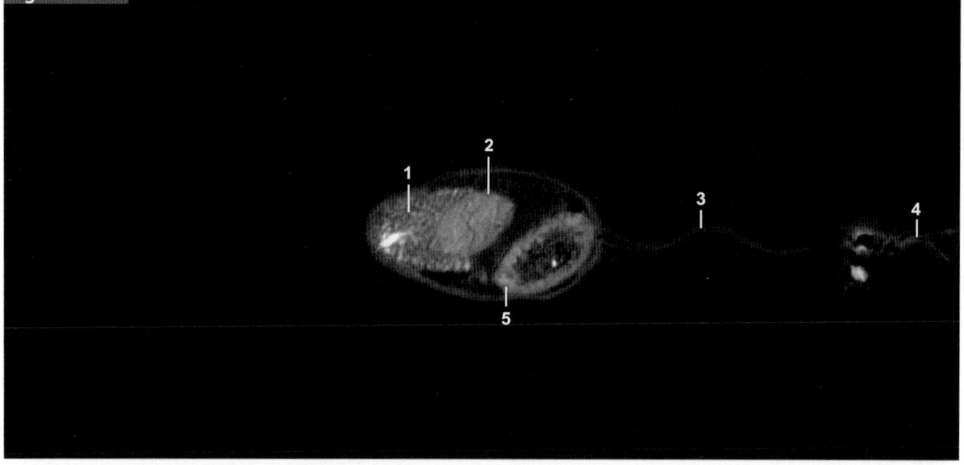

Figure 10.60

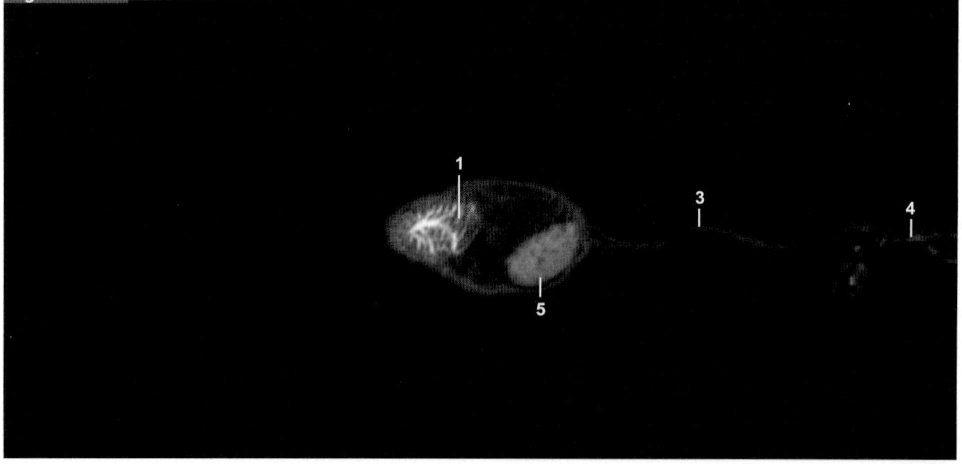

1. liver	**2.** intestinal bulb	**3.** median fin fold	**4.** anal fin	**5.** mid intestine	**6.** pectoral fin

Transverse Sections

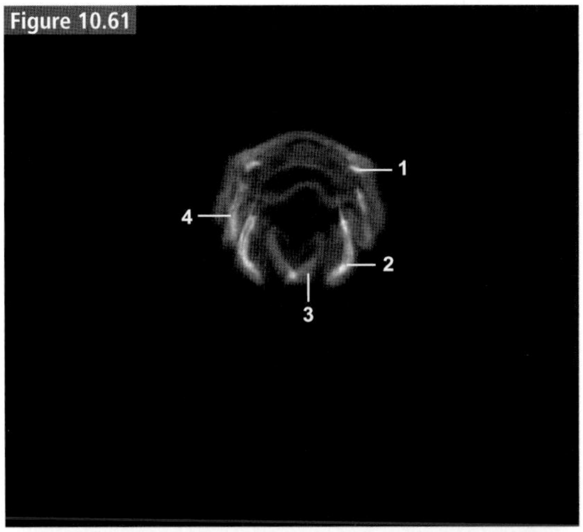

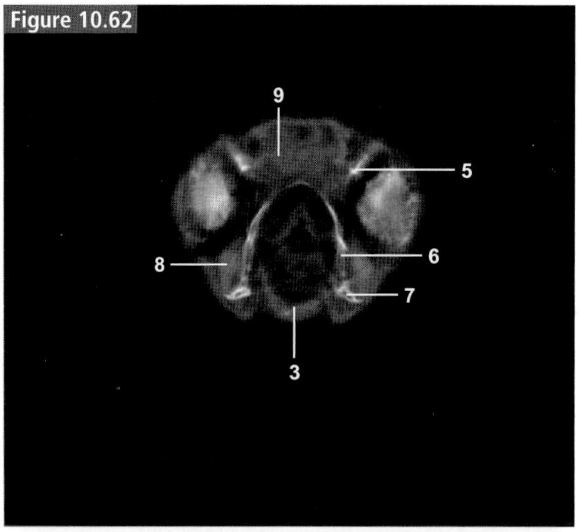

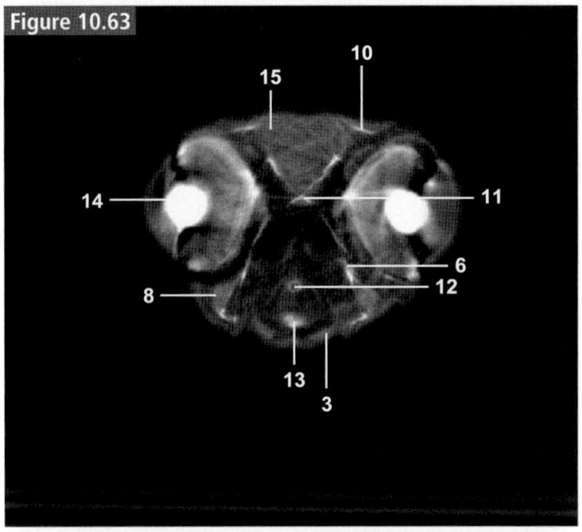

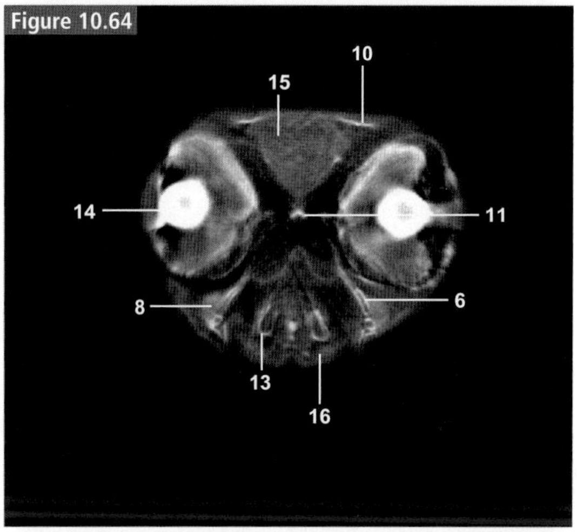

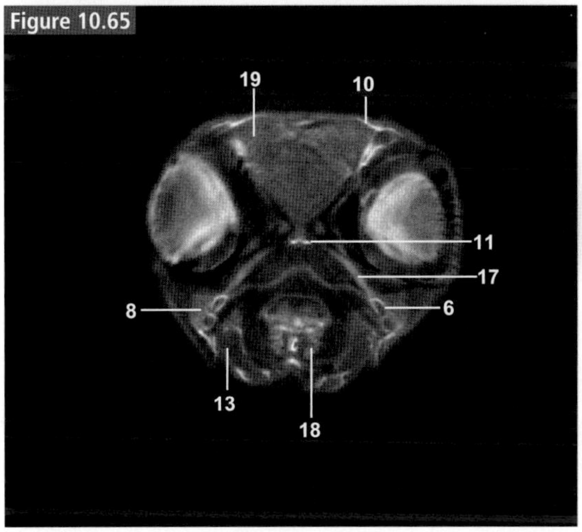

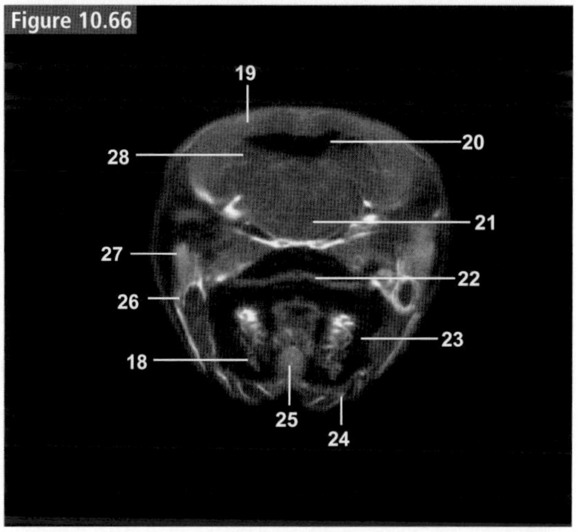

1. maxilla
2. dentary
3. intermandibularis
4. maxilla
5. lateral ethmoid
6. palatoquadrate
7. retroarticular
8. adductor mandibulae complex
9. olfactory bulb
10. frontal bone
11. parasphenoid
12. basihyal
13. ceratohyal
14. lens
15. telencephalon
16. interhyal
17. lateral rectus
18. gill filaments
19. optic tectum
20. tectal ventricle
21. hypothalamus
22. pharynx
23. opercular cavity
24. branchiostegal ray
25. sternohyoid
26. hyomandibula
27. dilator operculi
28. torus semicircularis

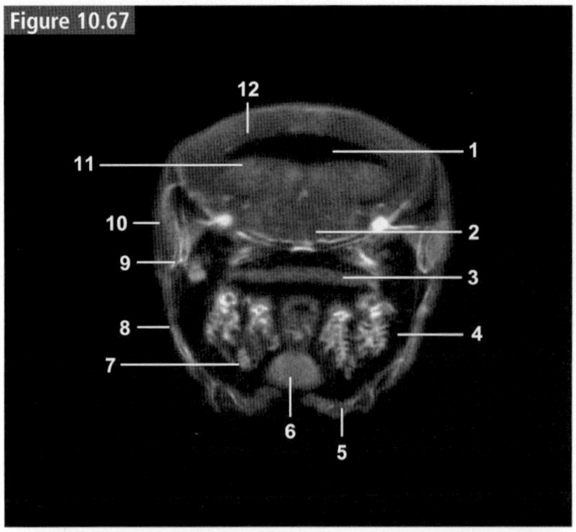

Figure 10.67

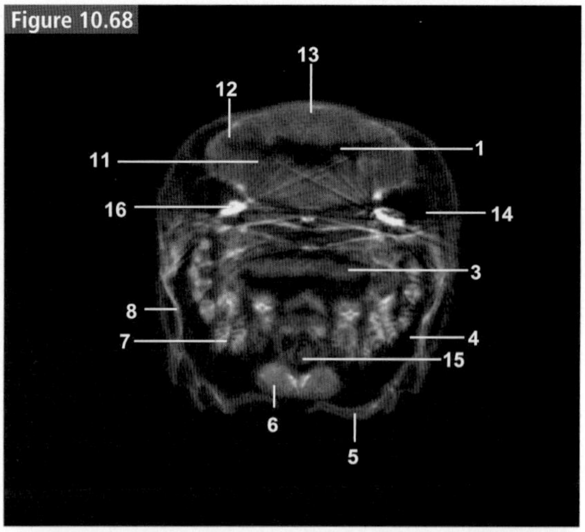

Figure 10.68

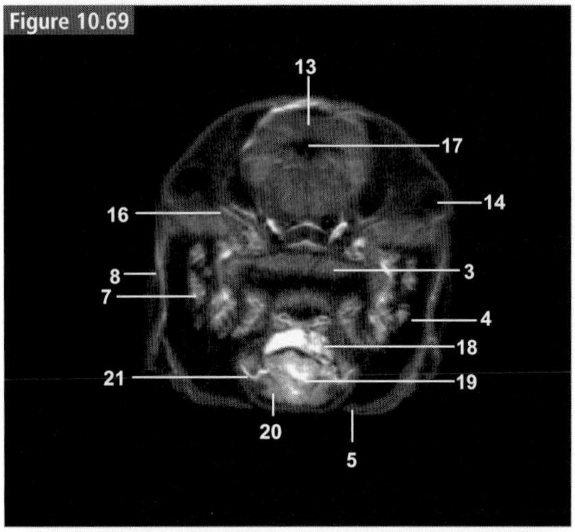

Figure 10.69

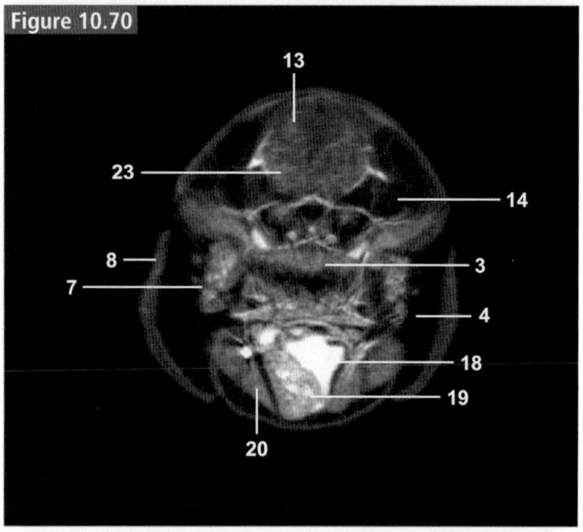

Figure 10.70

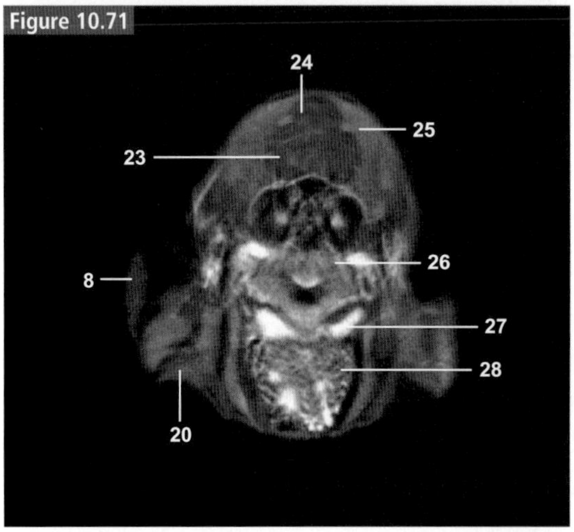

Figure 10.71

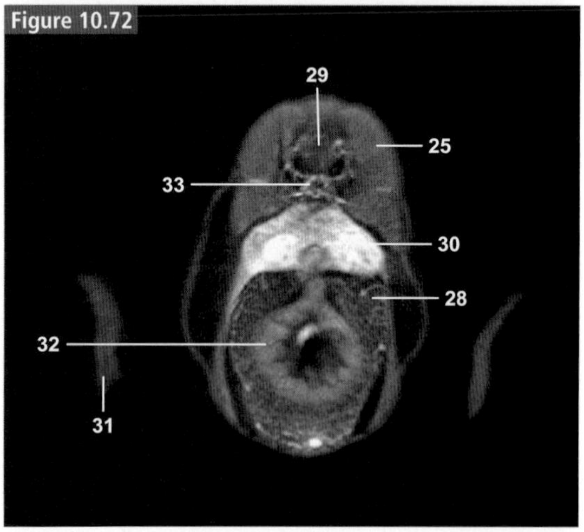

Figure 10.72

1. tectal ventricle
2. inferior lobe
3. pharynx
4. opercular cavity
5. branchiostegal ray
6. sternohyoid
7. gill filaments
8. opercle
9. hyomandibula
10. dilator operculi
11. torus semicircularis
12. optic tectum
13. cerebellum
14. semicircular canal
15. bulbus arteriosus
16. otolith
17. fourth ventricle
18. atrium
19. cardiac ventricle
20. pectoral fin musculature
21. cleithrum
22. granular eminence
23. medulla oblongata
24. vagal lobe
25. myotome
26. esophagus
27. common cardinal vein
28. liver
29. spinal cord
30. pronephros
31. pectoral fin
32. intestinal bulb
33. Weberian vertebra

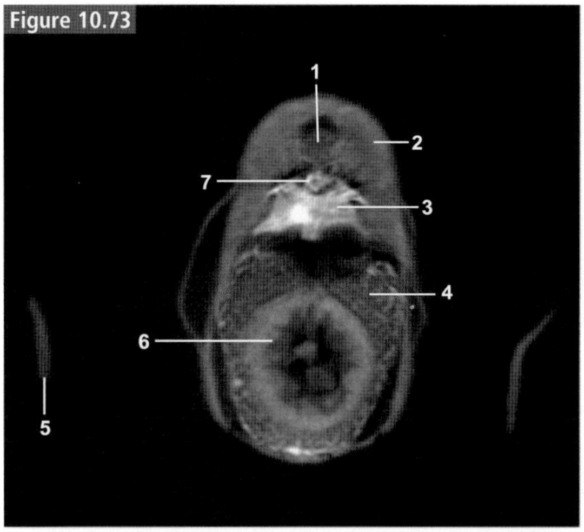

Figure 10.73

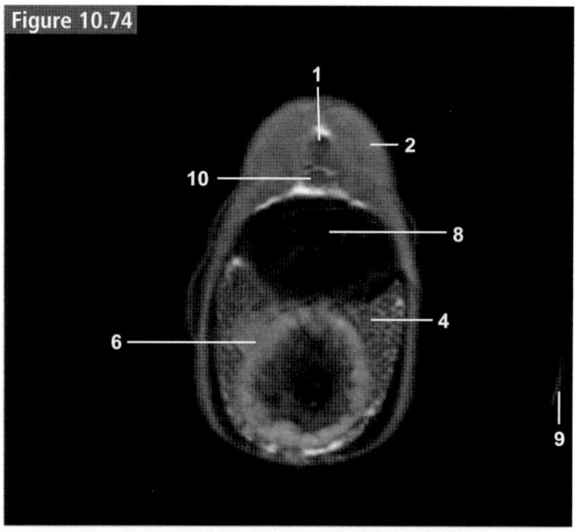

Figure 10.74

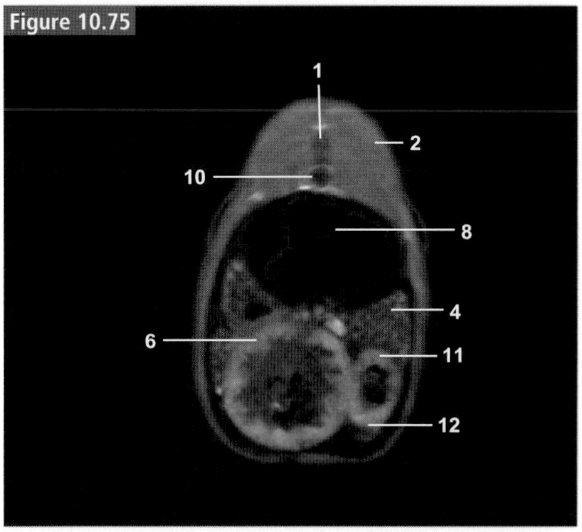

Figure 10.75

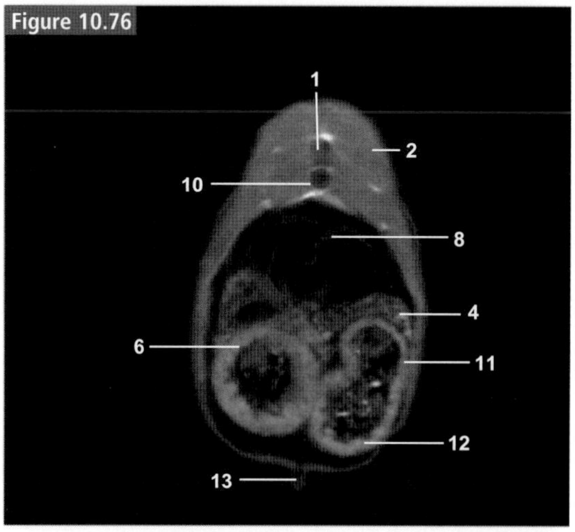

Figure 10.76

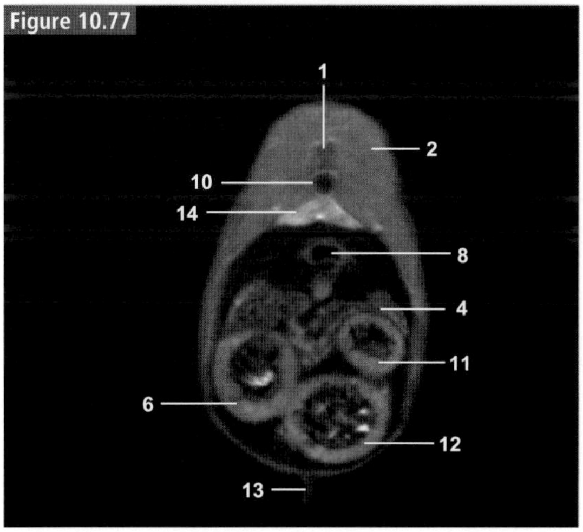

Figure 10.77

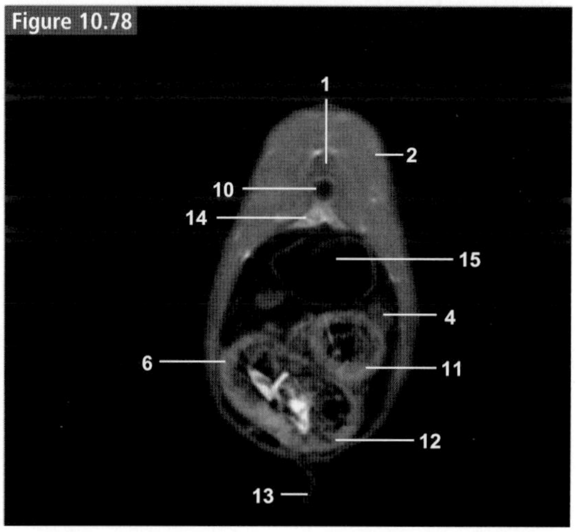

Figure 10.78

1. spinal cord
2. myotome
3. pronephros
4. liver
5. pectoral fin
6. intestinal bulb
7. Weberian vertebra
8. anterior chamber swim bladder
9. pectoral fin
10. precaudal vertebra
11. posterior intestine
12. mid intestine
13. median fin
14. mesonephros
15. posterior chamber swim bladder

Figure 10.79

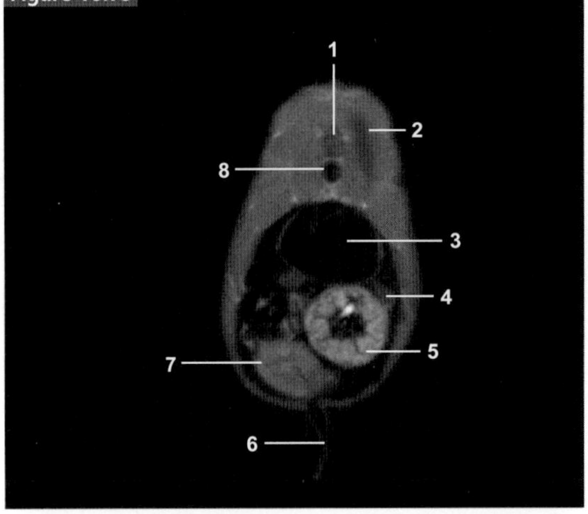

Figure 10.80

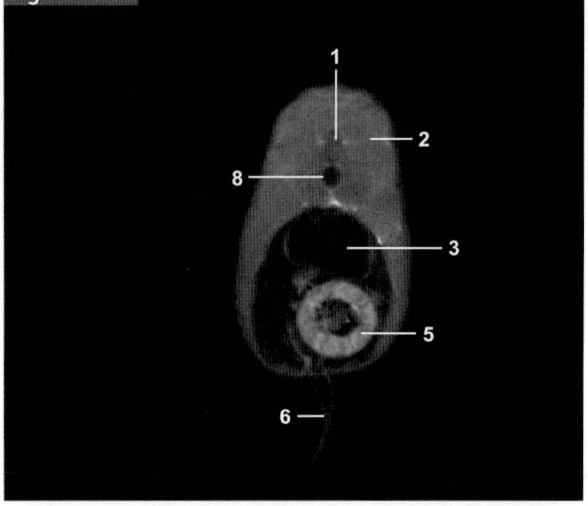

Figure 10.81

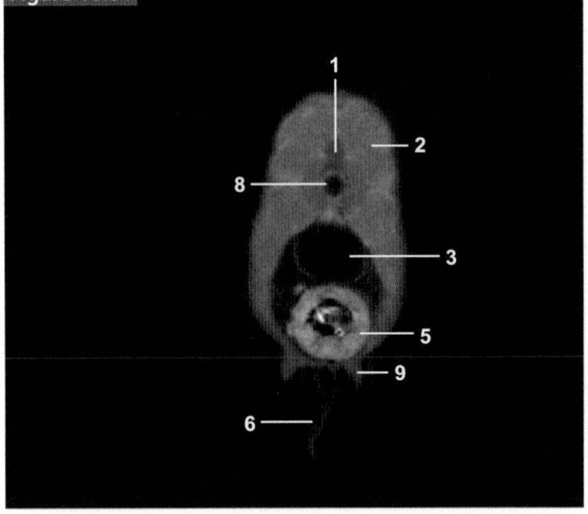

Figure 10.82

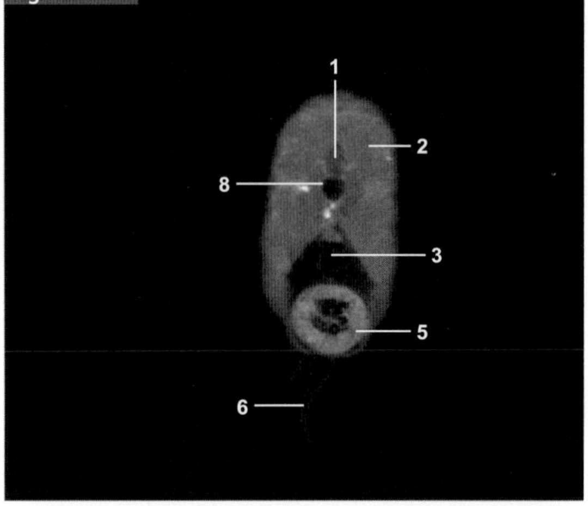

Figure 10.83

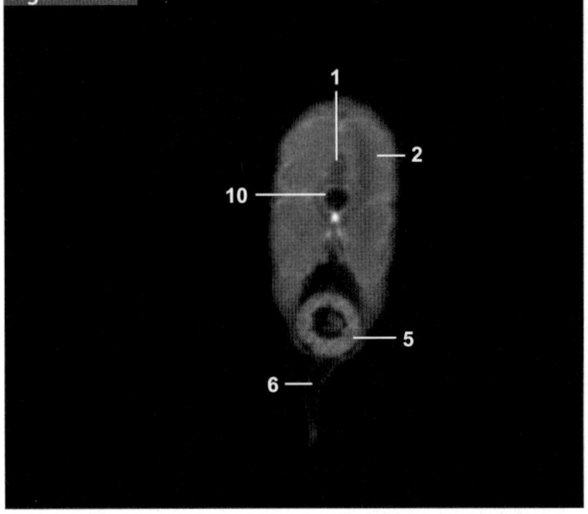

Figure 10.84

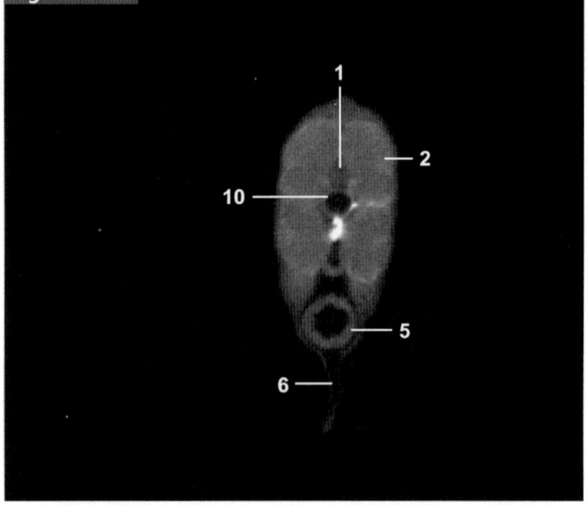

1. spinal cord	**3.** posterior chamber swim bladder	**5.** posterior intestine	**7.** intestinal bulb	**9.** pelvic fin
2. myotome	**4.** testis	**6.** median fin	**8.** precaudal vertebra	**10.** caudal vertebra

Figure 10.85

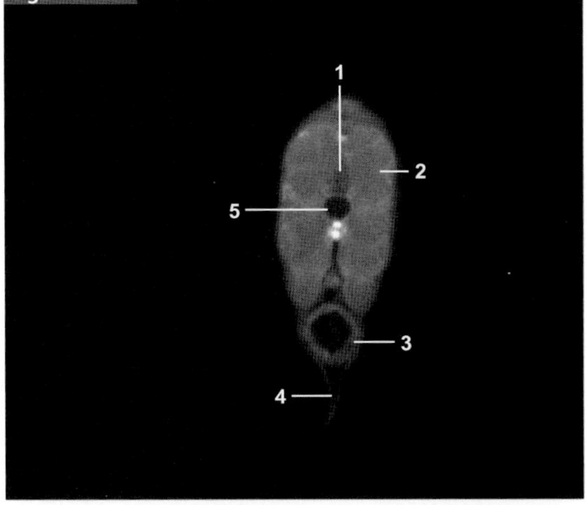

Figure 10.86

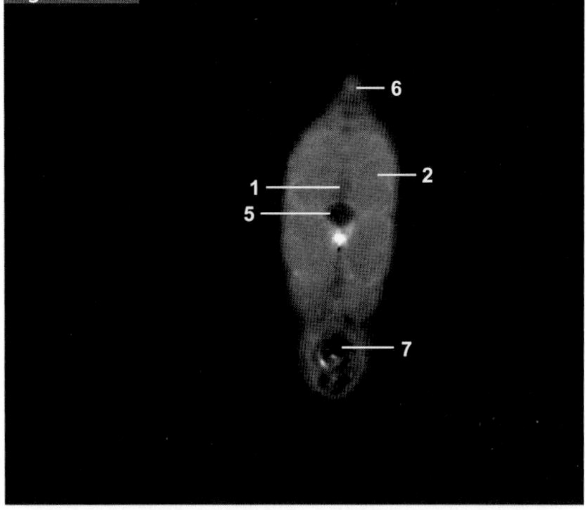

Figure 10.87

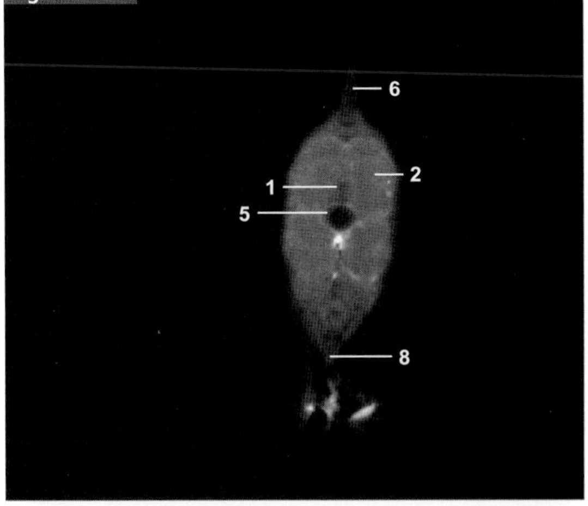

Figure 10.88

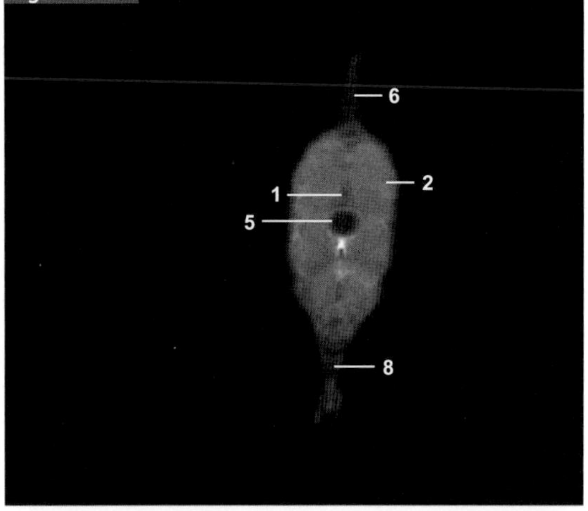

Figure 10.89

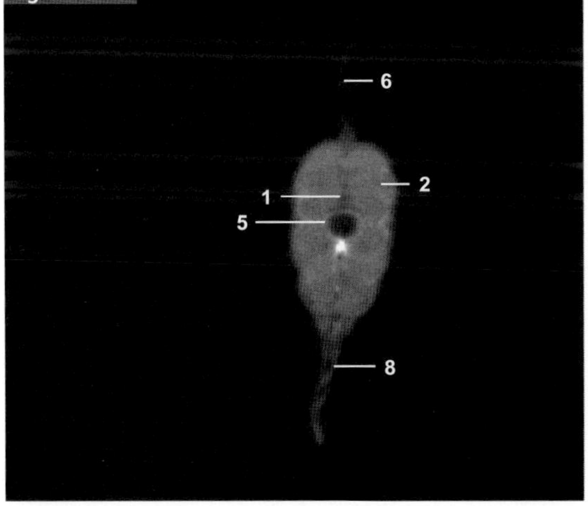

Figure 10.90

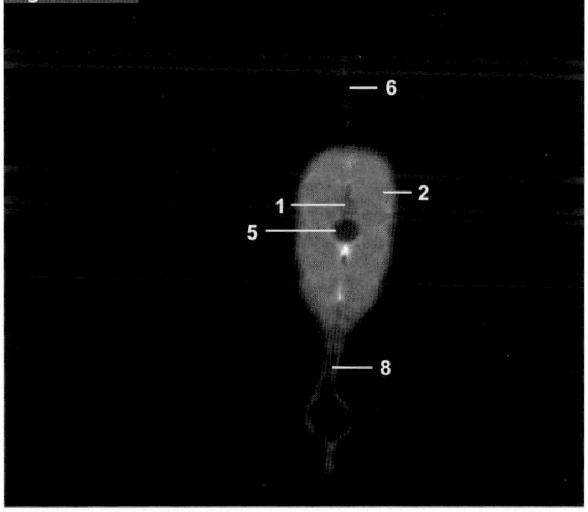

1. spinal cord	**3.** posterior intestine	**5.** caudal vertebra	**7.** cloacal chamber
2. myotome	**4.** median fin	**6.** dorsal fin	**8.** anal fin

CHAPTER 11

14 MM

Sagittal Sections

Figure 11.1

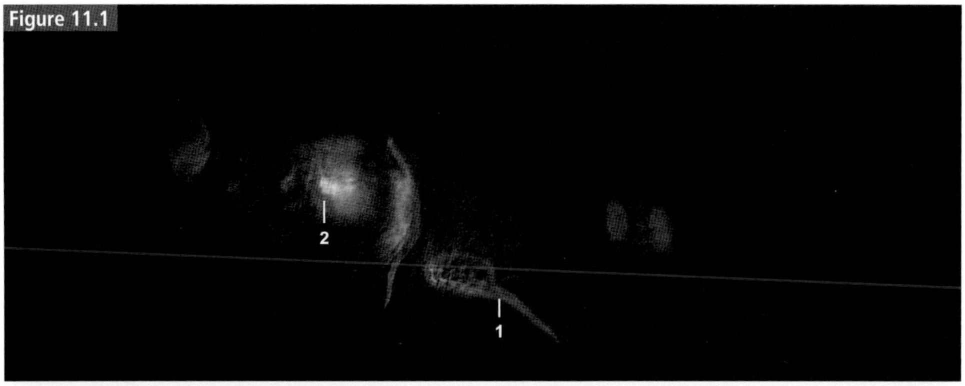

Figure 11.2

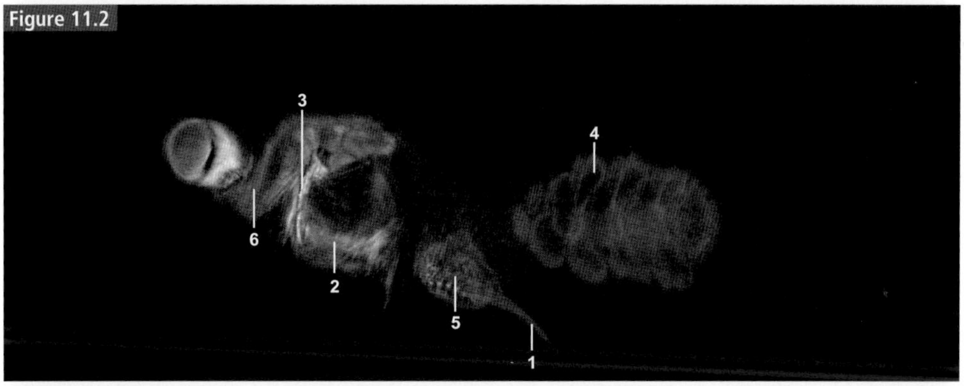

Figure 11.3

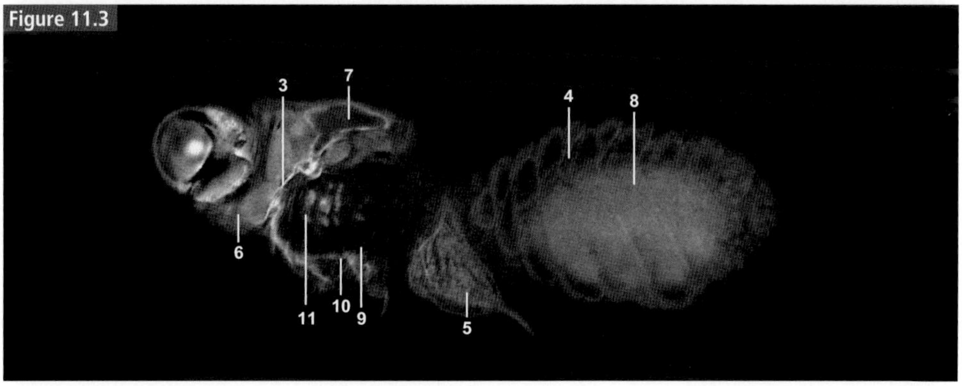

1. pectoral fin	4. scales	6. adductor mandibulae	8. myotome	10. branchiostegal ray
2. opercle	5. pectoral fin musculature	7. semicircular canal	9. opercular cavity	11. gill filaments
3. hyomandibula				

Atlas of Zebrafish Development. 10.1016/B978-0-12-374016-8.00011-X

Figure 11.4

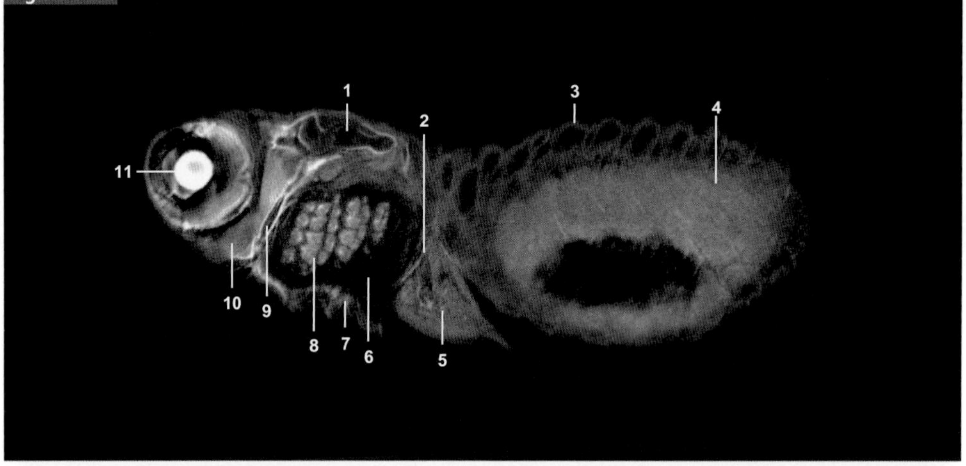

Figure 11.5

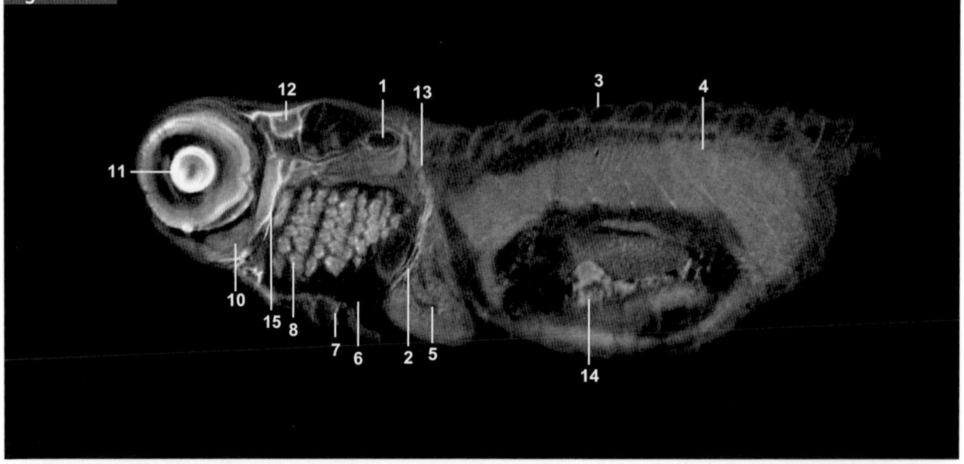

Figure 11.6

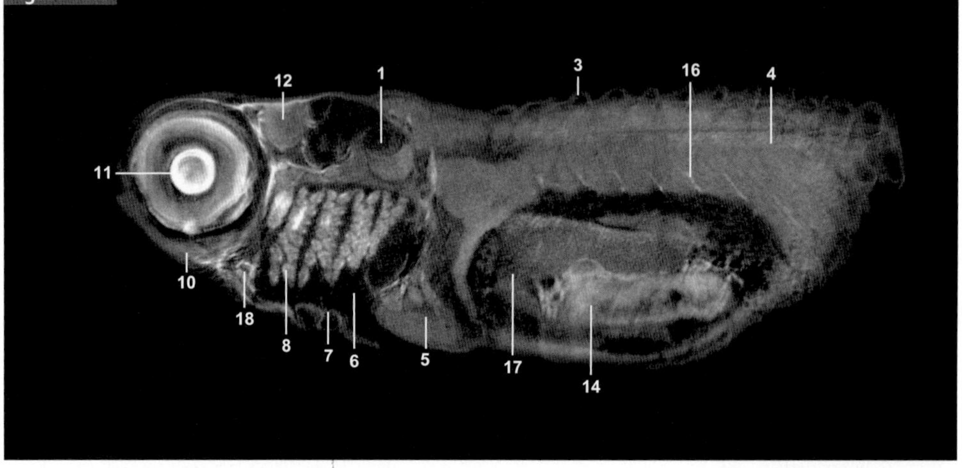

1. semicircular canal	**5.** pectoral fin musculature	**9.** hyomandibula	**13.** supracleithrum	**17.** liver
2. cleithrum	**6.** opercular cavity	**10.** adductor mandibulae	**14.** intestinal bulb	**18.** retroarticular
3. scales	**7.** branchiostegal ray	**11.** lens	**15.** preopercle	
4. myotome	**8.** gill filaments	**12.** optic tectum	**16.** rib	

Figure 11.7

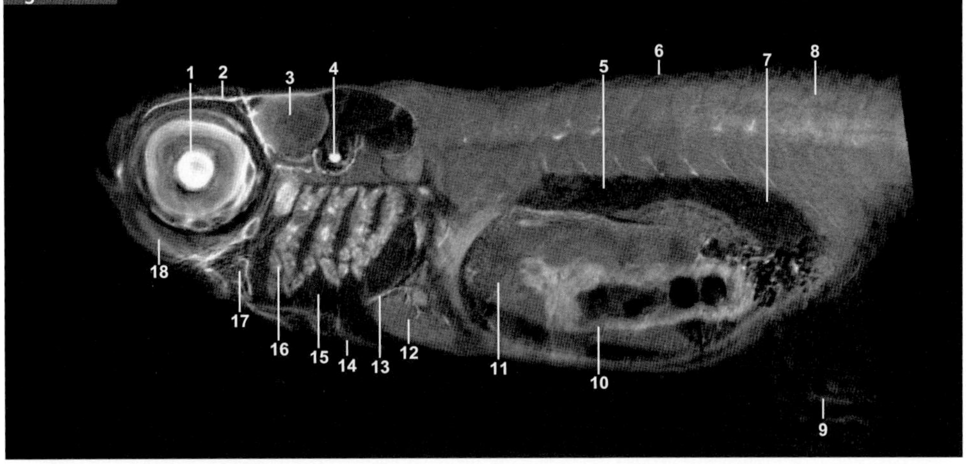

Figure 11.8

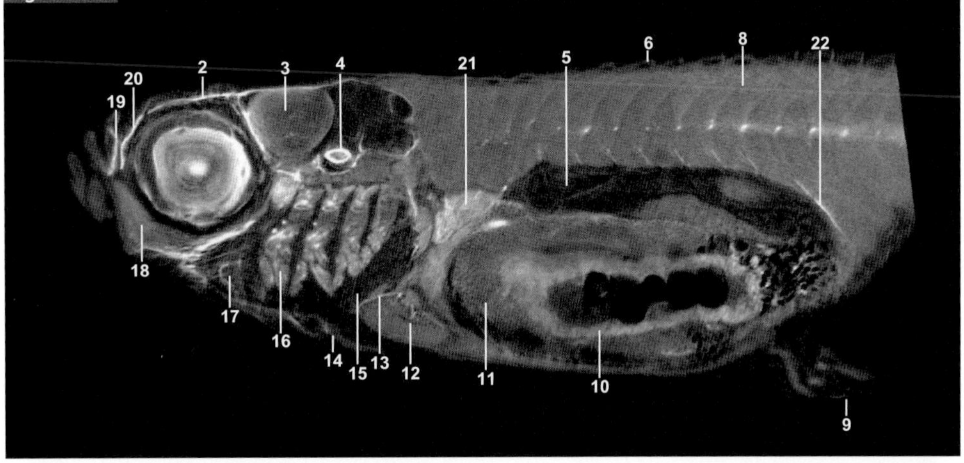

Figure 11.9

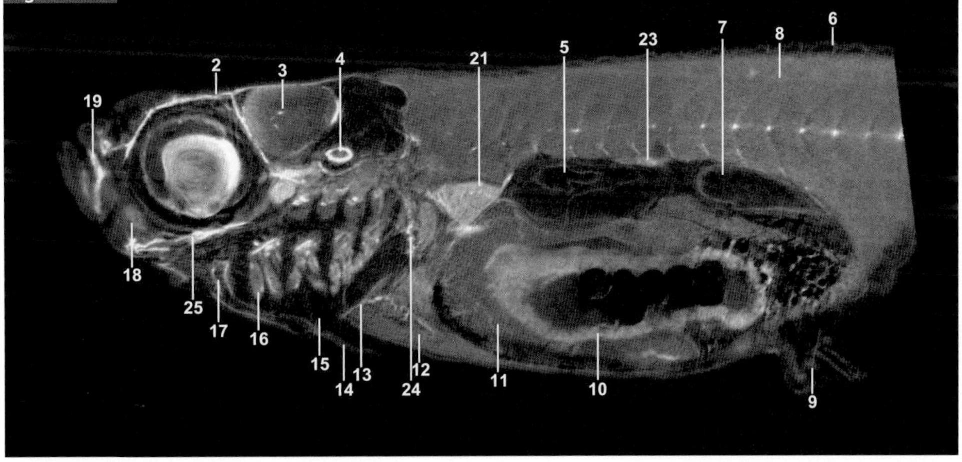

1. lens
2. frontal bone
3. optic tectum
4. otolith
5. anterior chamber swim bladder
6. scales
7. posterior chamber swim bladder
8. myotome
9. pelvic fin
10. intestinal bulb
11. liver
12. pectoral fin musculature
13. cleithrum
14. branchiostegal ray
15. opercular cavity
16. gill filaments
17. ceratohyal
18. adductor mandibulae
19. maxilla
20. lateral ethmoid
21. pronephros
22. rib
23. mesonephros
24. ceratobranchial 5 tooth
25. infraorbital

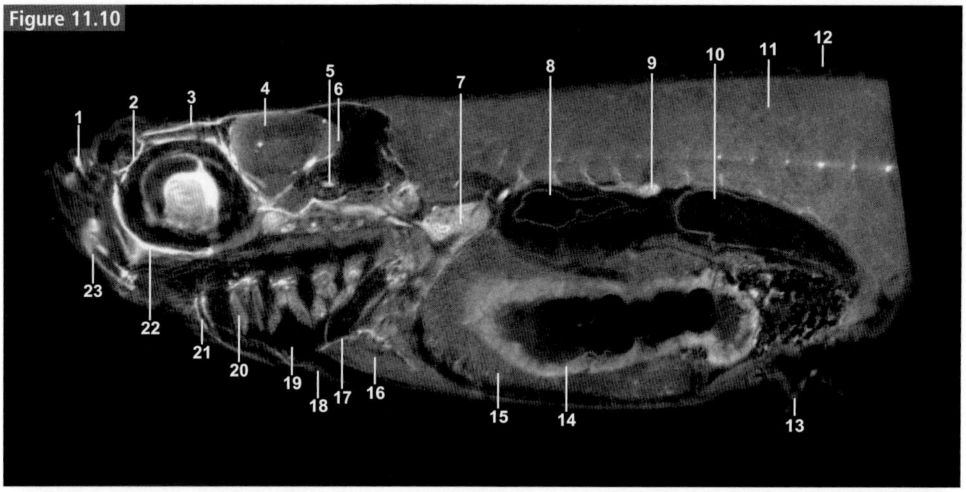

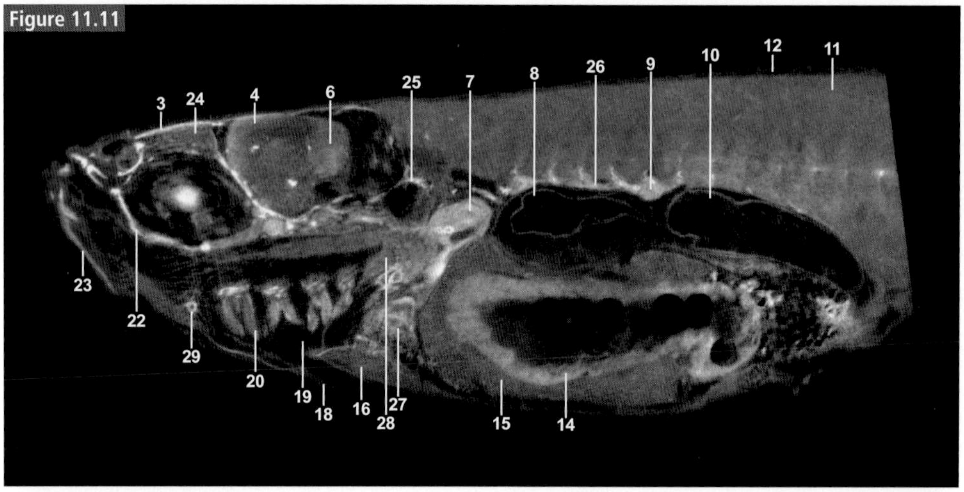

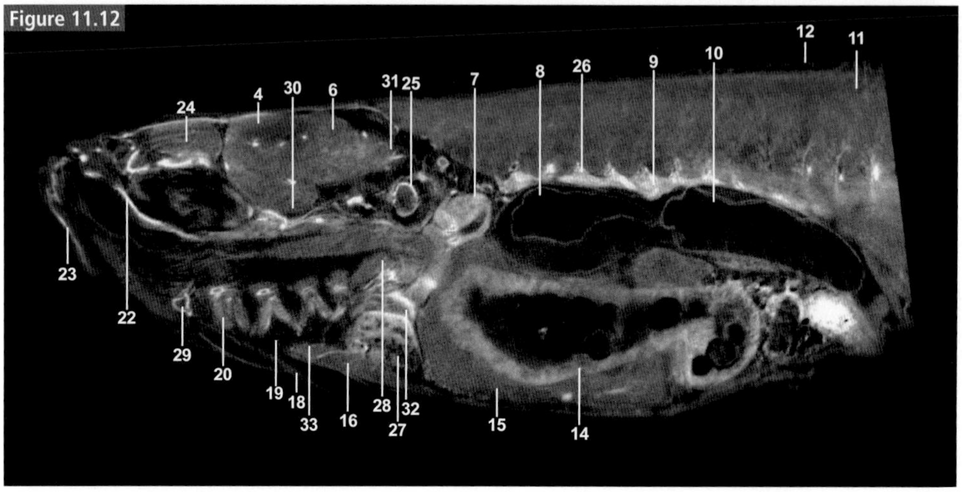

1. maxilla
2. lateral ethmoid
3. frontal bone
4. optic tectum
5. otolith
6. cerebellum
7. pronephros
8. anterior chamber swim bladder
9. mesonephros
10. posterior chamber swim bladder
11. myotome
12. scales
13. pelvic fin
14. intestinal bulb
15. liver
16. pectoral fin musculature
17. cleithrum
18. branchiostegal ray
19. opercular cavity
20. gill filaments
21. ceratohyal
22. infraorbital
23. dentary
24. telencephalon
25. lagena
26. mesonephric duct
27. cardiac ventricle
28. esophagus
29. retroarticular
30. hypothalamus
31. crista cerebellaris
32. atrium
33. sternohyoid

Figure 11.13

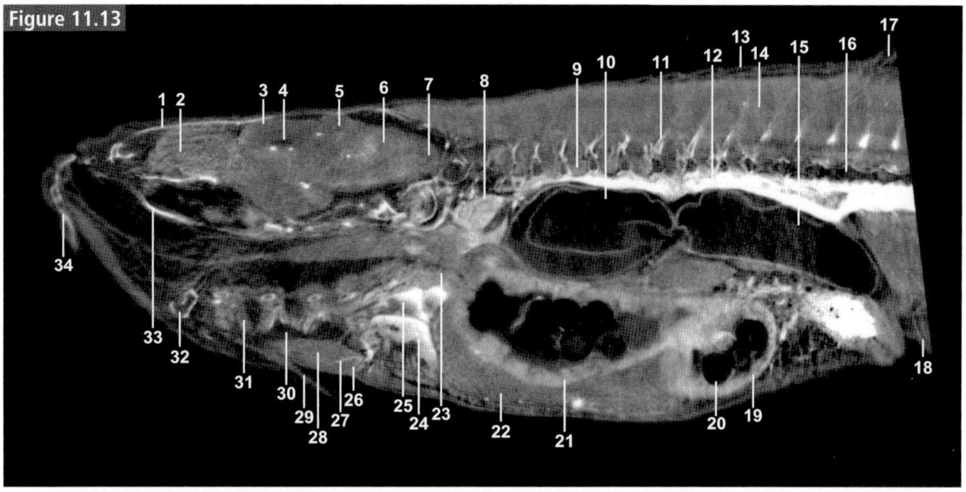

Figure 11.14

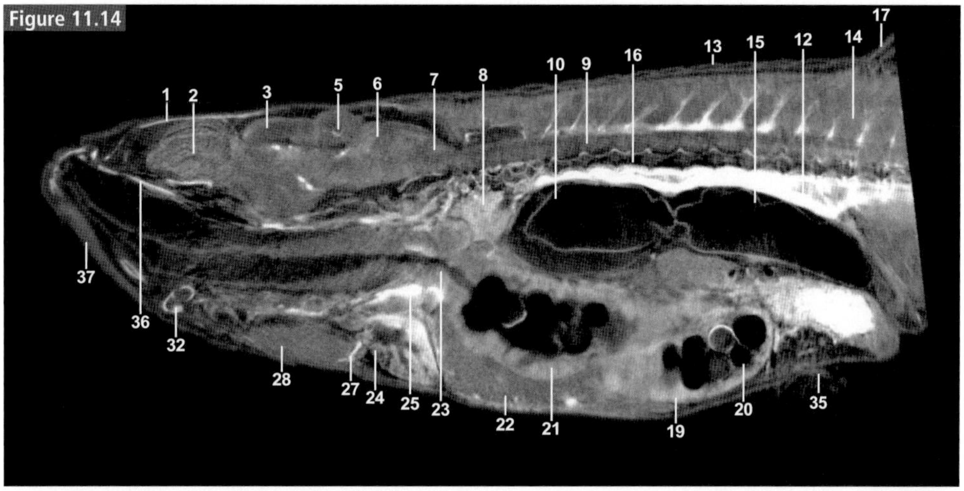

Figure 11.15

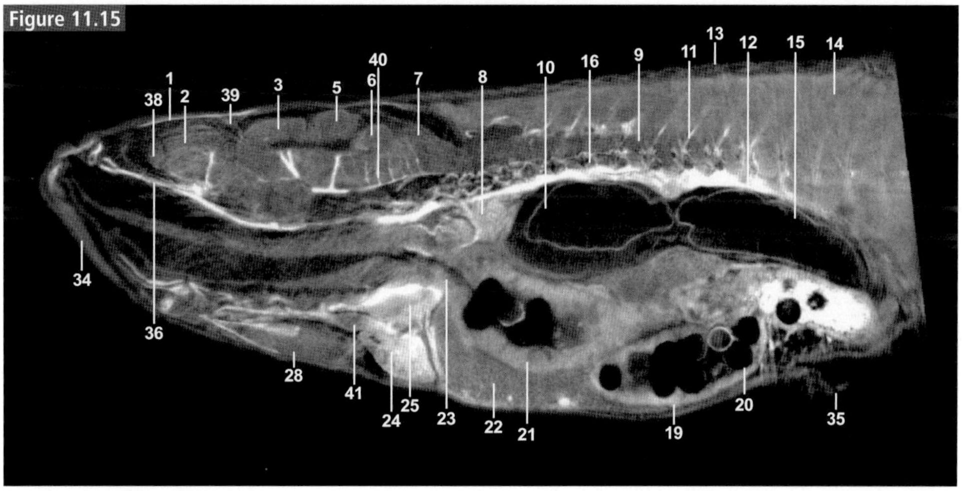

1. frontal bone	**10.** anterior chamber swim bladder	**17.** dorsal fin	**26.** pectoral fin musculature	**34.** dentary
2. telencephalon		**18.** anal fin	**27.** cleithrum	**35.** pelvic fin
3. optic tectum	**11.** neural spine	**19.** mid intestine	**28.** sternohyoid	**36.** parasphenoid
4. tectal ventricle	**12.** posterior cardinal vein	**20.** artemia cyst	**29.** branchiostegal ray	**37.** intermandibularis
5. cerebellum	**13.** scales	**21.** intestinal bulb	**30.** opercular cavity	**38.** olfactory bulb
6. crista cerebellaris	**14.** myotome	**22.** liver	**31.** gill filaments	**39.** epiphysis
7. facial lobe	**15.** posterior chamber swim bladder	**23.** esophagus	**32.** retroarticular	**40.** cranial vasculature
8. pronephros		**24.** cardiac ventricle	**33.** infraorbital	**41.** bulbus arteriosus
9. spinal cord	**16.** vertebral column	**25.** atrium		

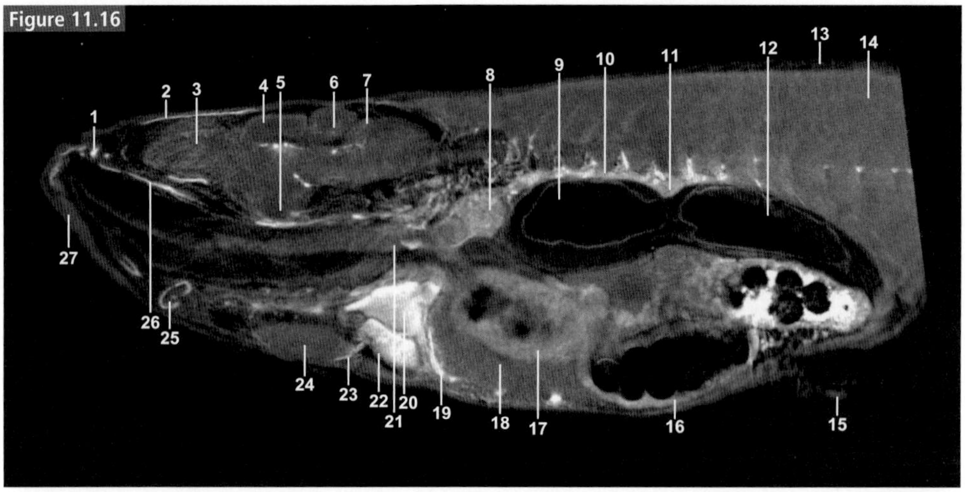

Figure 11.16

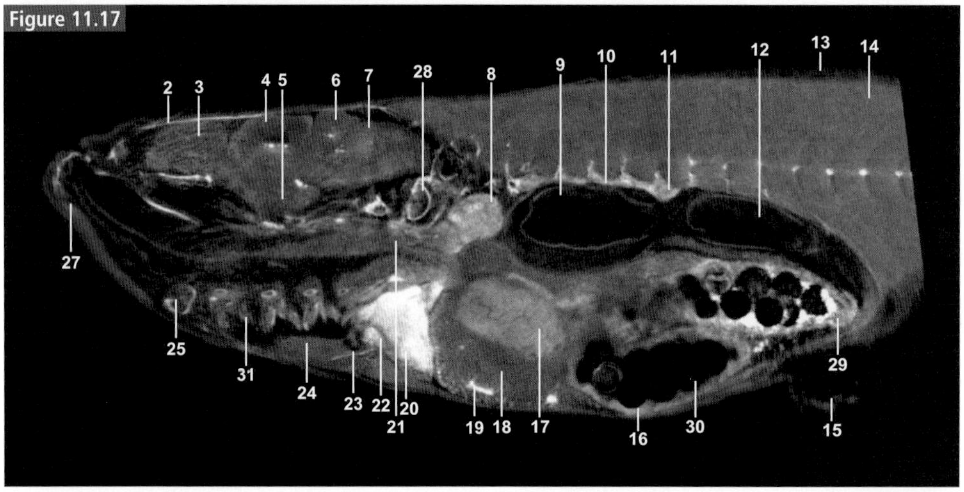

Figure 11.17

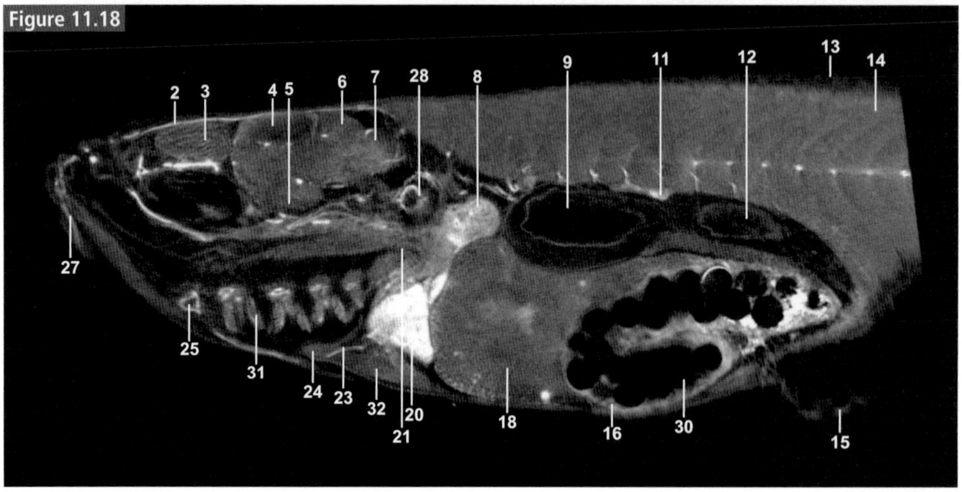

Figure 11.18

1. premaxilla	**8.** pronephros	**13.** scales	**20.** atrium	**27.** dentary
2. frontal bone	**9.** anterior chamber swim bladder	**14.** myotome	**21.** esophagus	**28.** lagena
3. telencephalon	**10.** mesonephric duct	**15.** pelvic fin	**22.** cardiac ventricle	**29.** posterior intestine
4. optic tectum	**11.** mesonephros	**16.** mid intestine	**23.** cleithrum	**30.** artemia cyst
5. hypothalamus	**12.** posterior chamber swim bladder	**17.** intestinal bulb	**24.** sternohyoid	**31.** gill filaments
6. cerebellum		**18.** liver	**25.** retroarticular	**32.** pectoral fin musculature
7. crista cerebellaris		**19.** hepatic portal vein	**26.** parasphenoid	

Figure 11.19

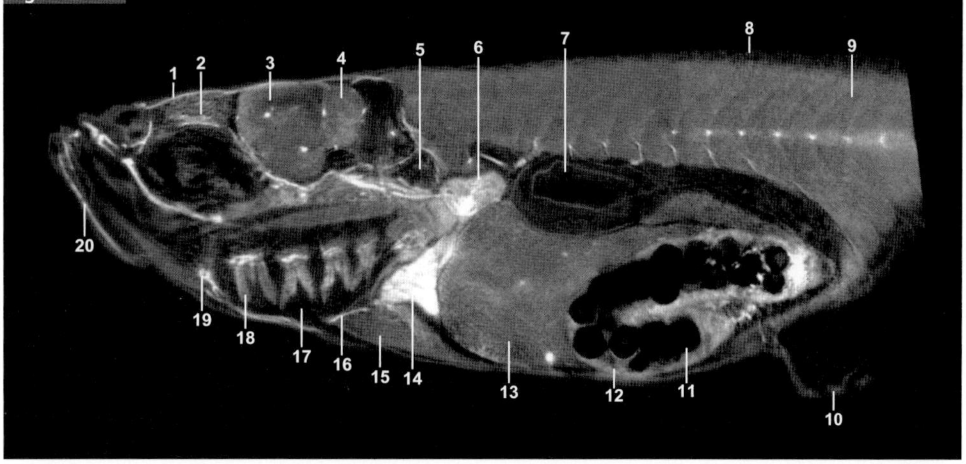

Figure 11.20

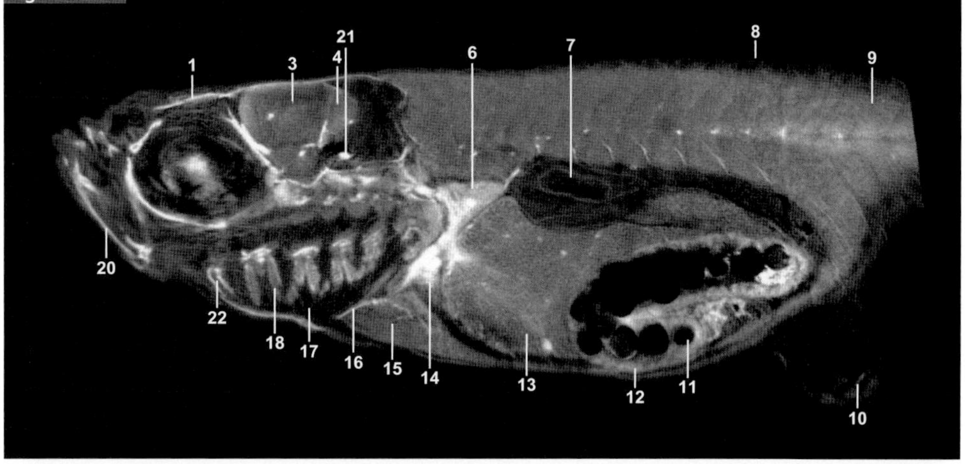

Figure 11.21

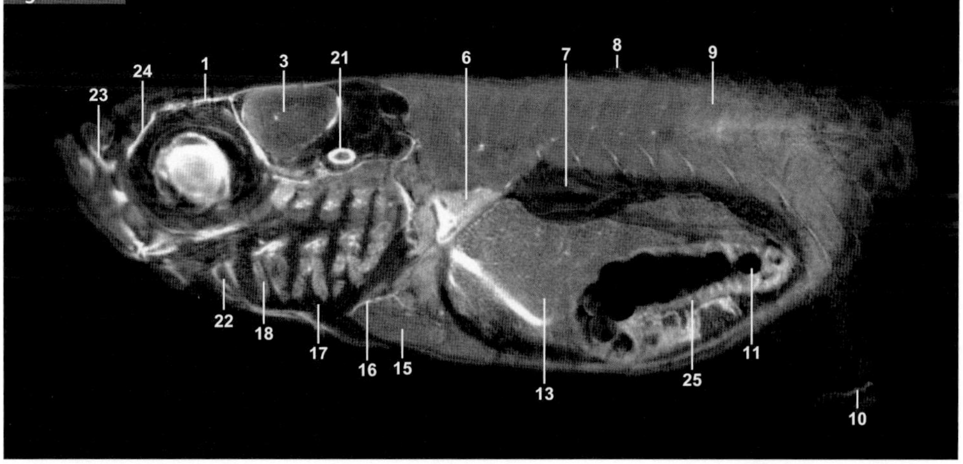

1. frontal bone
2. telencephalon
3. optic tectum
4. cerebellum
5. lagena
6. pronephros
7. anterior chamber swim bladder
8. scales
9. myotome
10. pelvic fin
11. artemia cyst
12. mid intestine
13. liver
14. atrium
15. pectoral fin musculature
16. cleithrum
17. opercular cavity
18. gill filaments
19. retroarticular
20. dentary
21. otolith
22. ceratohyal
23. maxilla
24. lateral ethmoid
25. posterior intestine

Figure 11.22

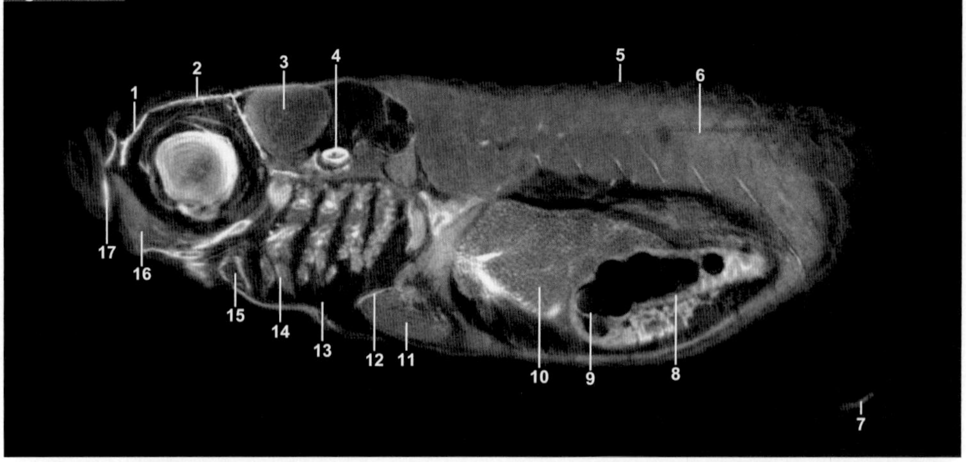

Figure 11.23

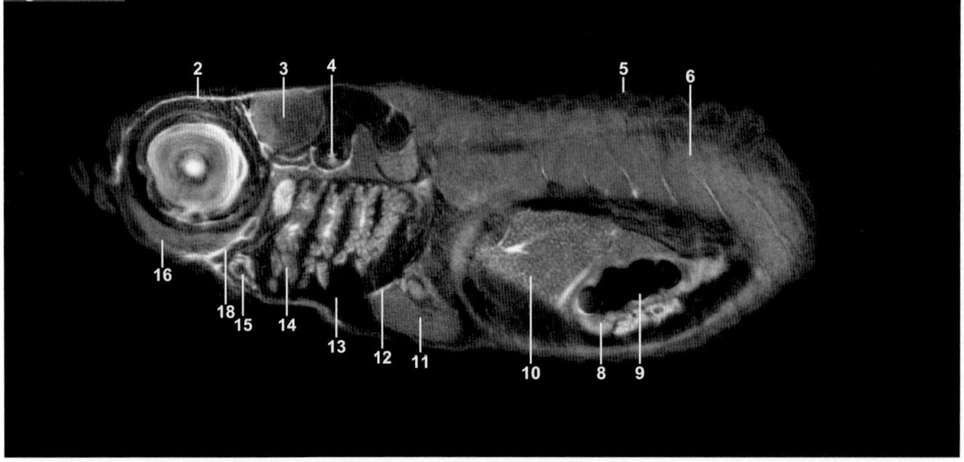

Figure 11.24

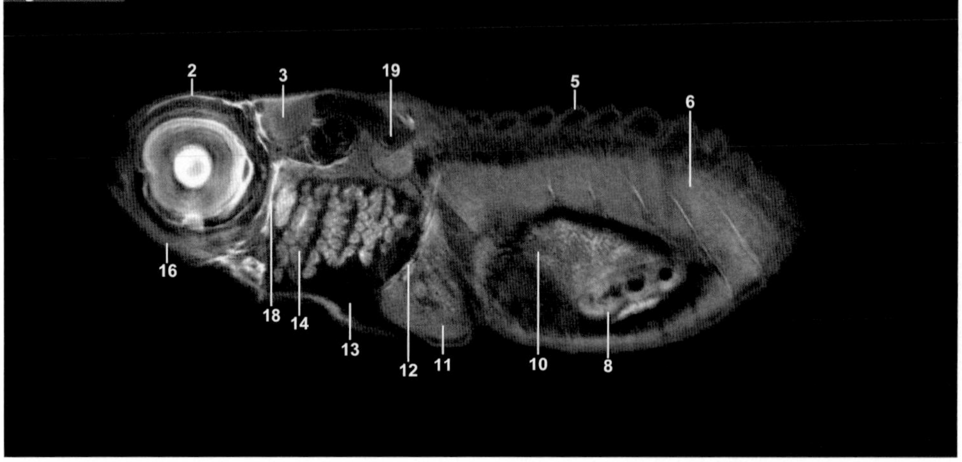

1. lateral ethmoid	**5.** scales	**9.** artemia cyst	**13.** opercular cavity	**17.** maxilla
2. frontal bone	**6.** myotome	**10.** liver	**14.** gill filaments	**18.** hyomandibula
3. optic tectum	**7.** pelvic fin	**11.** pectoral fin musculature	**15.** ceratohyal	**19.** semicircular canal
4. otolith	**8.** posterior intestine	**12.** cleithrum	**16.** adductor mandibulae	

Figure 11.25

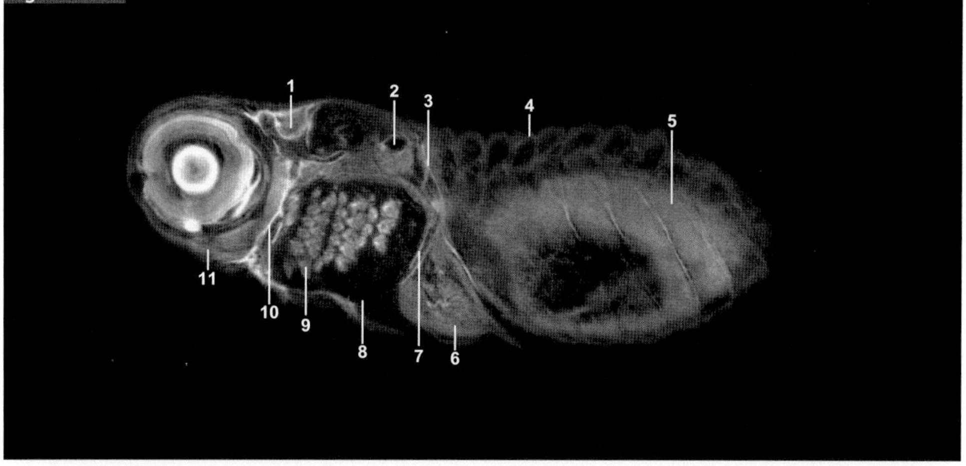

Figure 11.26

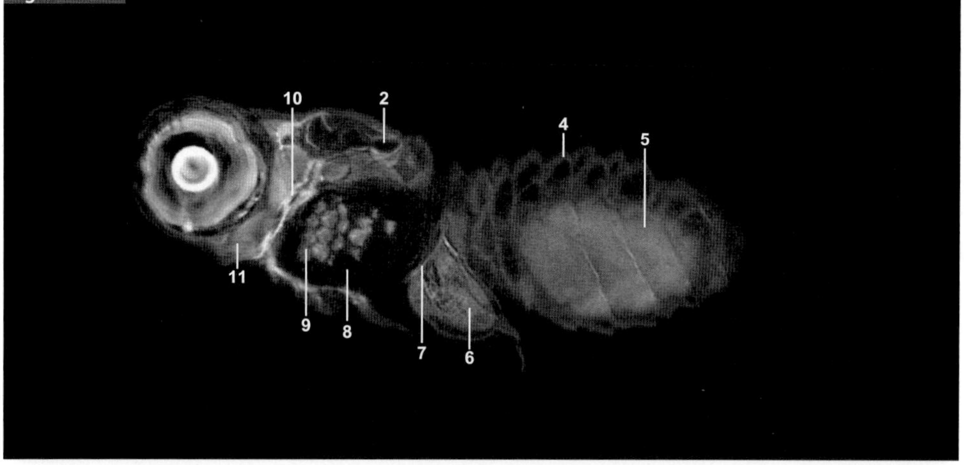

Figure 11.27

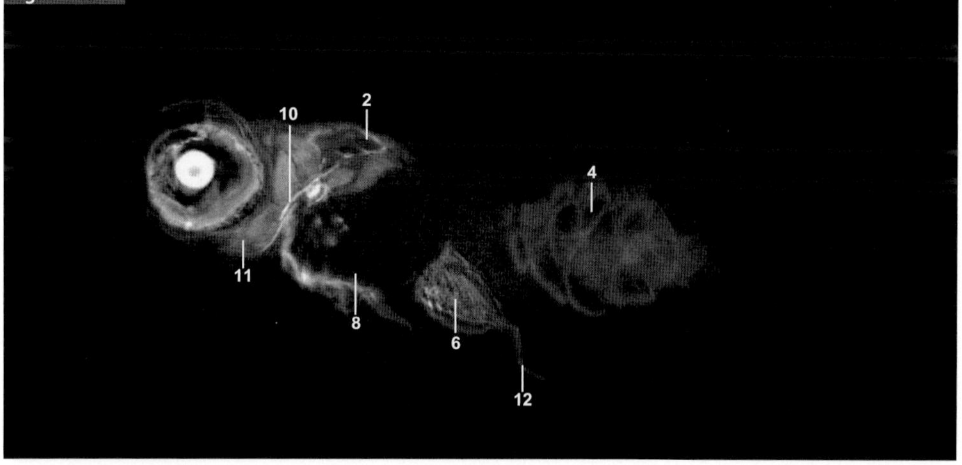

1. optic tectum	**4.** scales	**7.** cleithrum	**9.** gill filaments	**11.** adductor mandibulae
2. semicircular canal	**5.** myotome	**8.** opercular cavity	**10.** hyomandibula	**12.** pectoral fin
3. supracleithrum	**6.** pectoral fin musculature			

Figure 11.28

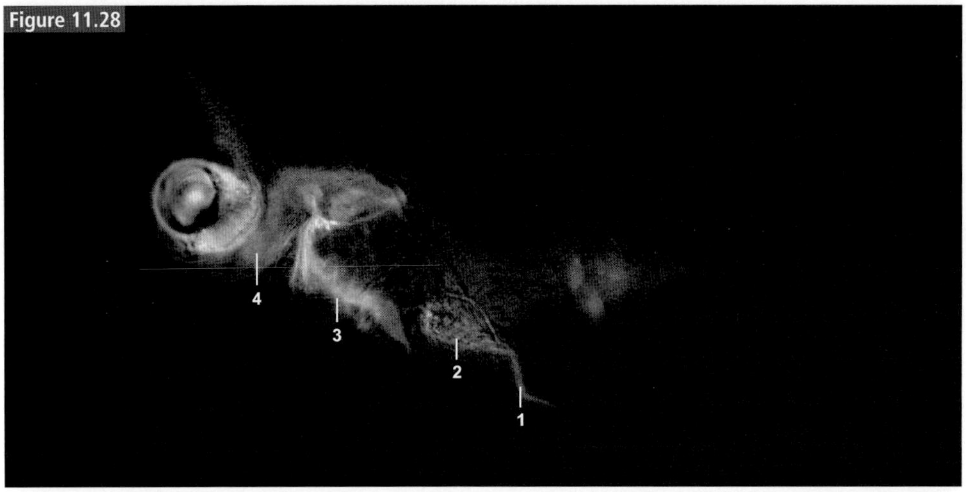

Figure 11.29

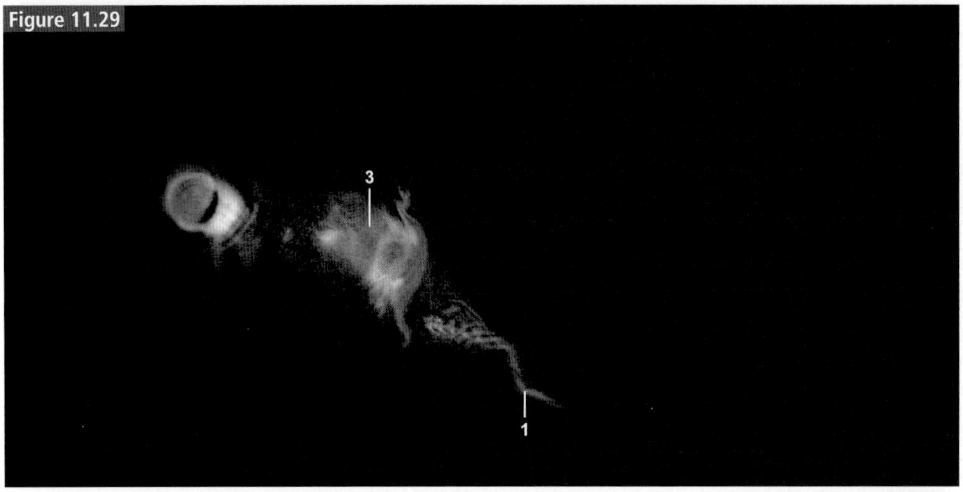

Figure 11.30

1. pectoral fin	**2.** pectoral fin musculature	**3.** opercle	**4.** adductor mandibulae

Coronal Sections

Figure 11.31

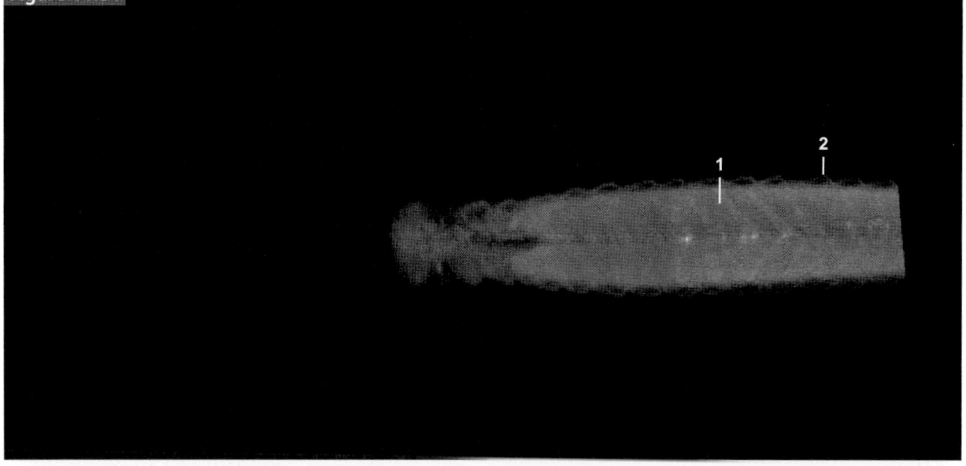

Figure 11.32

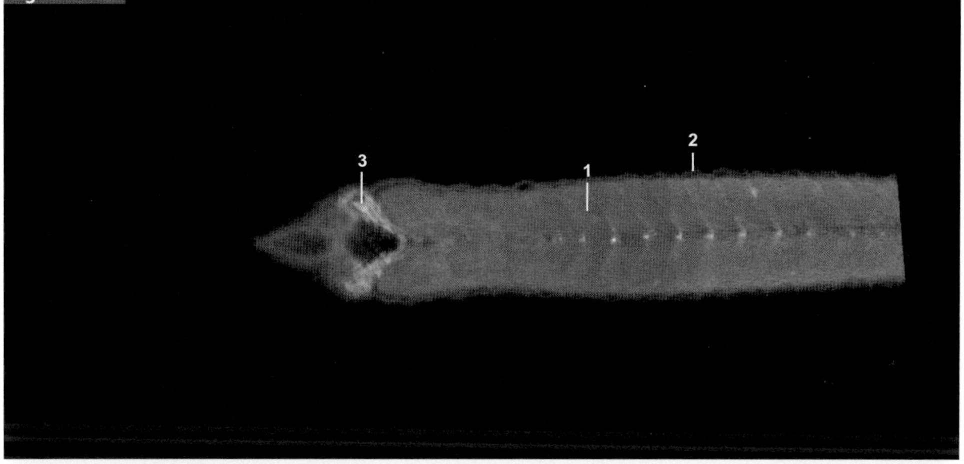

Figure 11.33

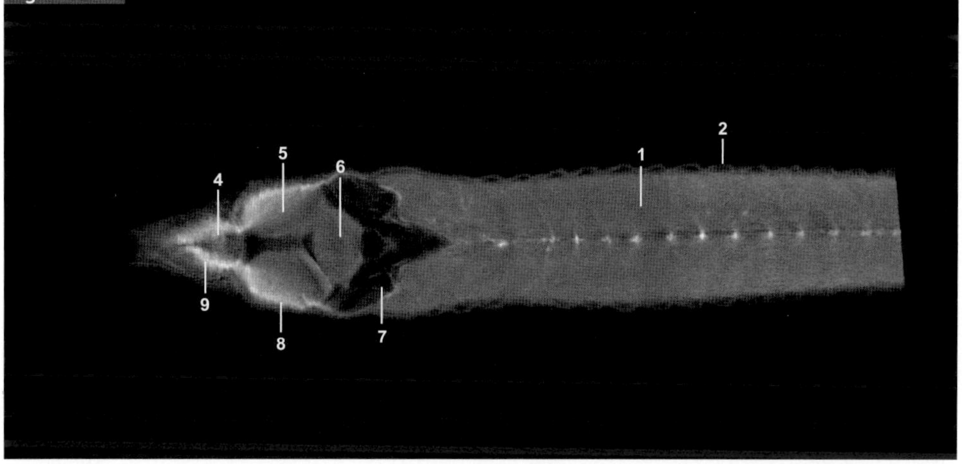

1. myotome	**3.** supraoccipital	**5.** optic tectum	**7.** semicircular canal	**9.** frontal bone
2. scales	**4.** telencephalon	**6.** cerebellum	**8.** parietal bone	

Figure 11.34

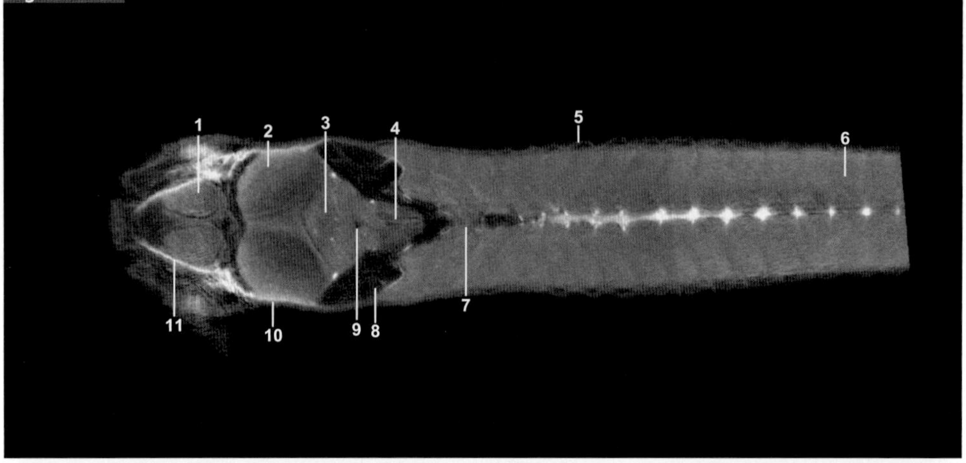

Figure 11.35

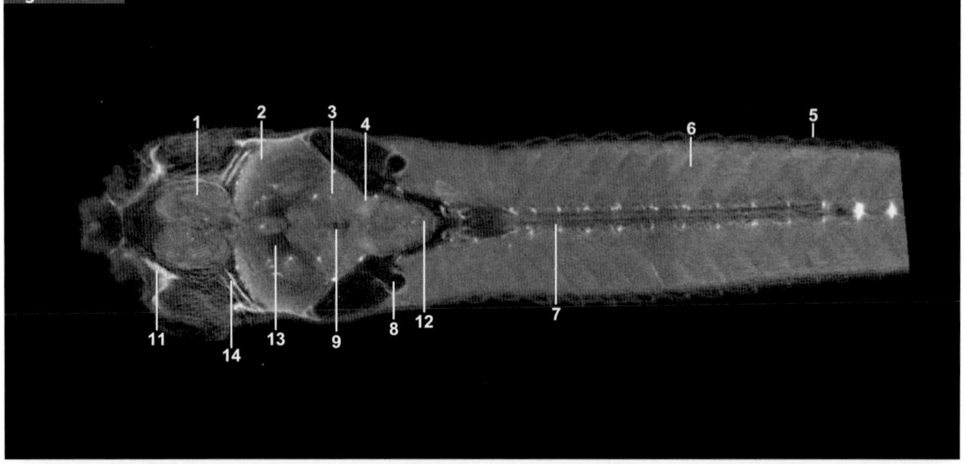

Figure 11.36

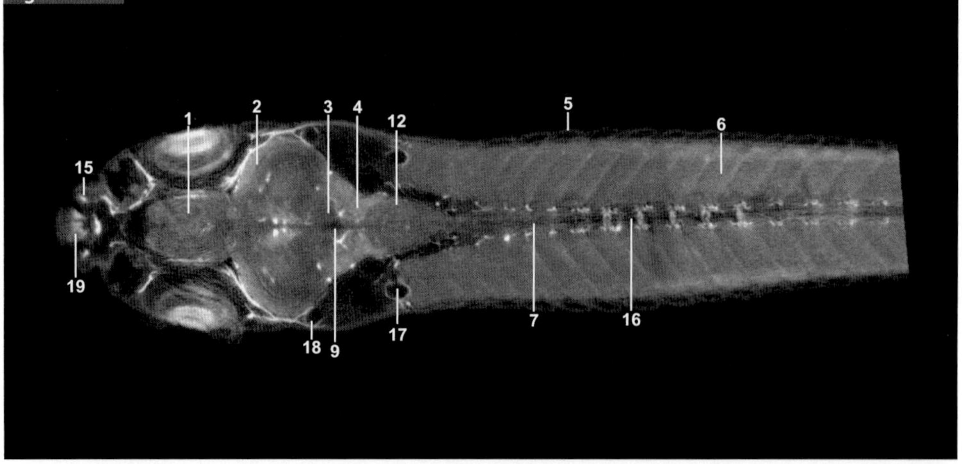

1. telencephalon	**5.** scales	**9.** fourth ventricle	**13.** tectal ventricle	**17.** posterior semicircular canal
2. optic tectum	**6.** myotome	**10.** parietal bone	**14.** infraorbital	**18.** anterior semicircular canal
3. cerebellum	**7.** spinal cord	**11.** frontal bone	**15.** maxilla	**19.** premaxilla
4. crista cerebellaris	**8.** semicircular canal	**12.** vagal lobe	**16.** precaudal vertebra	

Figure 11.37

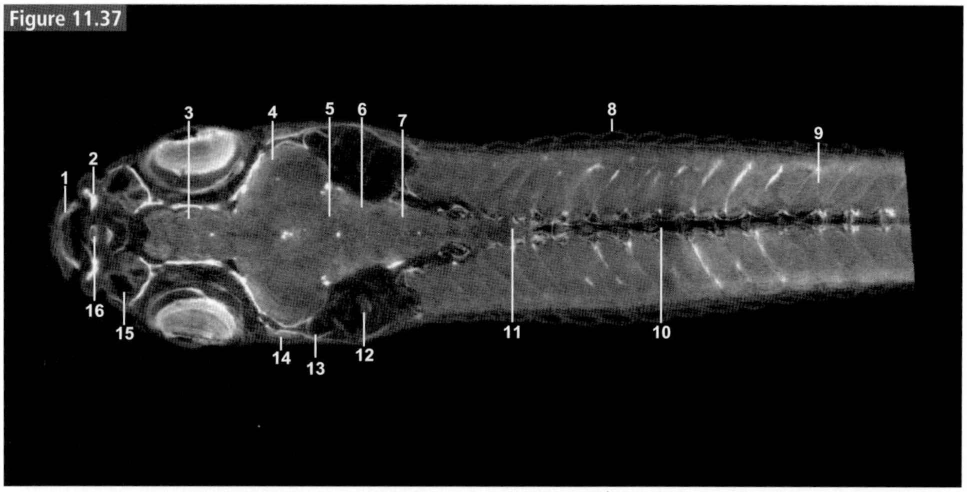

Figure 11.38

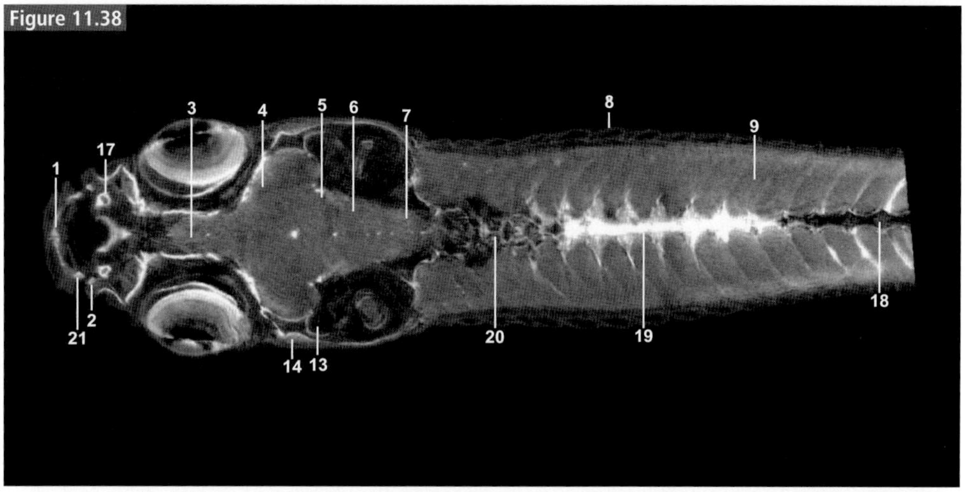

Figure 11.39

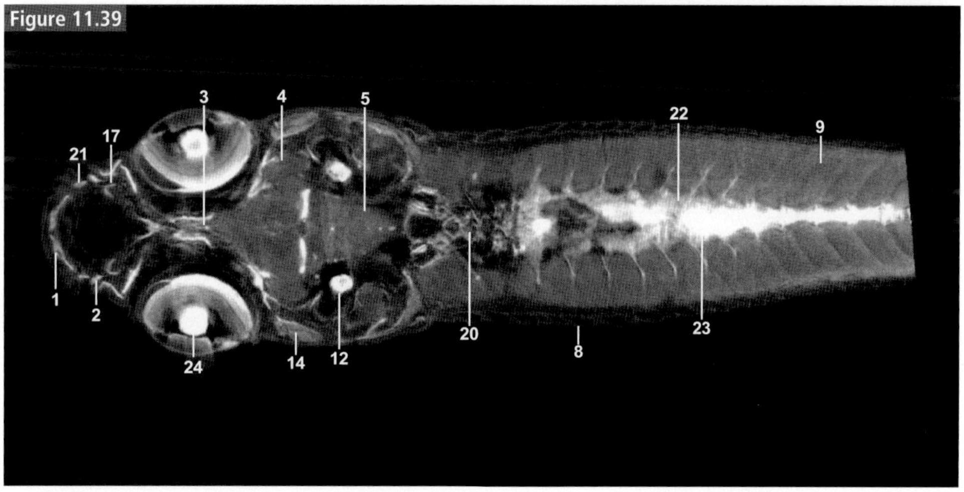

1. Meckel's cartilage
2. maxilla
3. telencephalon
4. optic tectum
5. cerebellum
6. crista cerebellaris
7. vagal lobe
8. scales
9. myotome
10. precaudal vertebra
11. spinal cord
12. otolith
13. anterior semicircular canal
14. dilator operculi
15. olfactory pit
16. kinethmoid
17. nasal bone
18. caudal vertebra
19. dorsal aorta
20. Weberian apparatus
21. premaxilla
22. mesonephros
23. posterior cardinal vein
24. lens

Figure 11.40

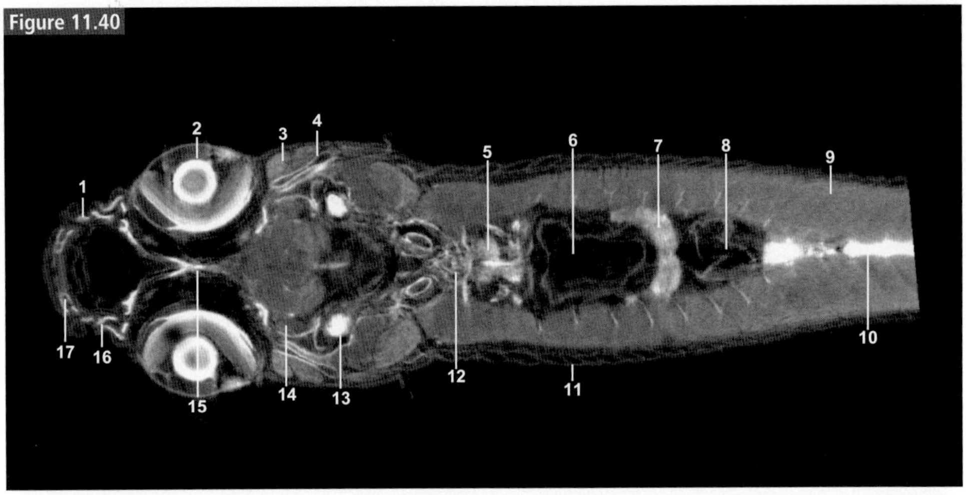

Figure 11.41

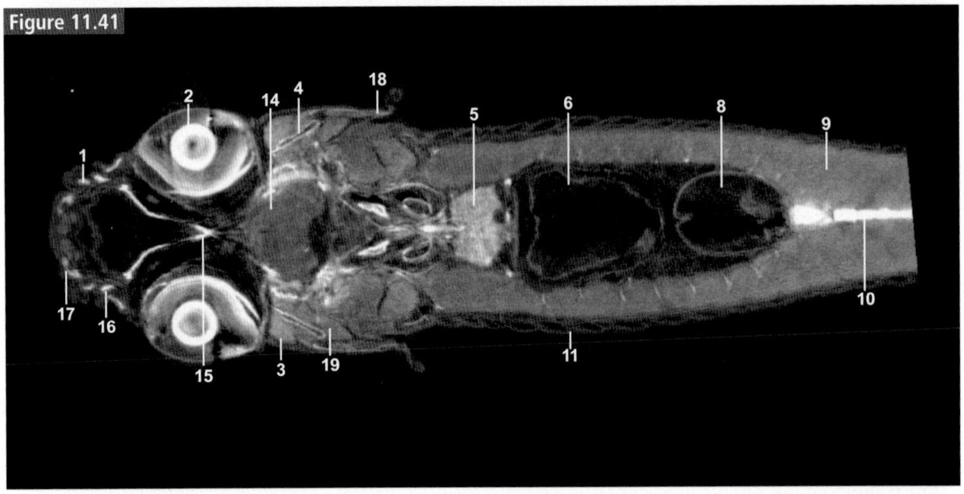

Figure 11.42

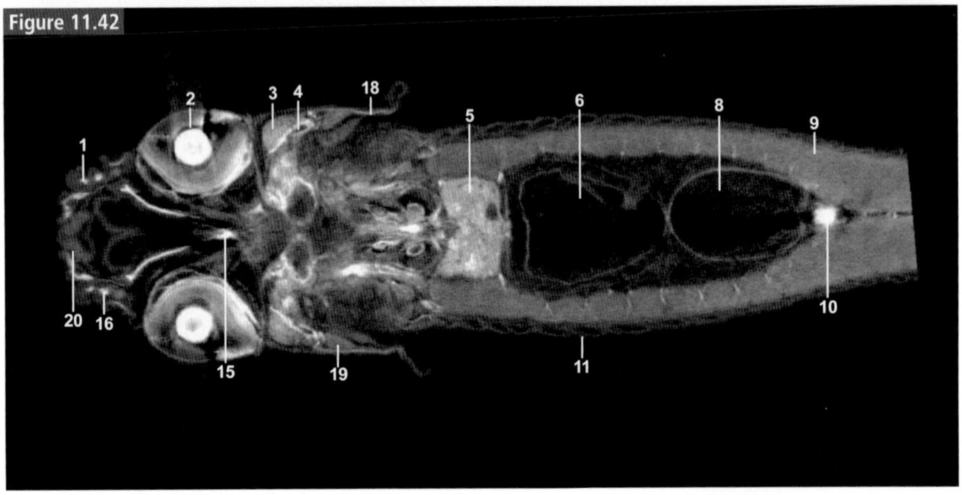

1. premaxilla
2. lens
3. dilator operculi
4. hyomandibula
5. pronephros
6. anterior chamber swim bladder
7. mesonephros
8. posterior chamber swim bladder
9. myotome
10. posterior cardinal vein
11. scales
12. Weberian apparatus
13. otolith
14. optic tectum
15. parasphenoid
16. maxilla
17. Meckel's cartilage
18. opercle
19. adductor operculi
20. intermandibularis

Figure 11.43

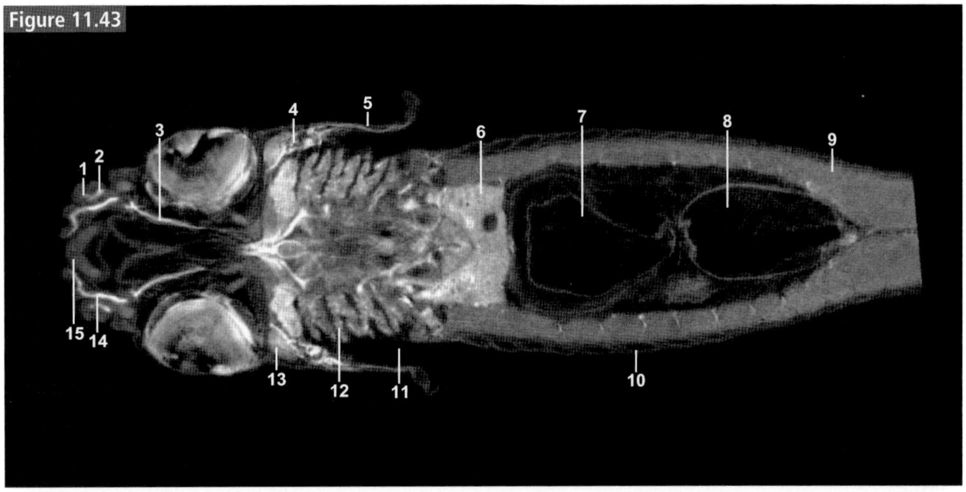

Figure 11.44

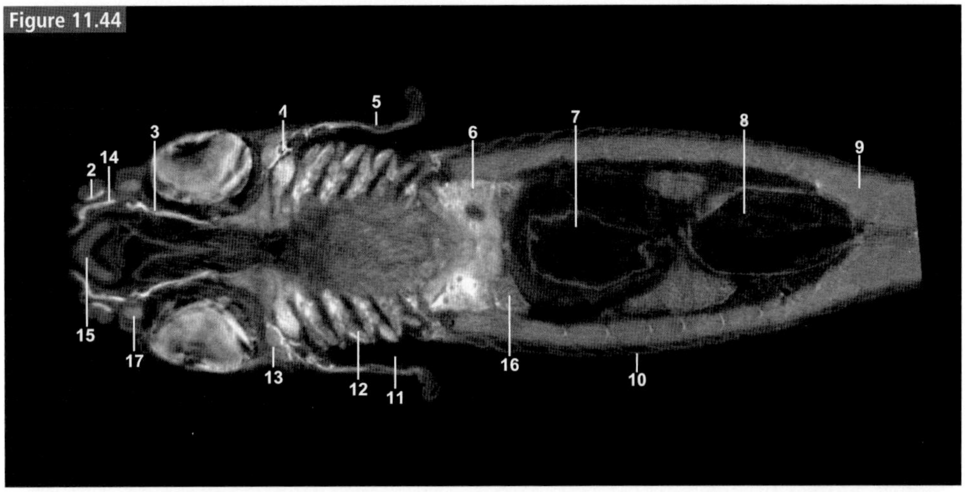

Figure 11.45

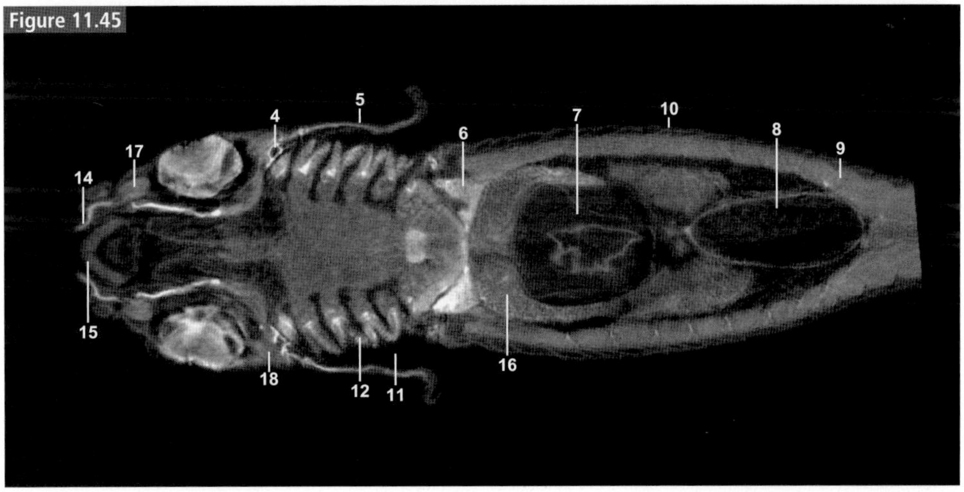

1. premaxilla
2. maxilla
3. palatoquadrate
4. hyomandibula
5. opercle
6. pronephros
7. anterior chamber swim bladder
8. posterior chamber swim bladder
9. myotome
10. scales
11. opercular cavity
12. gill filaments
13. dilator operculi
14. dentary
15. intermandibularis
16. liver
17. adductor mandibulae complex
18. adductor operculi

Figure 11.46

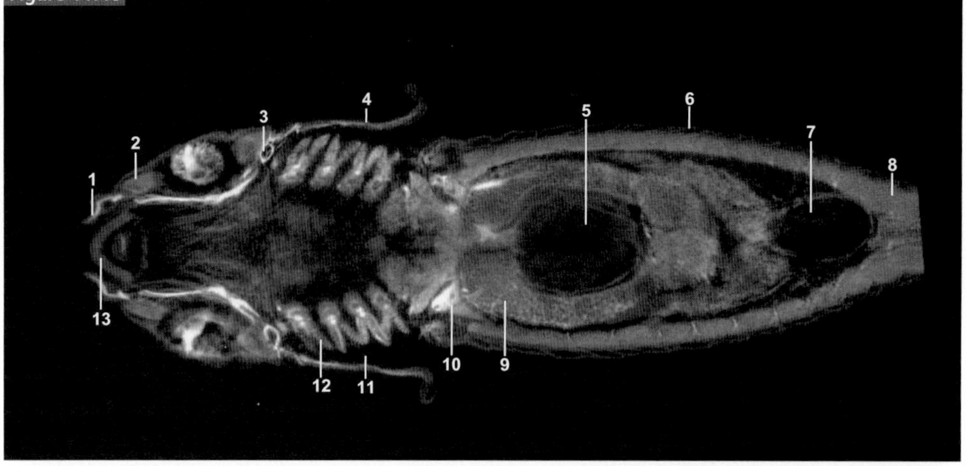

Figure 11.47

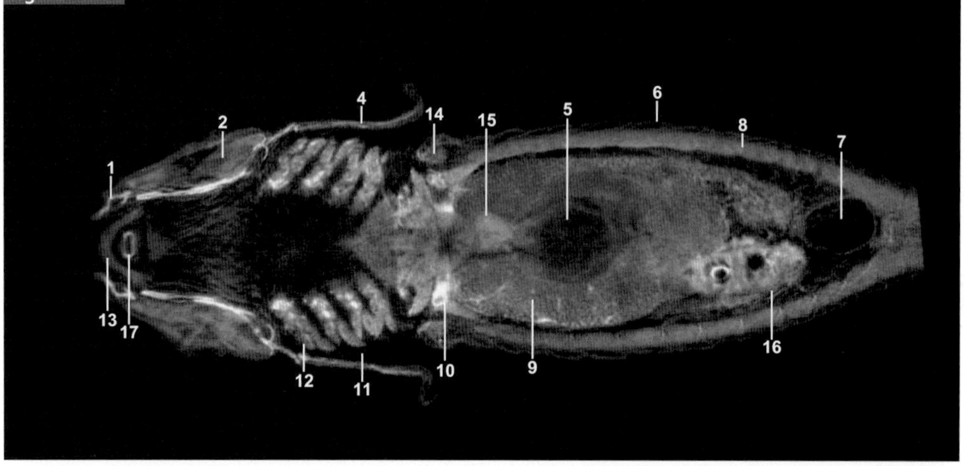

Figure 11.48

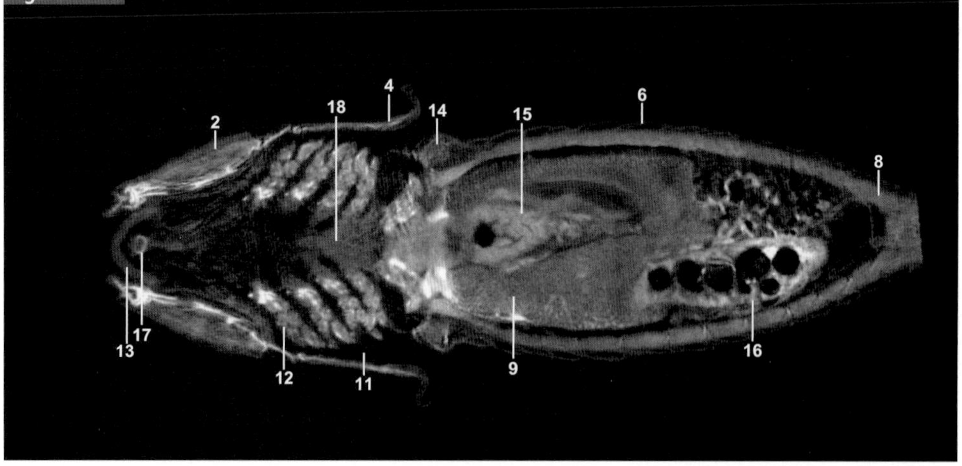

1. dentary
2. adductor mandibulae complex
3. hyomandibula
4. opercle
5. anterior chamber swim bladder
6. scales
7. posterior chamber swim bladder
8. myotome
9. liver
10. pronephros
11. opercular cavity
12. gill filaments
13. intermandibularis
14. pectoral fin musculature
15. intestinal bulb
16. posterior intestine
17. basihyal
18. basibranchial

Figure 11.49

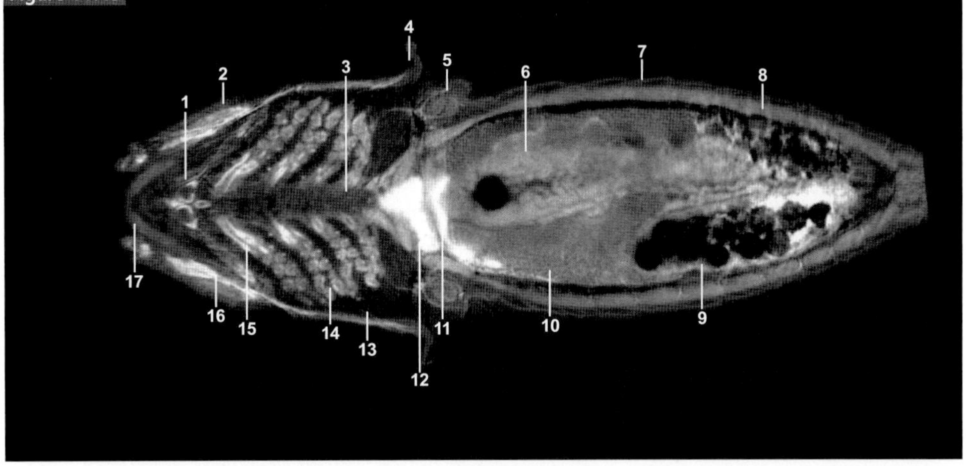

Figure 11.50

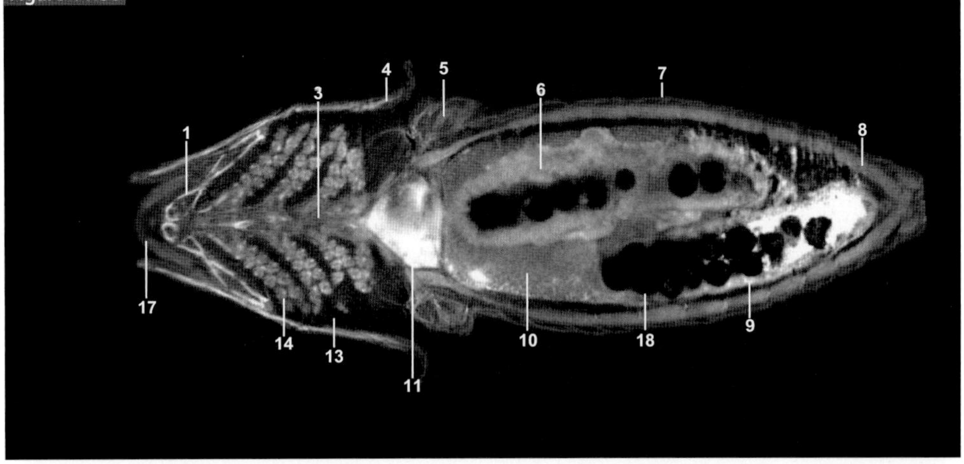

Figure 11.51

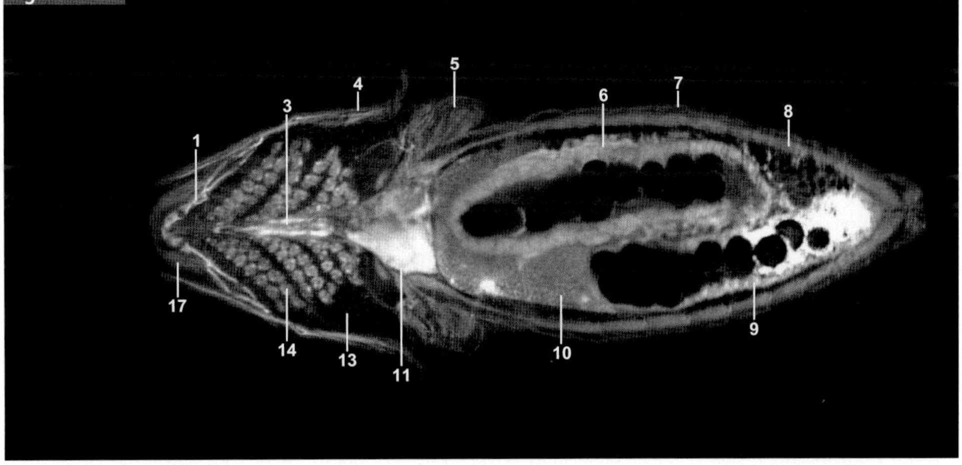

1. ceratohyal
2. adductor mandibulae complex
3. basibranchial
4. opercle
5. pectoral fin musculature
6. intestinal bulb
7. scales
8. myotome
9. posterior intestine
10. liver
11. atrium
12. sinus venosus
13. opercular cavity
14. gill filaments
15. aortic arch
16. ceratobranchial
17. intermandibularis
18. artemia cyst

Figure 11.52

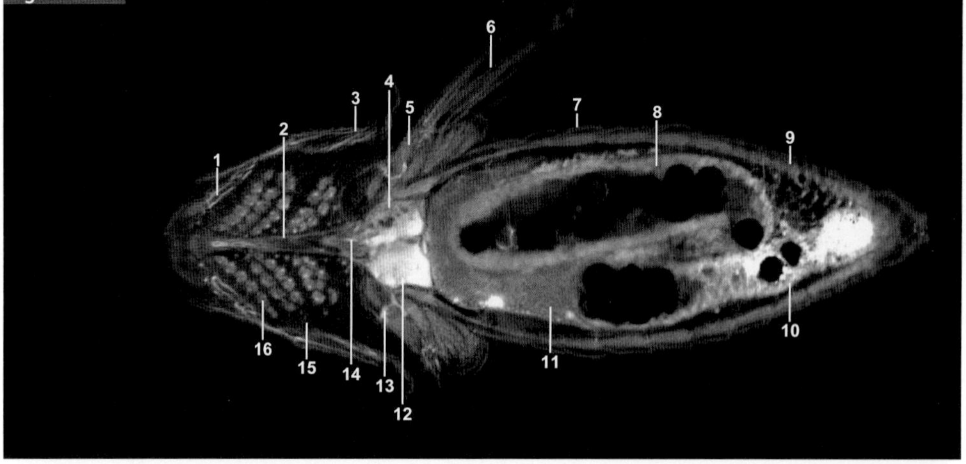

Figure 11.53

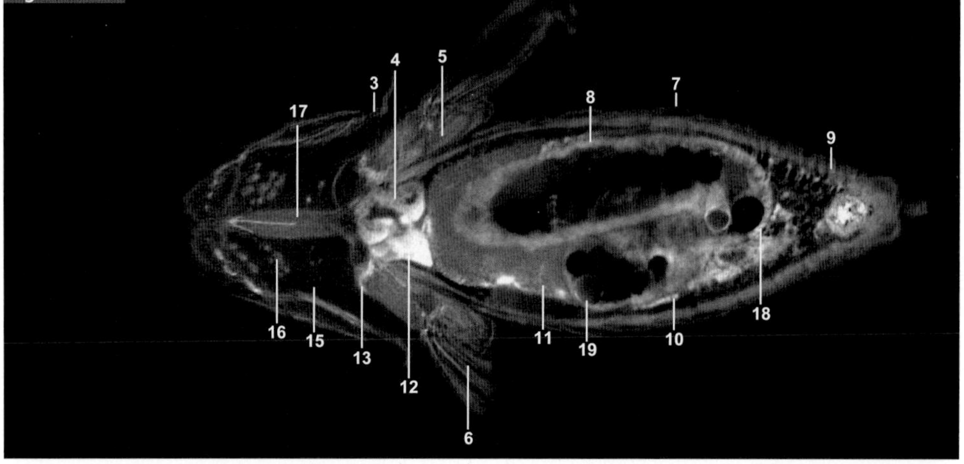

Figure 11.54

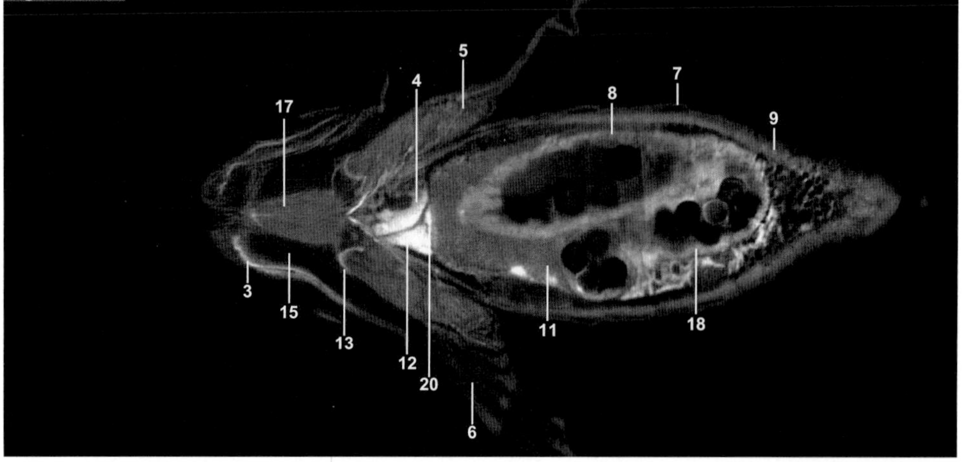

1. ceratohyal	**5.** pectoral fin musculature	**9.** myotome	**13.** cleithrum	**17.** sternohyoid
2. basibranchial	**6.** pectoral fin	**10.** posterior intestine	**14.** bulbus arteriosus	**18.** mid intestine
3. opercle	**7.** scales	**11.** liver	**15.** opercular cavity	**19.** artemia cyst
4. cardiac ventricle	**8.** intestinal bulb	**12.** atrium	**16.** gill filaments	**20.** sinus venosus

Figure 11.55

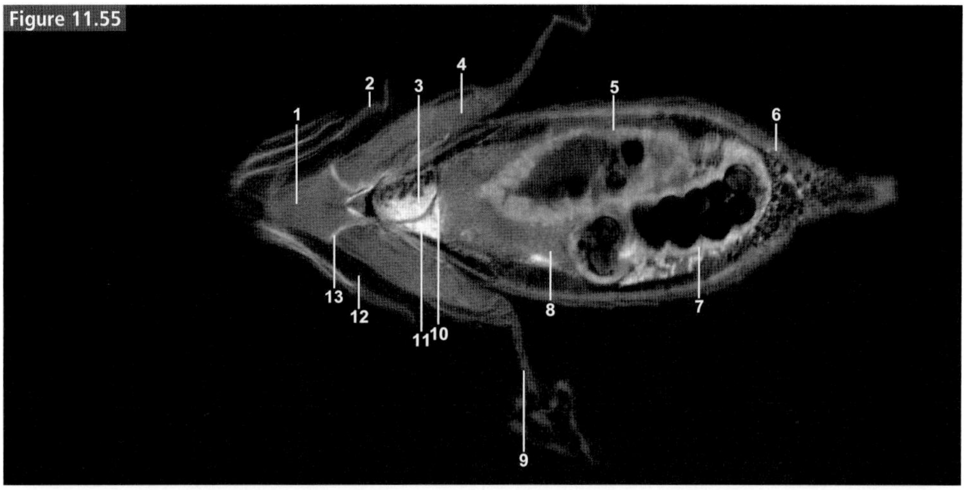

Figure 11.56

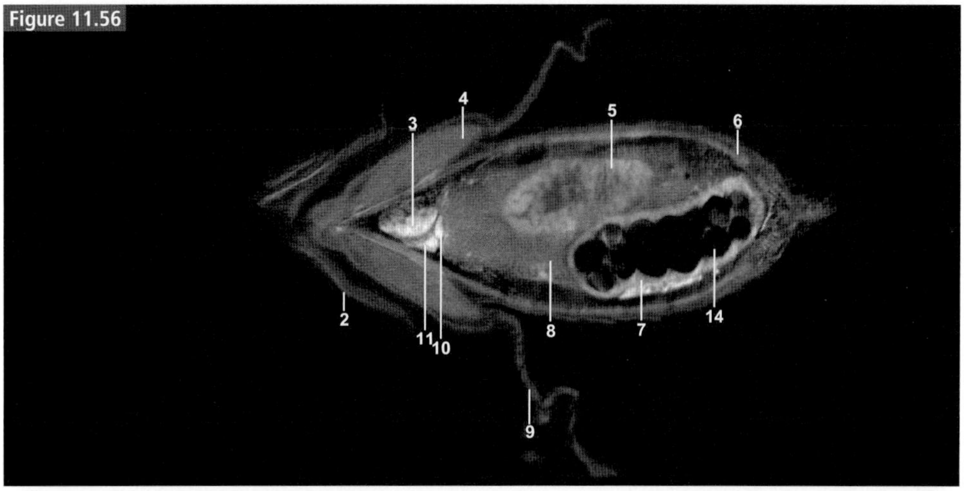

Figure 11.57

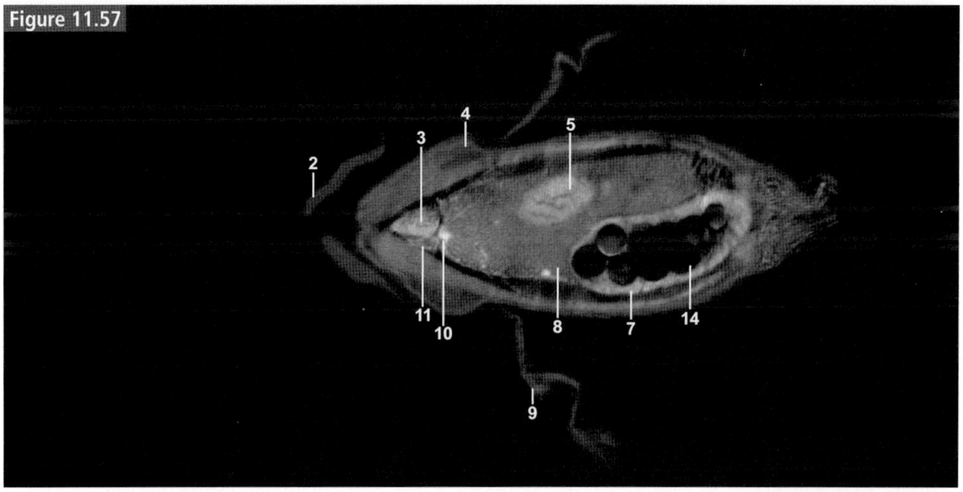

1. sternohyoid	**4.** pectoral fin musculature	**7.** mid intestine	**10.** sinus venosus	**13.** cleithrum
2. opercle	**5.** intestinal bulb	**8.** liver	**11.** atrium	**14.** artemia cyst
3. cardiac ventricle	**6.** myotome	**9.** pectoral fin	**12.** opercular cavity	

Figure 11.58

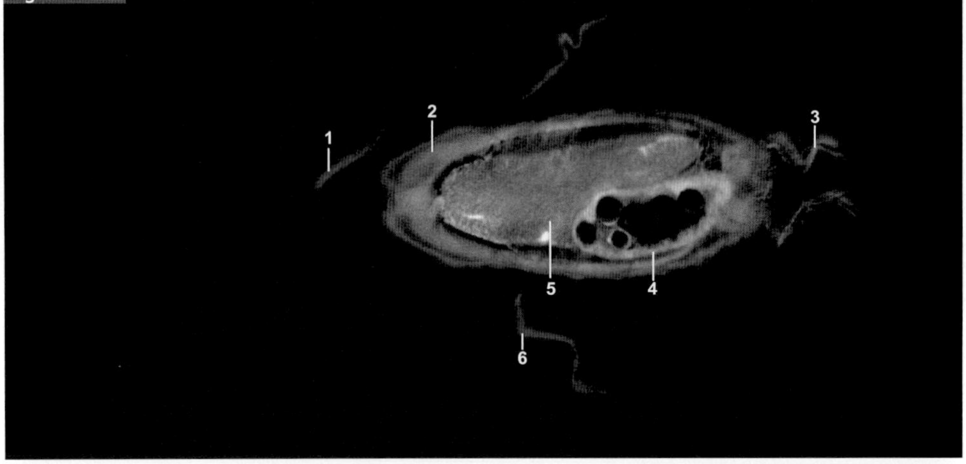

Figure 11.59

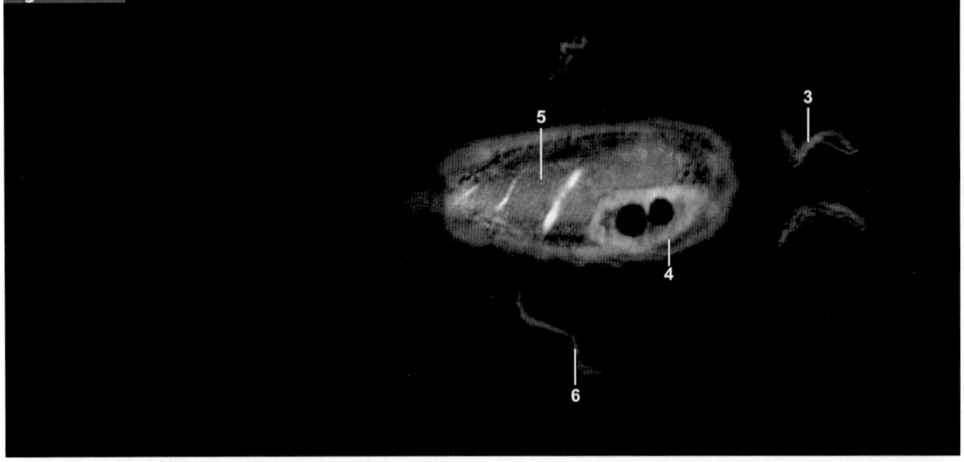

Figure 11.60

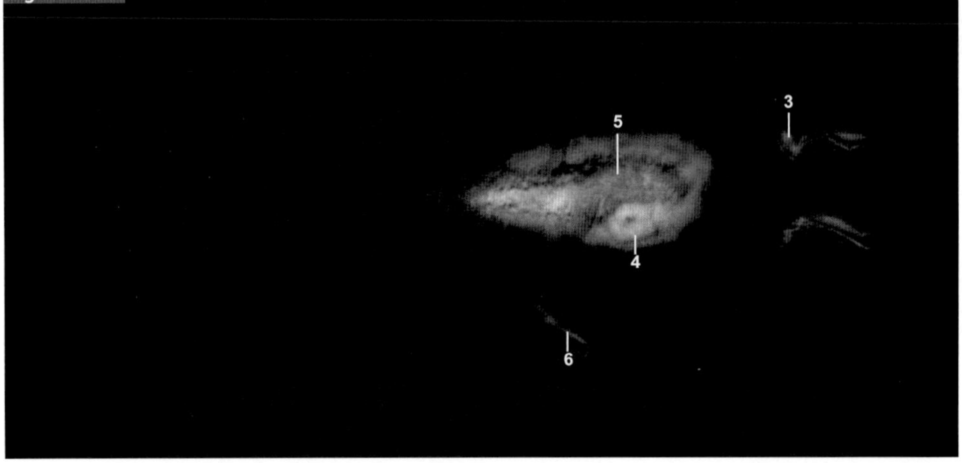

1. opercle	**2.** pectoral fin musculature	**3.** pelvic fin	**4.** mid intestine	**5.** liver	**6.** pectoral fin

Transverse Sections

Figure 11.61

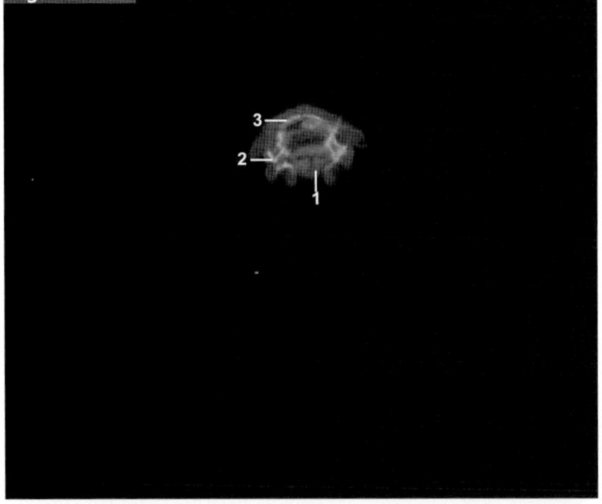

Figure 11.62

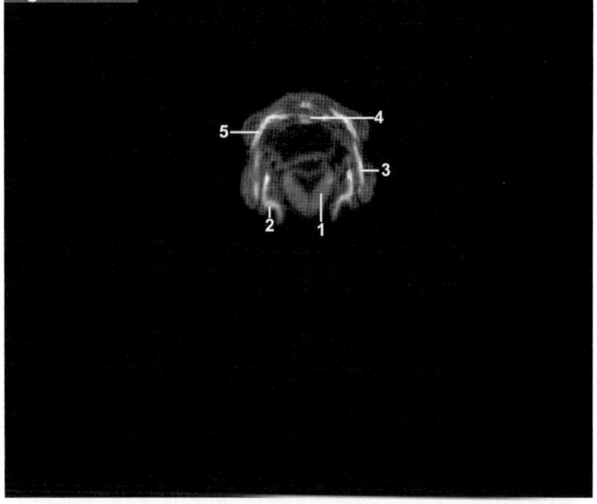

Figure 11.63

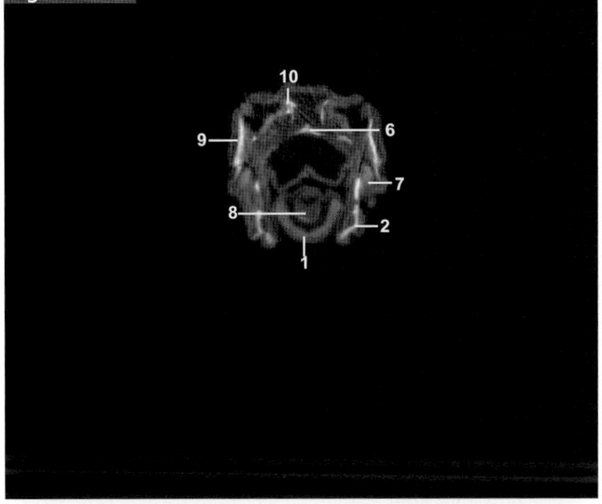

Figure 11.64

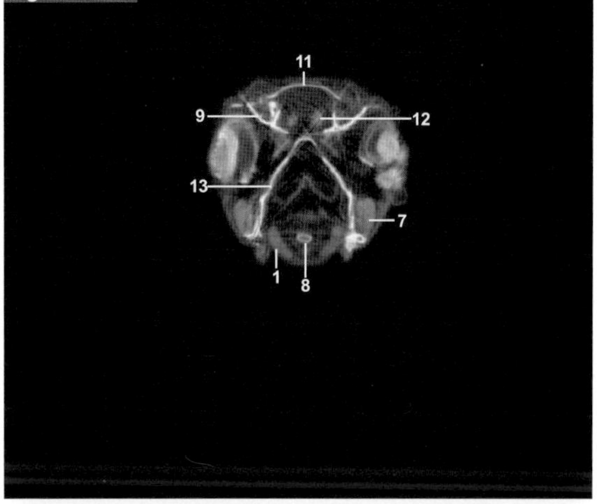

Figure 11.65

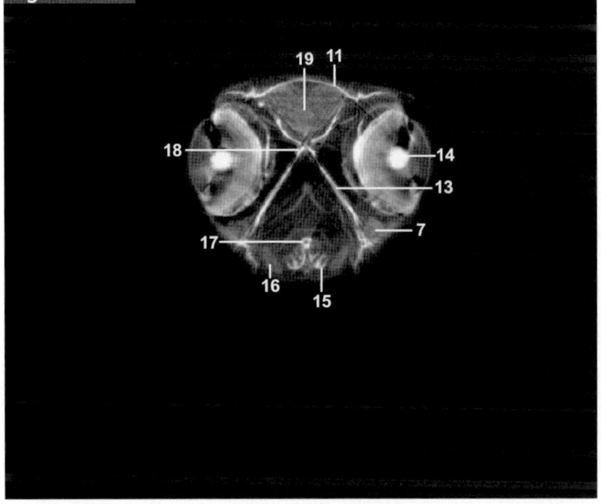

Figure 11.66

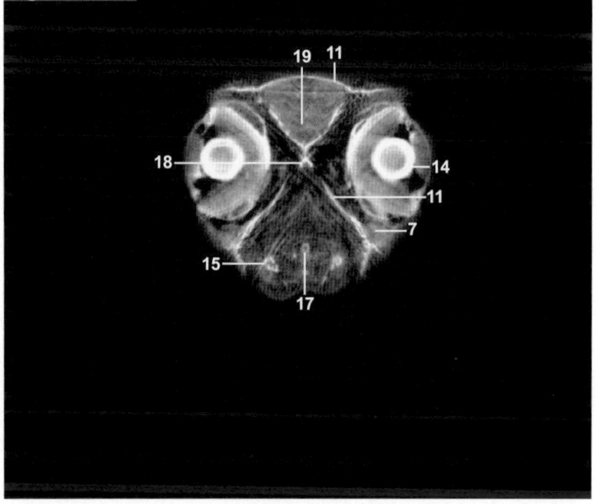

1. intermandibularis	**5.** maxilla	**8.** basihyal	**12.** olfactory bulb	**16.** interhyal
2. dentary	**6.** supraethmoid	**9.** lateral ethmoid	**13.** palatoquadrate	**17.** basibranchial
3. premaxilla	**7.** adductor mandibulae complex	**10.** hyosymplectic	**14.** lens	**18.** parasphenoid
4. kinethmoid		**11.** frontal bone	**15.** ceratohyal	**19.** telencephalon

Figure 11.67

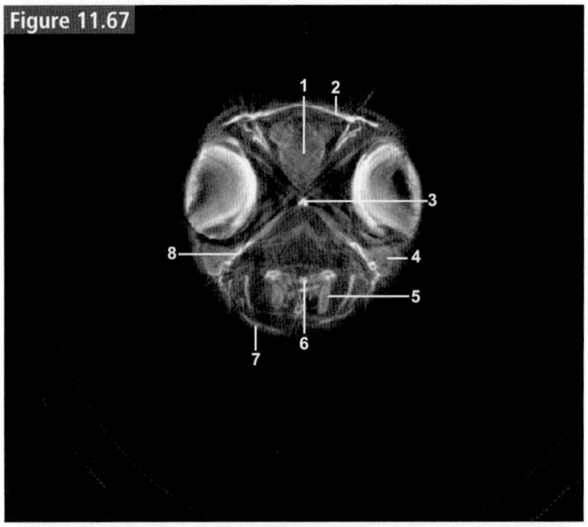

Figure 11.68

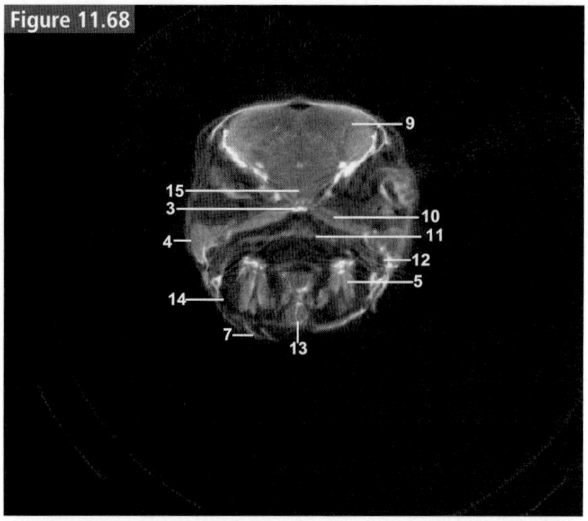

Figure 11.69

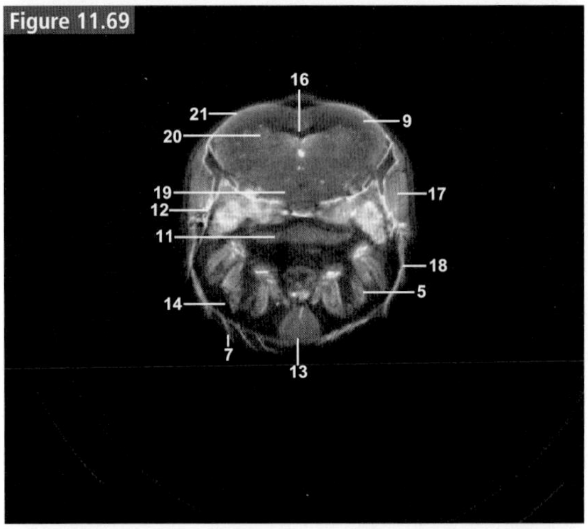

Figure 11.70

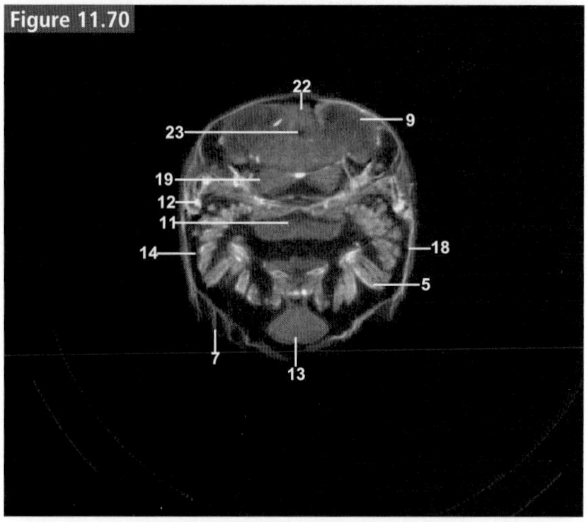

Figure 11.71

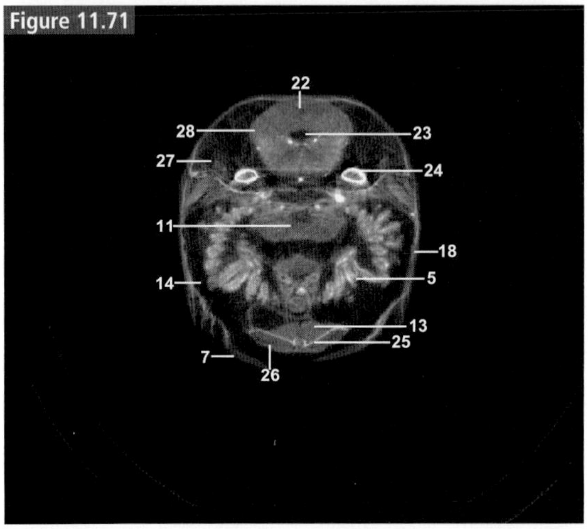

Figure 11.72

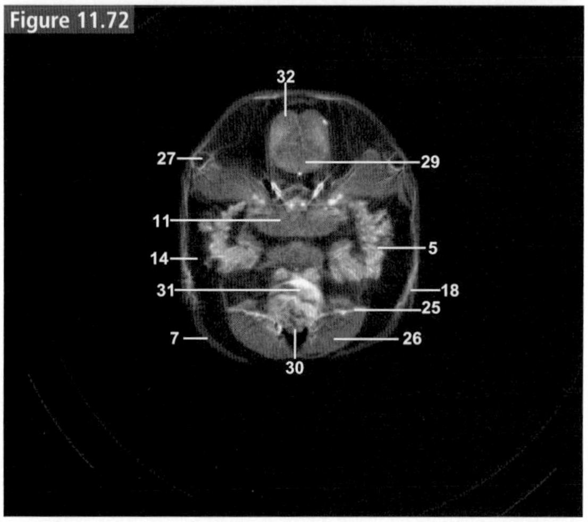

1. telencephalon
2. frontal bone
3. parasphenoid
4. adductor mandibulae complex
5. gill filaments
6. basibranchial
7. branchiostegal ray
8. palatoquadrate
9. optic tectum
10. lateral rectus
11. pharynx
12. hyomandibula
13. sternohyoid
14. opercular cavity
15. hypothalamus
16. tectal ventricle
17. dilator operculi
18. opercle
19. inferior lobe
20. torus semicircularis
21. parietal bone
22. cerebellum
23. fourth ventricle
24. otolith
25. cleithrum
26. pectoral fin musculature
27. semicircular canal
28. granular eminence
29. medulla oblongata
30. cardiac ventricle
31. atrium
32. crista cerebellaris

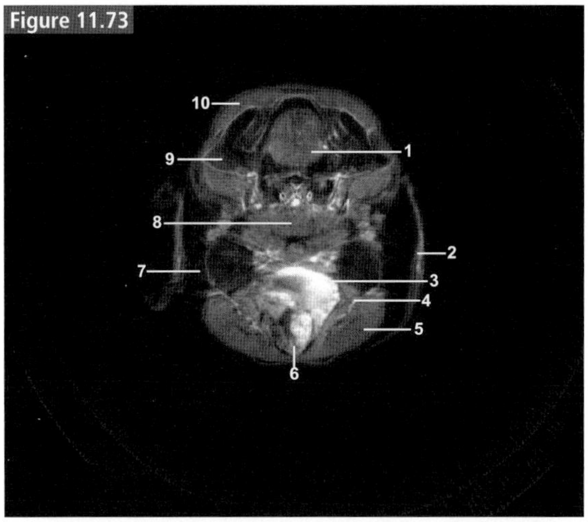

Figure 11.73

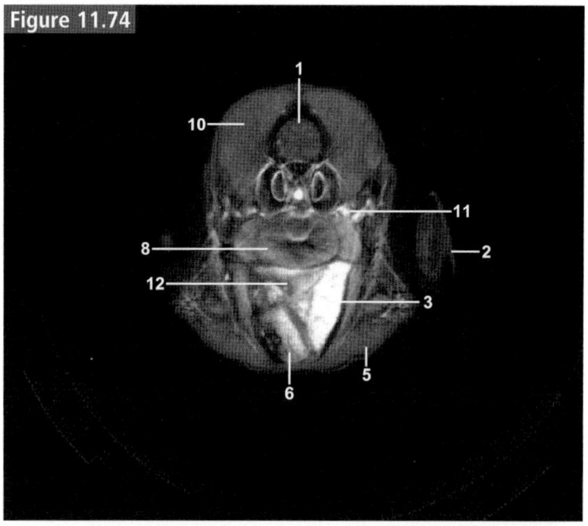

Figure 11.74

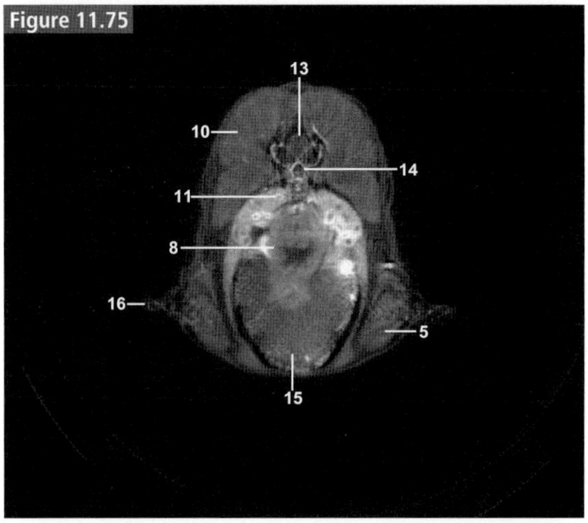

Figure 11.75

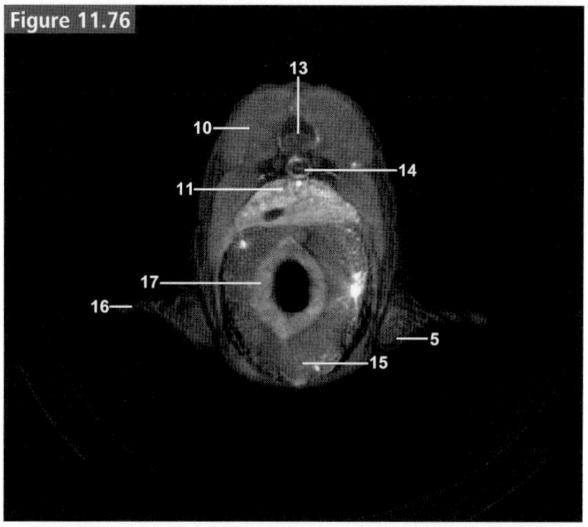

Figure 11.76

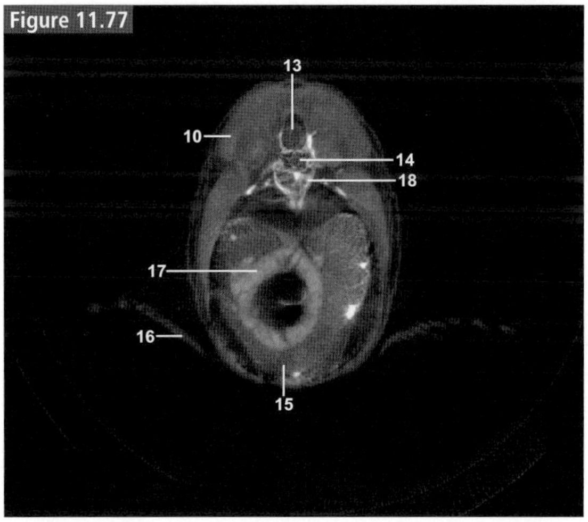

Figure 11.77

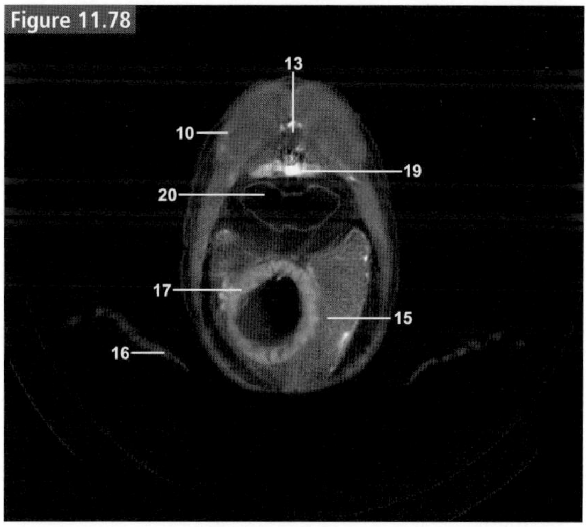

Figure 11.78

1. medulla oblongata
2. opercle
3. atrium
4. cleithrum
5. pectoral fin musculature
6. cardiac ventricle
7. opercular cavity
8. esophagus
9. semicircular canal
10. myotome
11. pronephros
12. sinus venosus
13. spinal cord
14. Weberian vertebra
15. liver
16. pectoral fin
17. intestinal bulb
18. Weberian apparatus
19. posterior cardinal vein
20. anterior chamber swim bladder

Figure 11.79

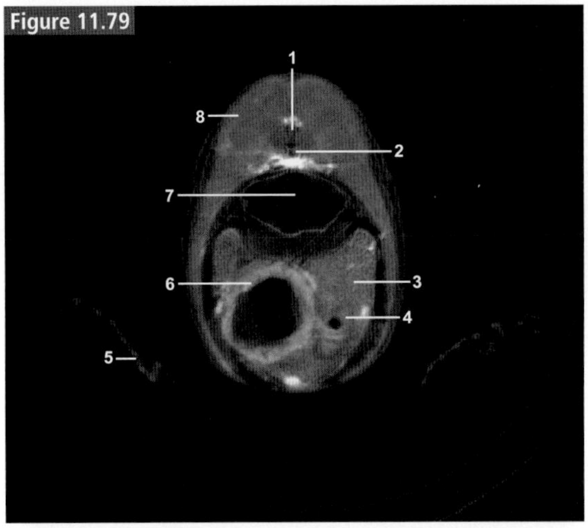

Figure 11.80

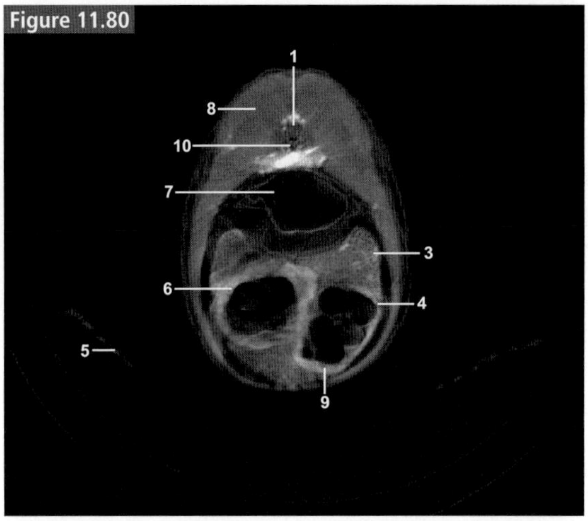

Figure 11.81

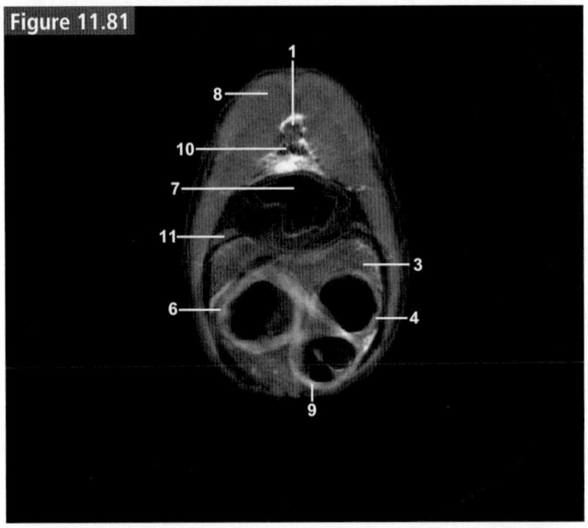

Figure 11.82

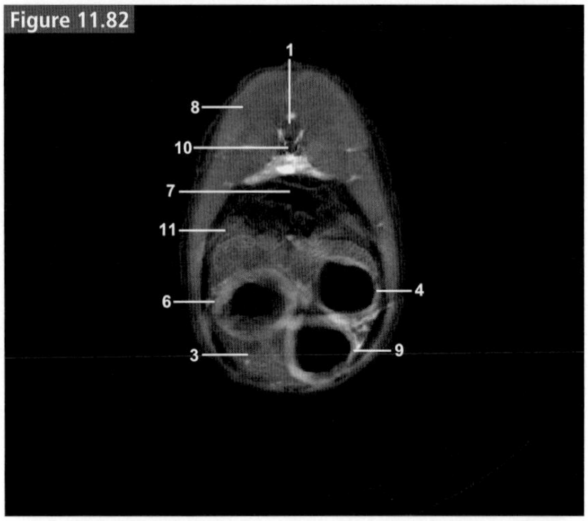

Figure 11.83

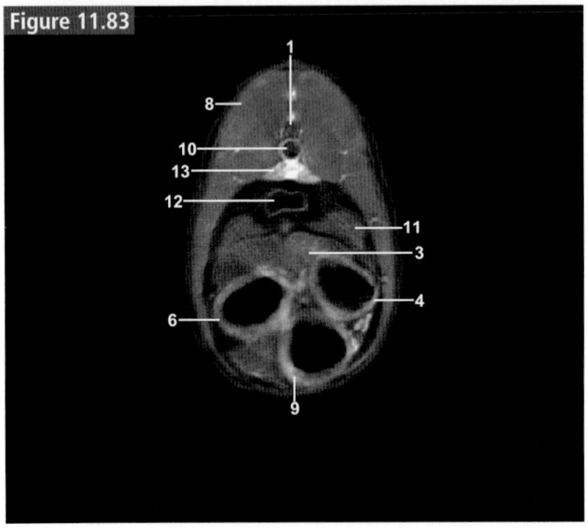

Figure 11.84

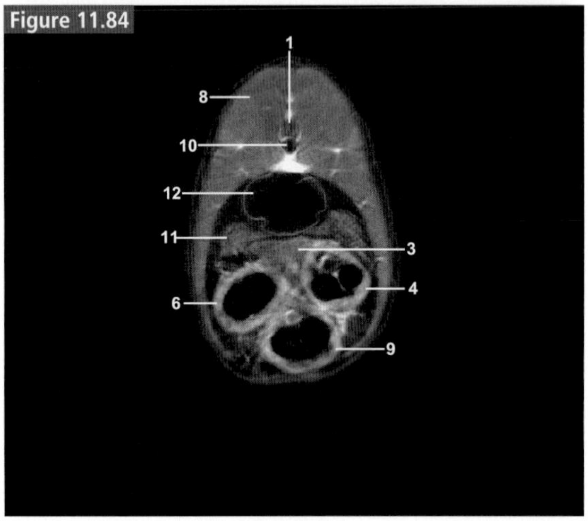

1. spinal cord	**4.** posterior intestine	**7.** anterior chamber swim bladder	**9.** mid intestine	**12.** posterior chamber swim bladder
2. vertebral column	**5.** pectoral fin	**8.** myotome	**10.** precaudal vertebra	**13.** mesonephros
3. liver	**6.** intestinal bulb		**11.** testis	

Figure 11.85

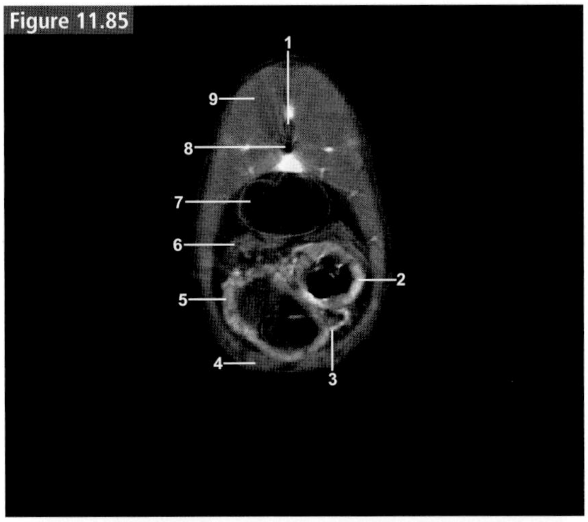

Figure 11.86

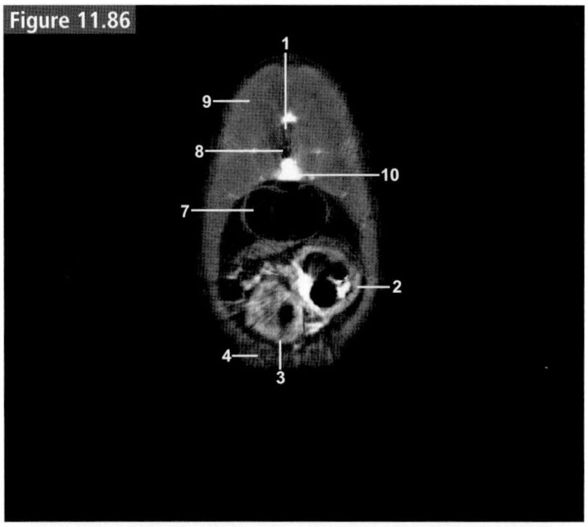

Figure 11.87

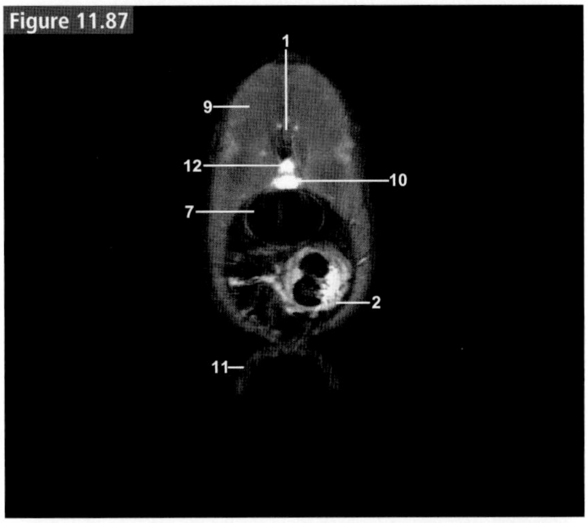

Figure 11.88

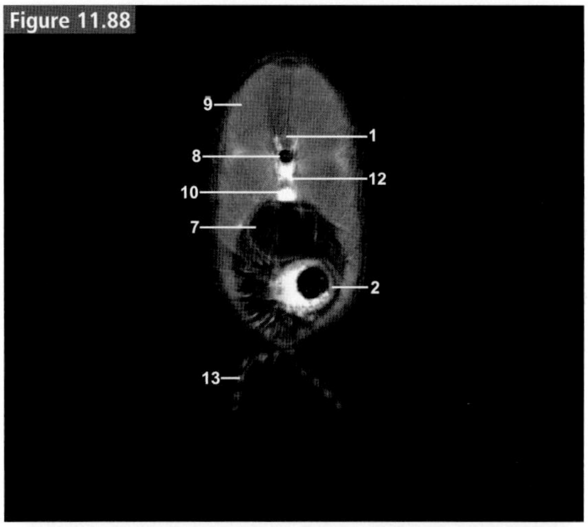

Figure 11.89

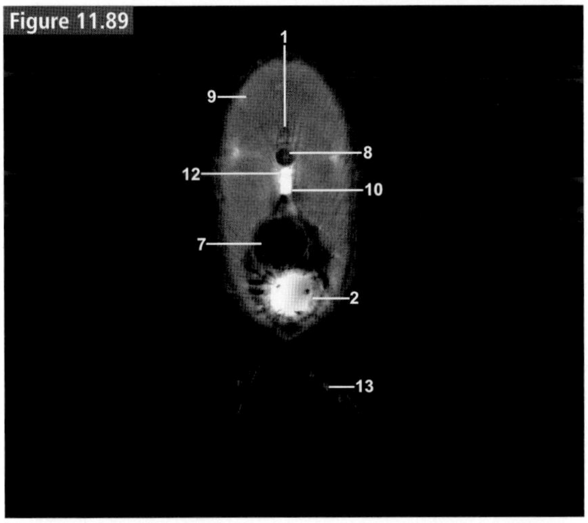

Figure 11.90

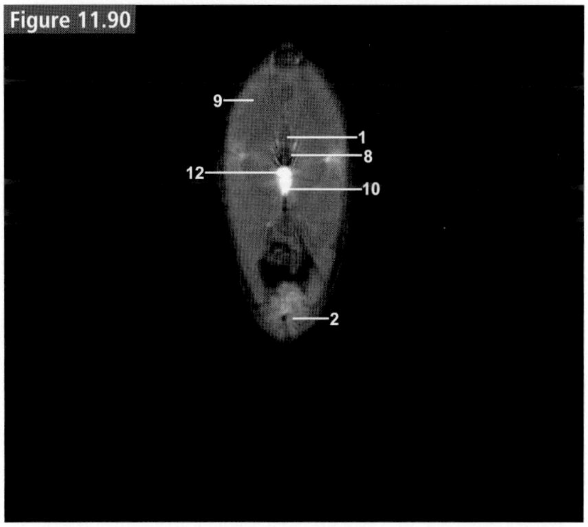

1. spinal cord	**4.** pelvic fin musculature	**7.** posterior chamber swim bladder	**9.** myotome	**12.** caudal artery
2. posterior intestine	**5.** intestinal bulb	**8.** precaudal vertebra	**10.** posterior cardinal vein	**13.** pelvic fin lepidotrichia
3. mid intestine	**6.** testis		**11.** pelvic fin	

PART III

THE ADULT ZEBRAFISH

CHAPTER 12

16 MM

Sagittal Sections

Figure 12.1

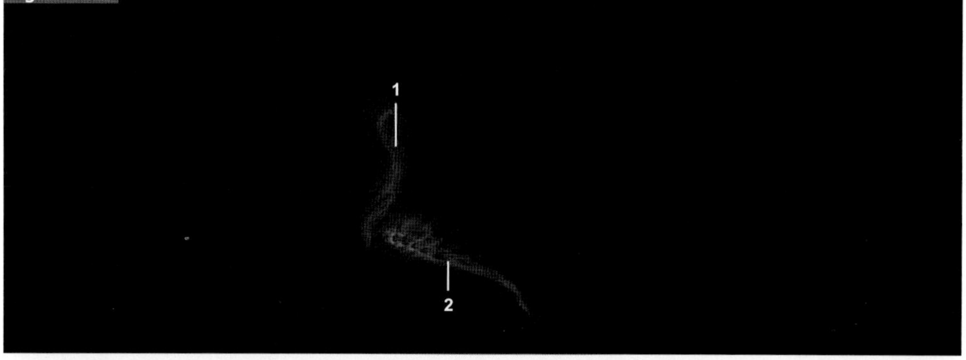

Figure 12.2

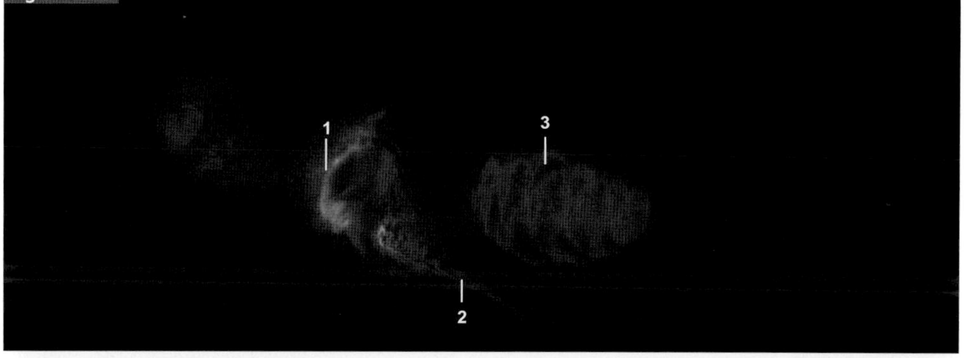

Figure 12.3

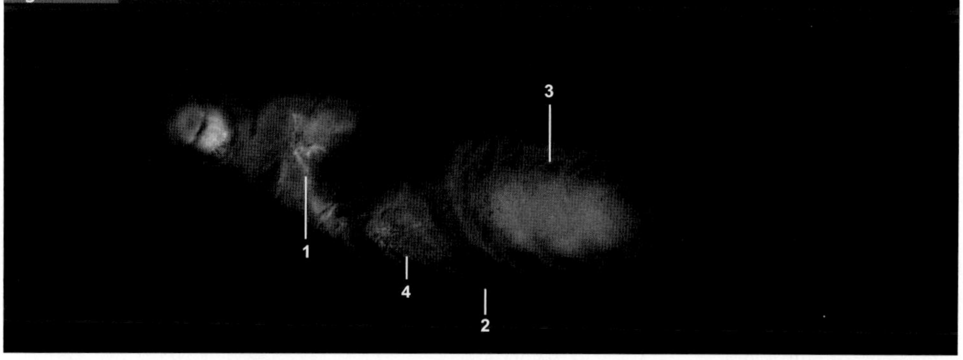

1. opercle	**2.** pectoral fin	**3.** scales	**4.** pectoral fin musculature

Atlas of Zebrafish Development. 10.1016/B978-0-12-374016-8.00012-1

Figure 12.4

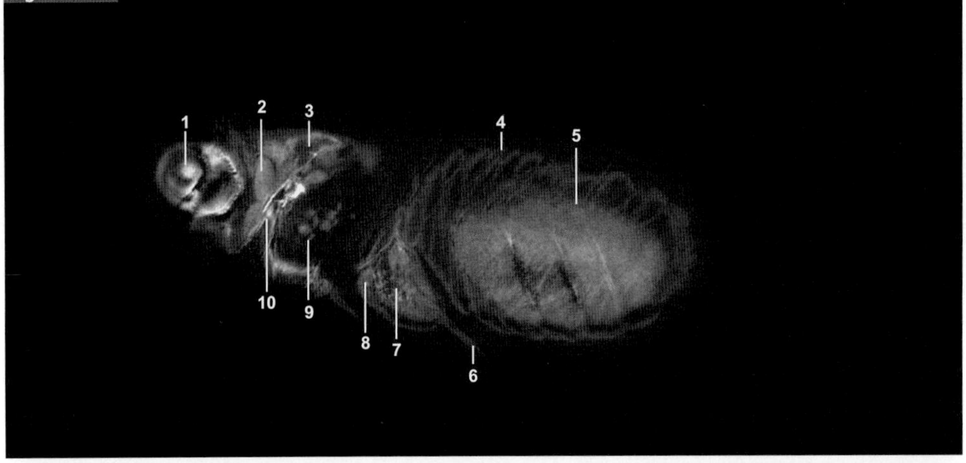

Figure 12.5

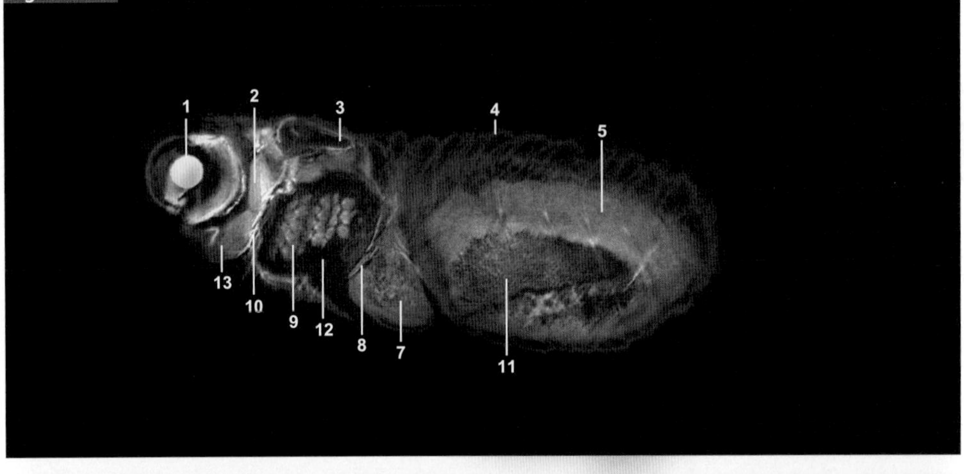

Figure 12.6

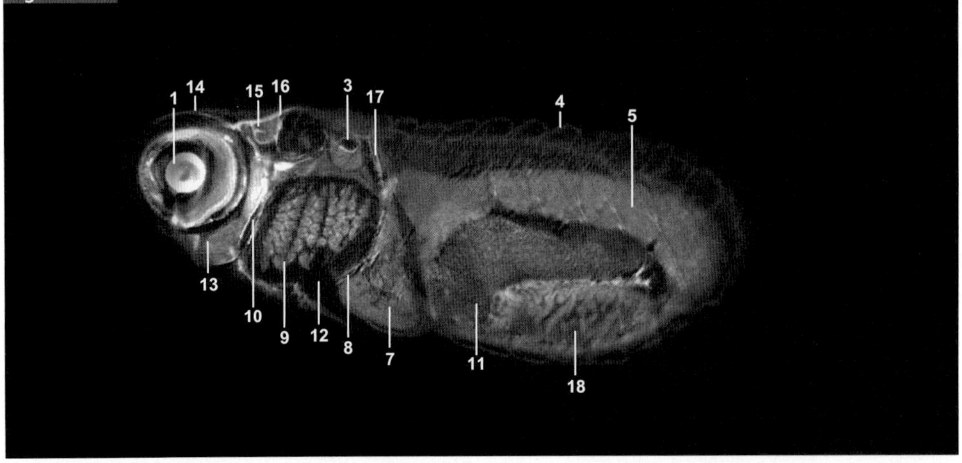

1. lens	**5.** myotome	**8.** cleithrum	**12.** opercular cavity	**15.** optic tectum
2. dilator operculi	**6.** pectoral fin	**9.** gill filaments	**13.** adductor mandibulae complex	**16.** parietal bone
3. semicircular canal	**7.** pectoral fin musculature	**10.** hyomandibula	**14.** frontal bone	**17.** supracleithrum
4. scales		**11.** pancreas		**18.** intestinal bulb

Figure 12.7

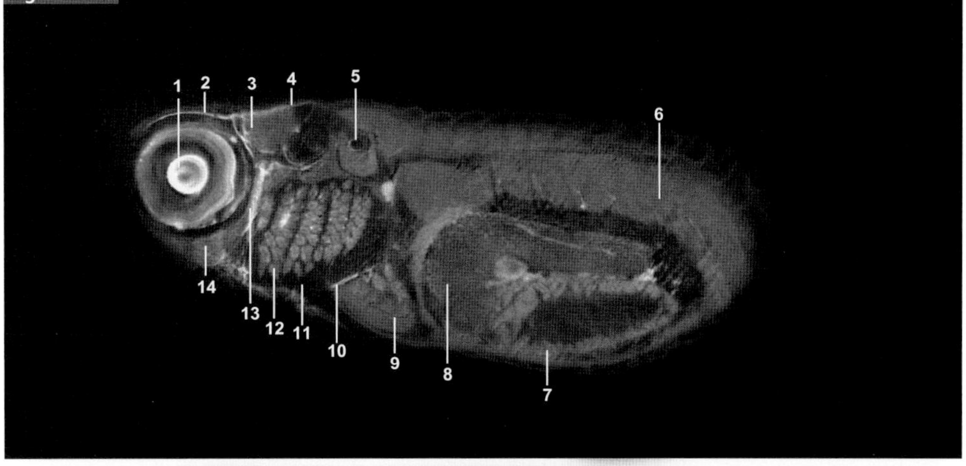

Figure 12.8

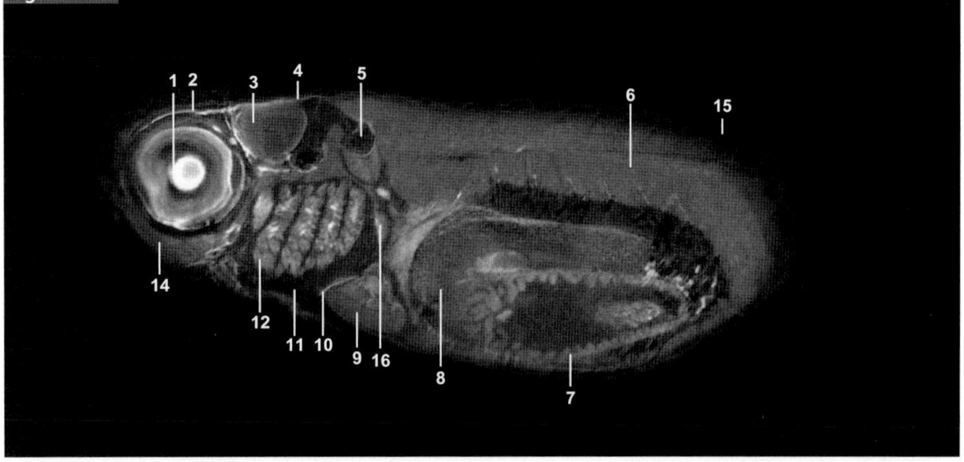

Figure 12.9

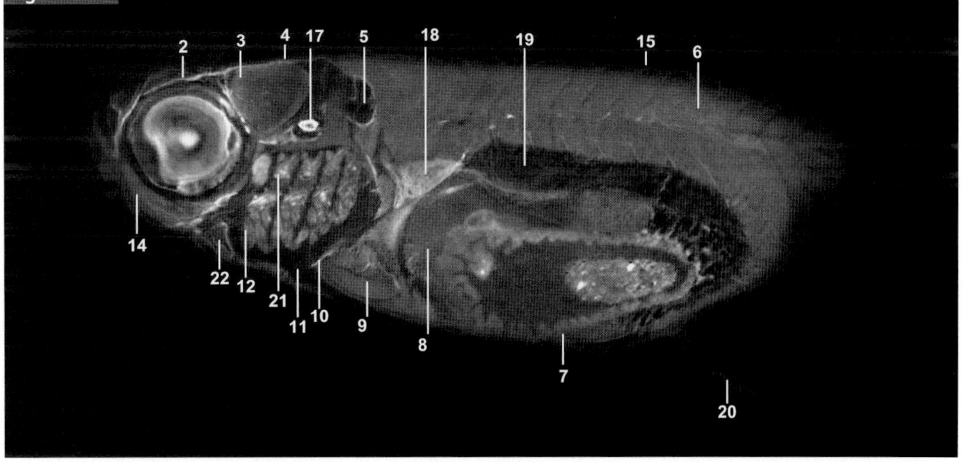

1. lens	**7.** intestinal bulb	**13.** hyomandibula	**18.** pronephros
2. frontal bone	**8.** pancreas	**14.** adductor mandibulae complex	**19.** anterior chamber swim bladder
3. optic tectum	**9.** pectoral fin musculature	**15.** scales	**20.** pelvic fin
4. parietal bone	**10.** cleithrum	**16.** ceratobranchial 5	**21.** aortic arch
5. semicircular canal	**11.** opercular cavity	**17.** otolith	**22.** ceratohyal
6. myotome	**12.** gill filaments		

Figure 12.10

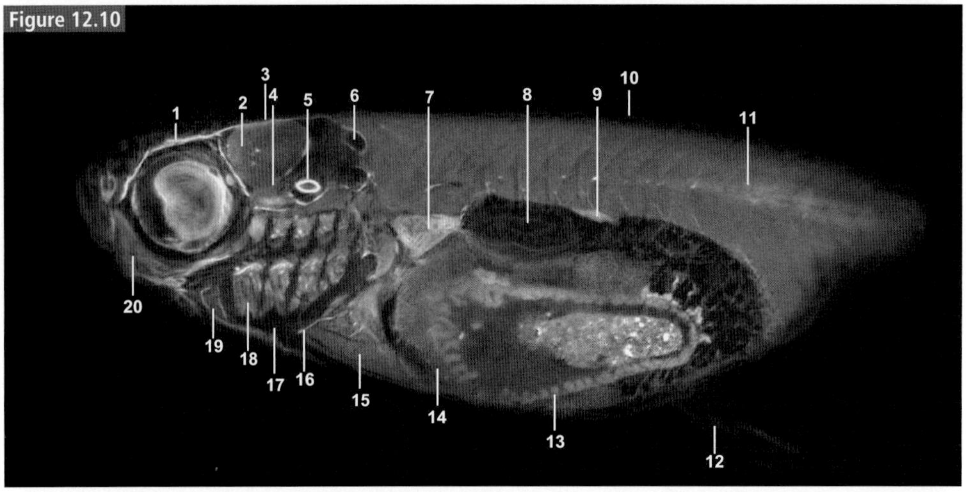

Figure 12.11

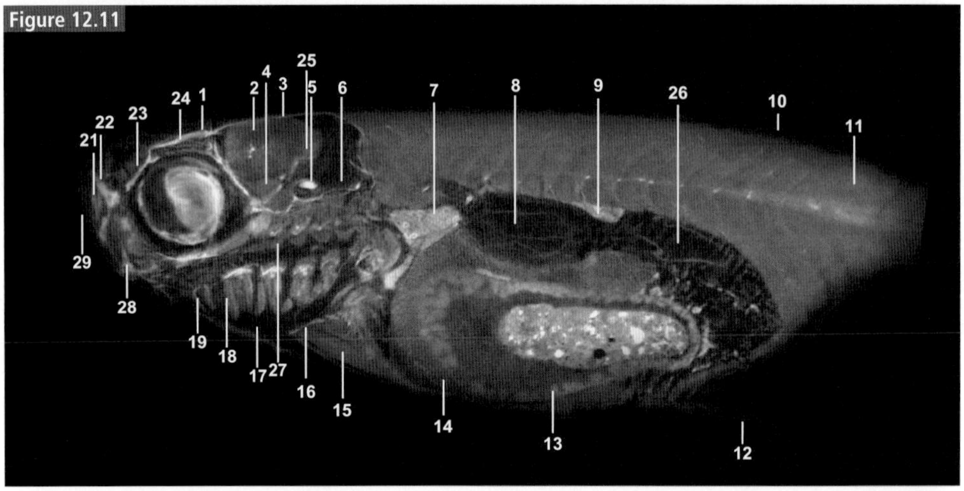

Figure 12.12

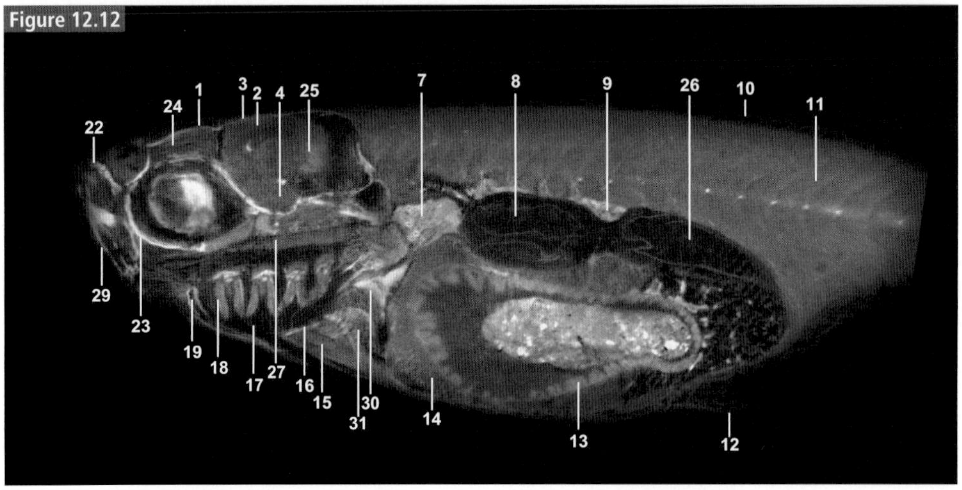

1. frontal bone
2. optic tectum
3. parietal bone
4. inferior lobe
5. otolith
6. semicircular canal
7. pronephros
8. anterior chamber swim bladder
9. mesonephros
10. scales
11. myotome
12. pelvic fin
13. intestinal bulb
14. liver
15. pectoral fin musculature
16. cleithrum
17. opercular cavity
18. gill filaments
19. ceratohyal
20. adductor mandibulae complex
21. premaxilla
22. maxilla
23. infraorbital
24. telencephalon
25. granular eminence
26. posterior chamber swim bladder
27. pharynx
28. retroarticular
29. dentary
30. atrium
31. cardiac ventricle

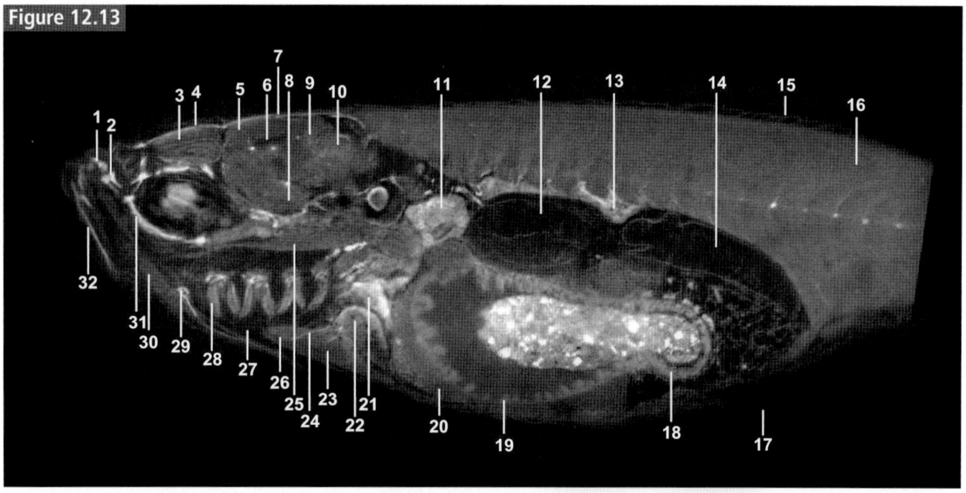

Figure 12.13

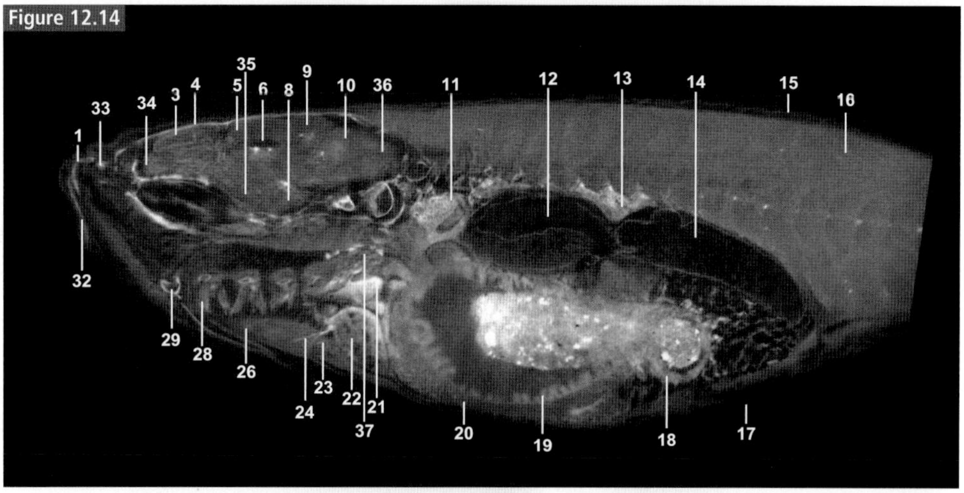

Figure 12.14

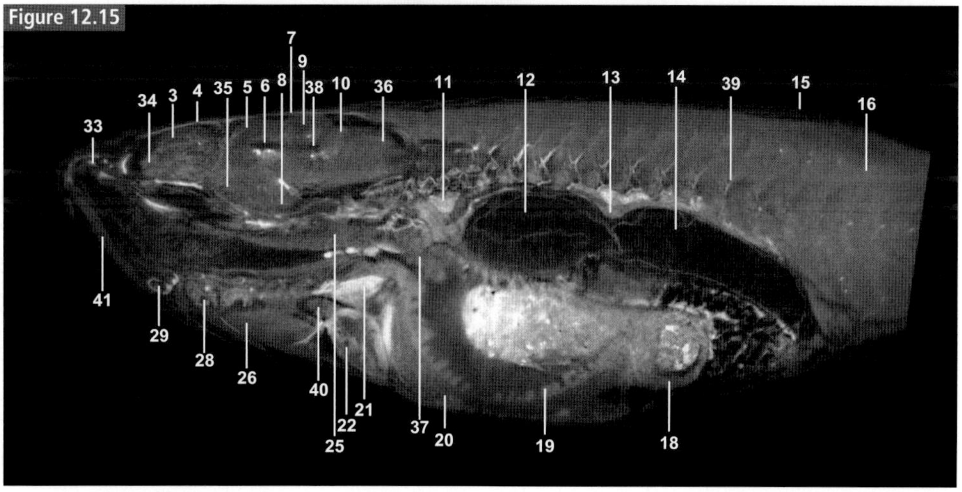

Figure 12.15

1. premaxilla	**10.** crista cerebellaris	**17.** pelvic fin	**26.** sternohyoid	**34.** olfactory bulb
2. nasal bone	**11.** pronephros	**18.** mid intestine	**27.** opercular cavity	**35.** hypothalamus
3. telencephalon	**12.** anterior chamber swim bladder	**19.** intestinal bulb	**28.** gill filaments	**36.** vagal lobe
4. frontal bone	**13.** mesonephros	**20.** liver	**29.** ceratohyal	**37.** esophagus
5. optic tectum	**14.** posterior chamber swim bladder	**21.** atrium	**30.** interhyal	**38.** fourth ventricle
6. tectal ventricle	**15.** scales	**22.** cardiac ventricle	**31.** infraorbital I	**39.** neural spine
7. parietal bone	**16.** myotome	**23.** pectoral fin musculature	**32.** dentary	**40.** bulbus arteriosus
8. inferior lobe		**24.** cleithrum	**33.** maxilla	**41.** intermandibularis
9. cerebellum		**25.** pharynx		

Figure 12.16

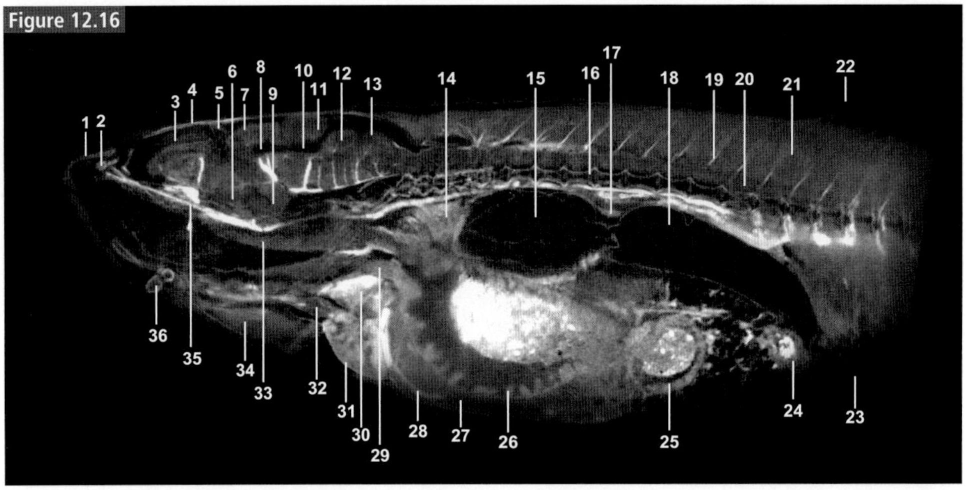

Figure 12.17

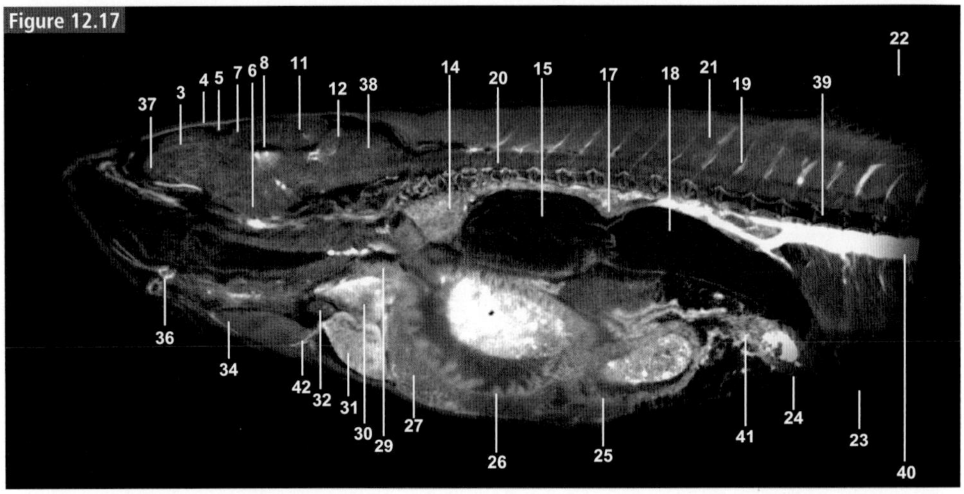

Figure 12.18

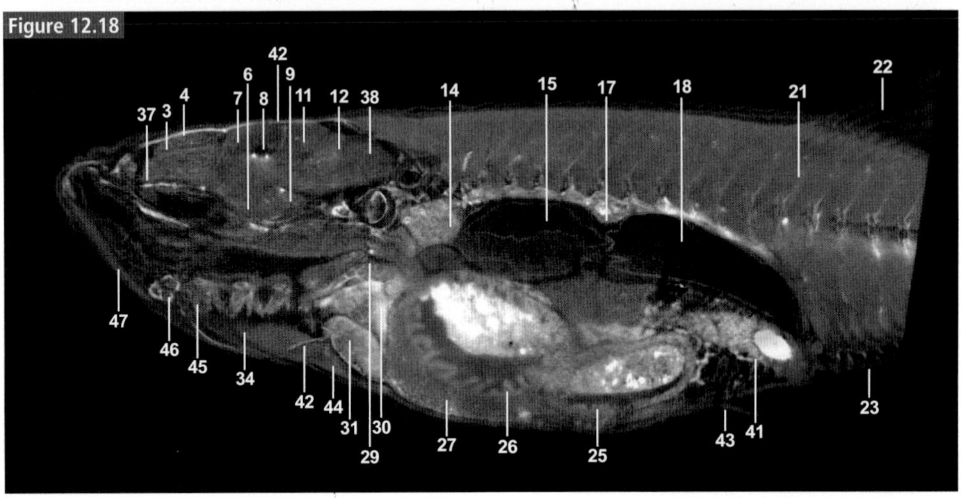

1. premaxilla
2. vomer
3. telencephalon
4. frontal bone
5. epiphysis
6. hypothalamus
7. optic tectum
8. tectal ventricle
9. inferior lobe
10. fourth ventricle
11. cerebellum
12. crista cerebellaris
13. facial lobe
14. pronephros
15. anterior chamber swim bladder
16. precaudal vertebra
17. mesonephros
18. posterior chamber swim bladder
19. neural spine
20. spinal cord
21. myotome
22. dorsal fin
23. anal fin
24. cloacal chamber
25. mid intestine
26. intestinal bulb
27. liver
28. hepatic portal vein
29. esophagus
30. atrium
31. cardiac ventricle
32. bulbus arteriosus
33. pharynx
34. sternohyoid
35. parasphenoid
36. ceratohyal
37. olfactory bulb
38. vagal lobe
39. vertebral column
40. caudal vein
41. posterior intestine
42. cleithrum
43. pelvic fin
44. pectoral fin musculature
45. gill filaments
46. retroarticular
47. intermandibularis

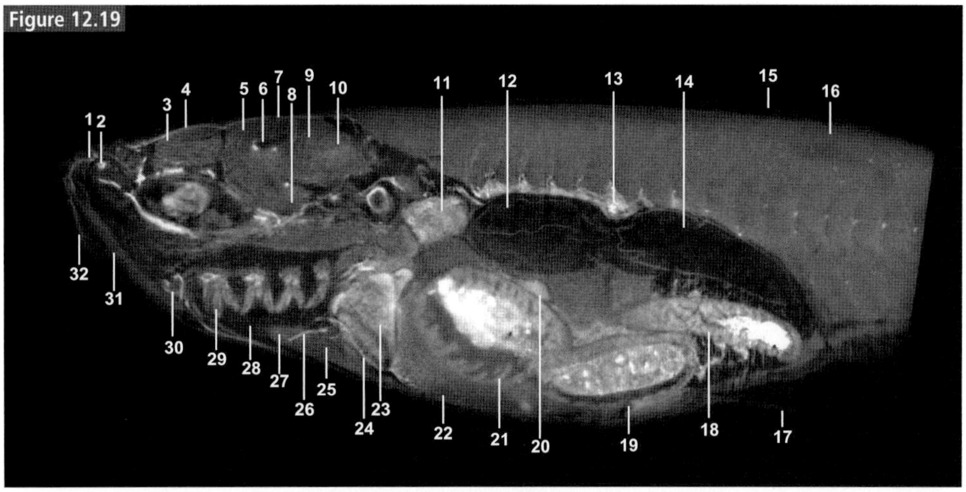

Figure 12.19

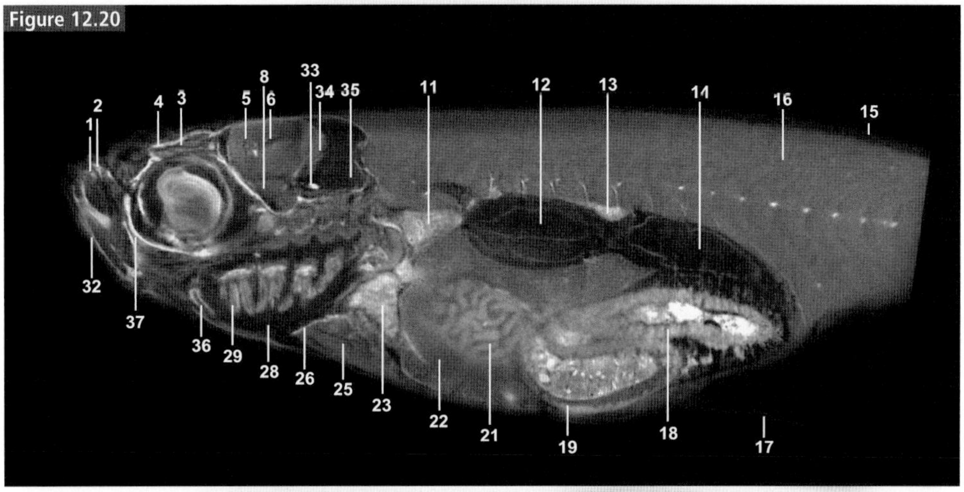

Figure 12.20

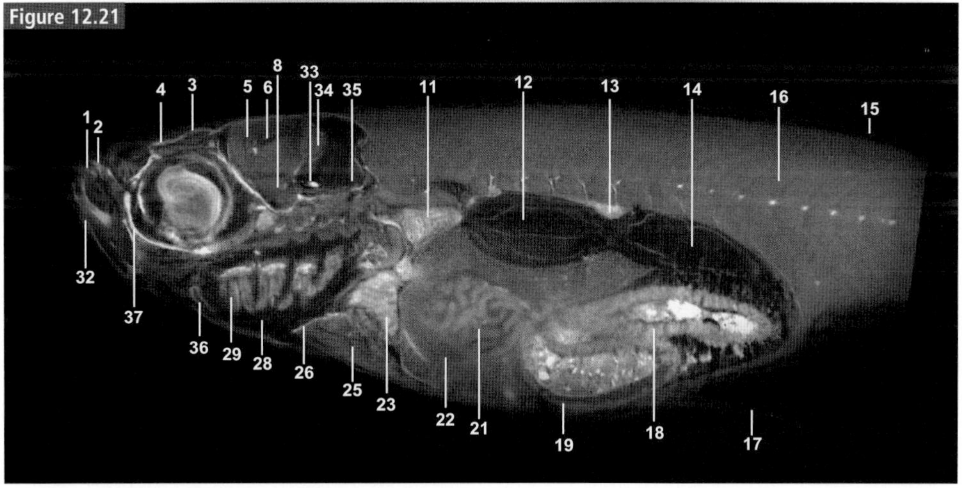

Figure 12.21

1. premaxilla
2. maxilla
3. telencephalon
4. frontal bone
5. optic tectum
6. tectal ventricle
7. parietal bone
8. inferior lobe
9. cerebellum
10. crista cerebellaris
11. pronephros
12. anterior chamber swim bladder
13. mesonephros
14. posterior chamber swim bladder
15. scales
16. myotome
17. pelvic fin
18. posterior intestine
19. mid intestine
20. spleen
21. intestinal bulb
22. liver
23. atrium
24. cardiac ventricle
25. pectoral fin musculature
26. cleithrum
27. sternohyoid
28. opercular cavity
29. gill filaments
30. retroarticular
31. intermandibularis
32. dentary
33. otolith
34. eminentia granularis
35. semicircular canal
36. ceratohyal
37. infraorbital I

Figure 12.22

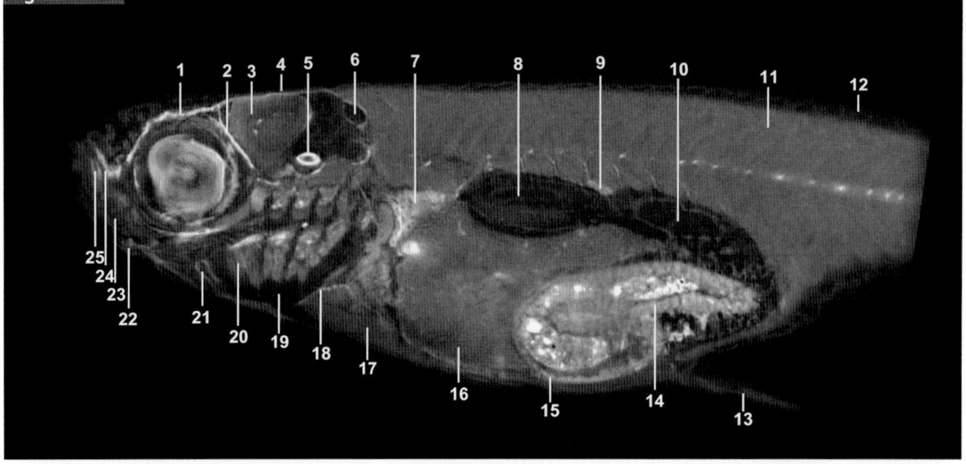

Figure 12.23

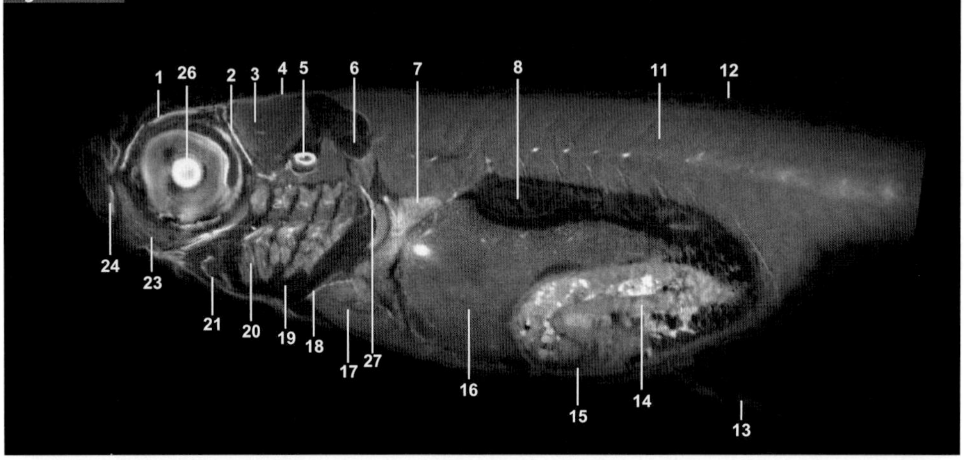

Figure 12.24

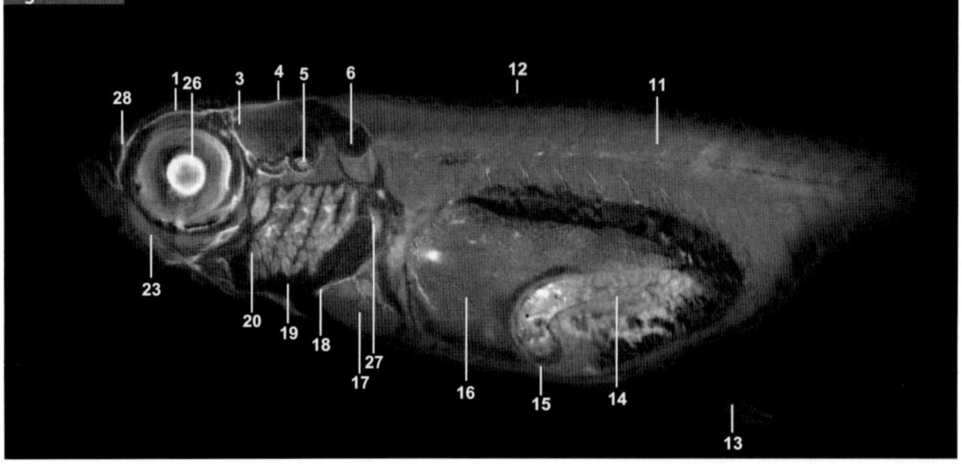

1. frontal bone
2. infraorbital
3. optic tectum
4. parietal bone
5. otolith
6. semicircular canal
7. pronephros
8. anterior chamber swim bladder
9. mesonephros
10. posterior chamber swim bladder
11. myotome
12. scales
13. pelvic fin
14. posterior intestine
15. mid intestine
16. liver
17. pectoral fin musculature
18. cleithrum
19. opercular cavity
20. gill filaments
21. ceratohyal
22. retroarticular
23. adductor mandibulae complex
24. maxilla
25. premaxilla
26. lens
27. ceratobranchial 5
28. lateral ethmoid

Figure 12.25

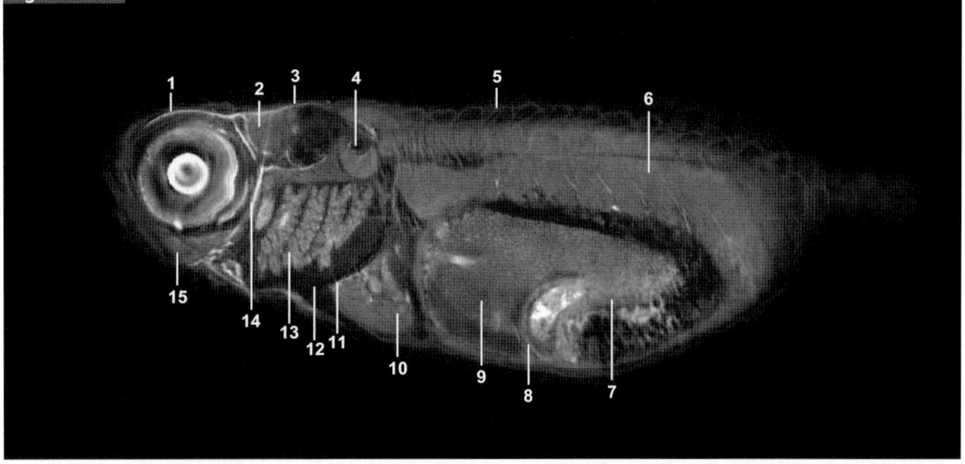

Figure 12.26

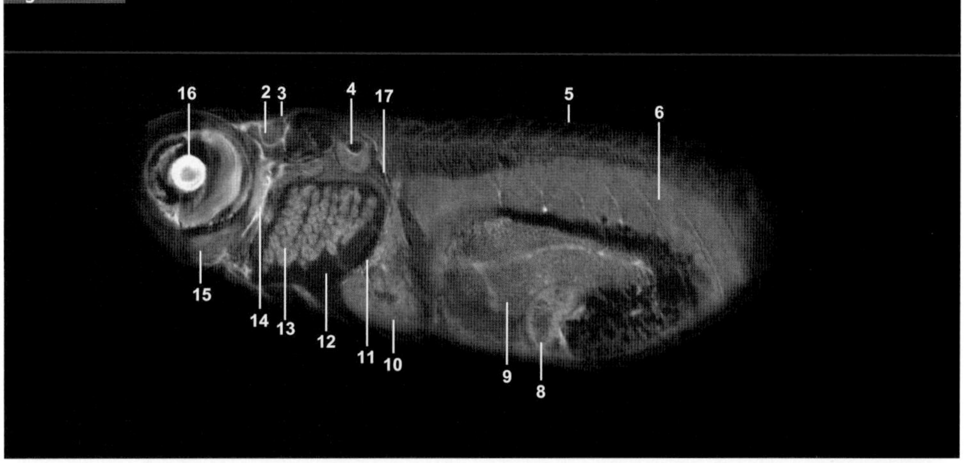

Figure 12.27

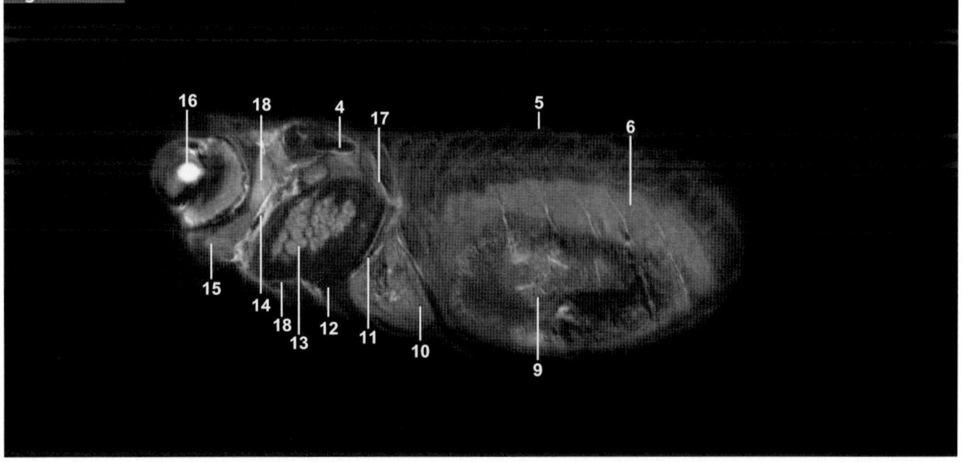

1. frontal bone
2. optic tectum
3. parietal bone
4. semicircular canal
5. scales
6. myotome
7. posterior intestine
8. mid intestine
9. liver
10. pectoral fin musculature
11. cleithrum
12. opercular cavity
13. gill filaments
14. hyomandibula
15. adductor mandibulae complex
16. lens
17. supracleithrum
18. dilator operculi
19. branchiostegal ray

Figure 12.28

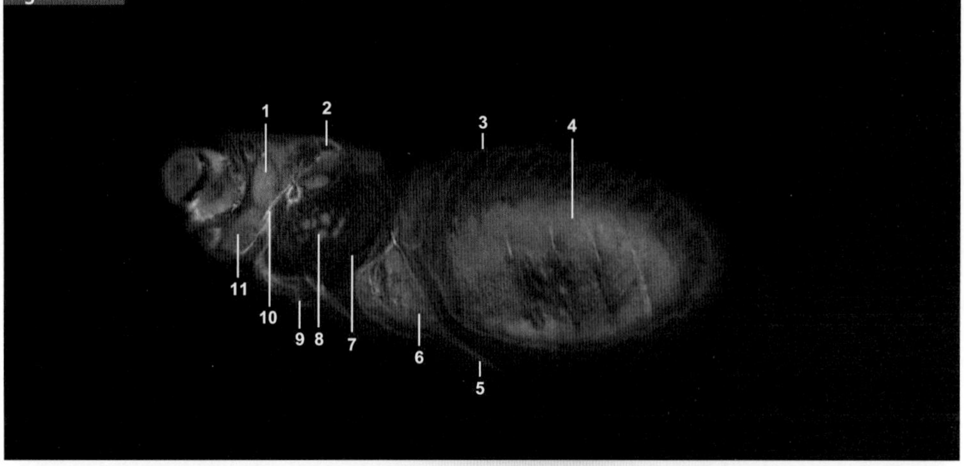

Figure 12.29

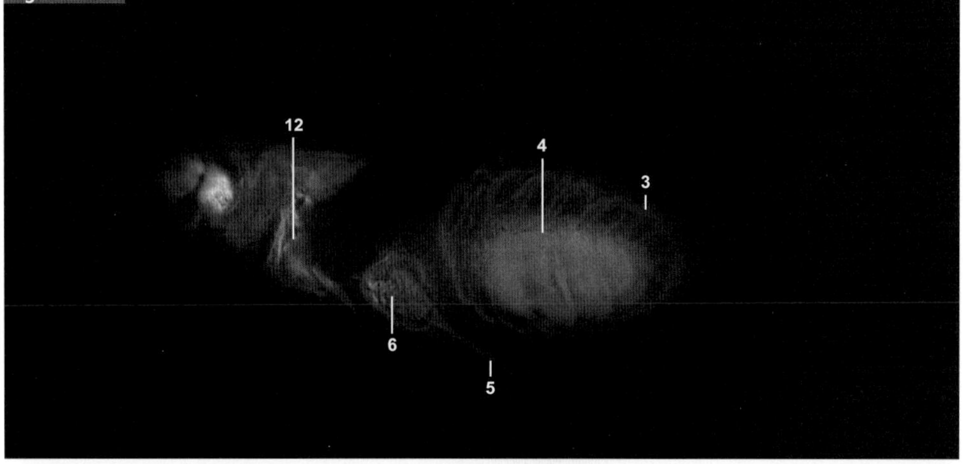

Figure 12.30

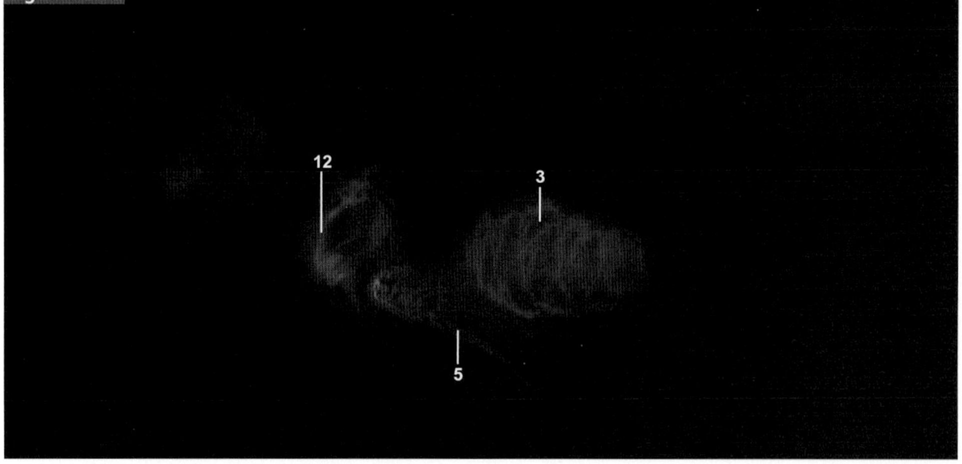

1. dilator operculi	**5.** pectoral fin	**9.** branchiostegal ray
2. semicircular canal	**6.** pectoral fin musculature	**10.** hyomandibula
3. scales	**7.** opercular cavity	**11.** adductor mandibulae complex
4. myotome	**8.** gill filaments	**12.** opercle

Coronal Sections

Figure 12.31

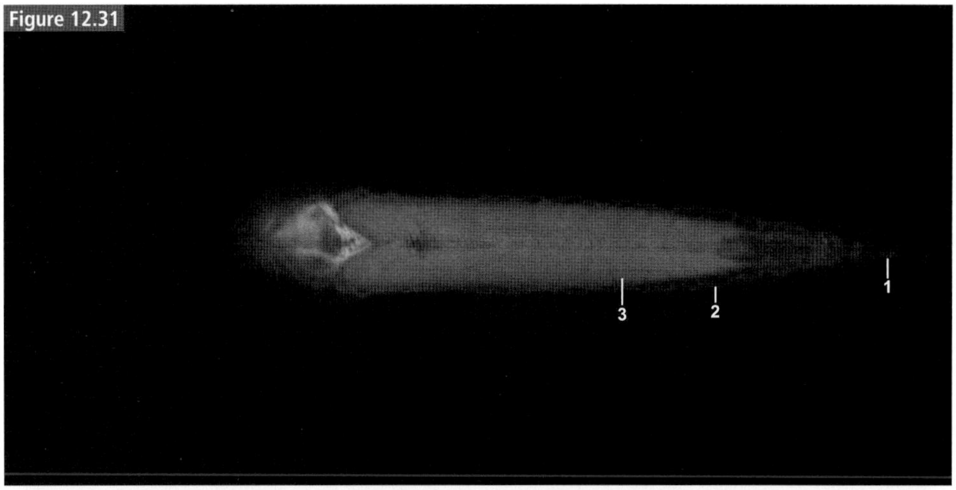

Figure 12.32

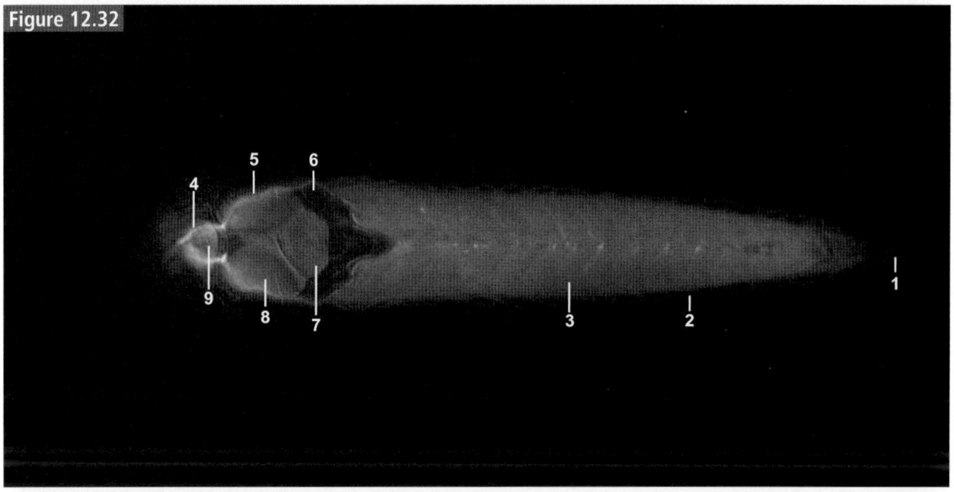

Figure 12.33

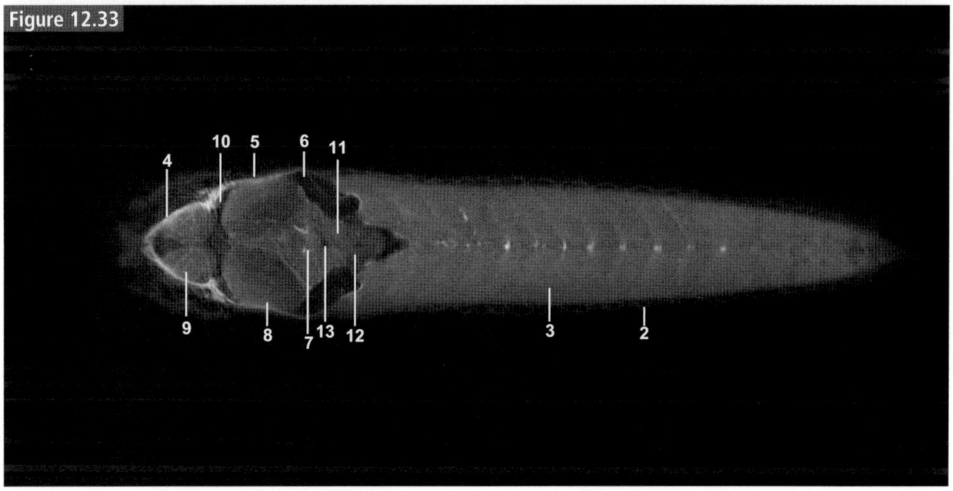

1. dorsal fin
2. scales
3. myotome
4. frontal bone
5. parietal bone
6. semicircular canal
7. cerebellum
8. optic tectum
9. telencephalon
10. telencephalic ventricle
11. crista cerebellaris
12. vagal lobe
13. fourth ventricle

Figure 12.34

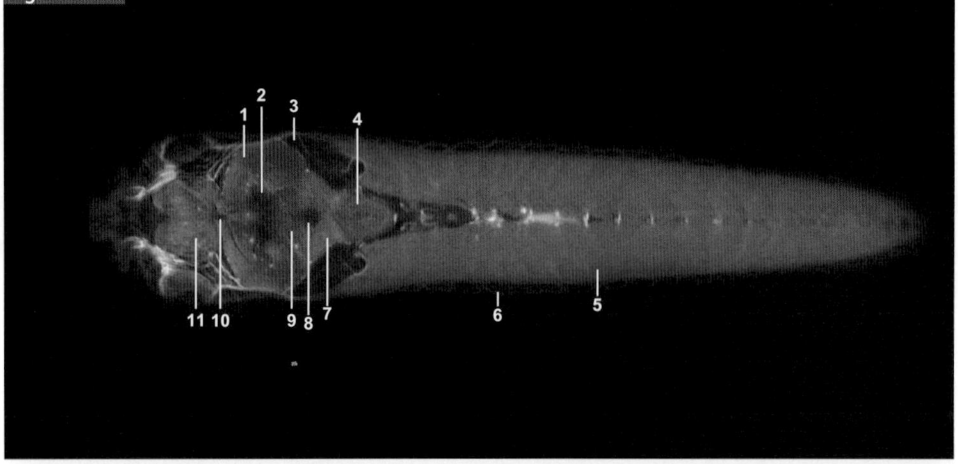

Figure 12.35

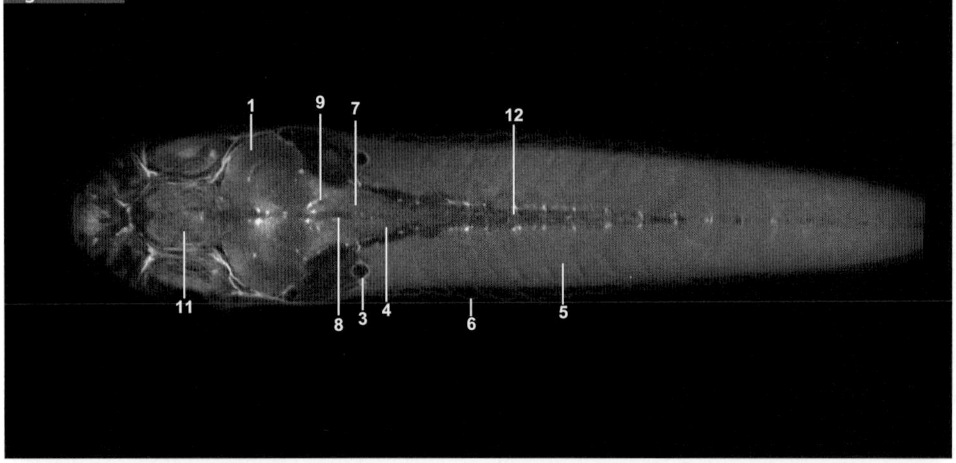

Figure 12.36

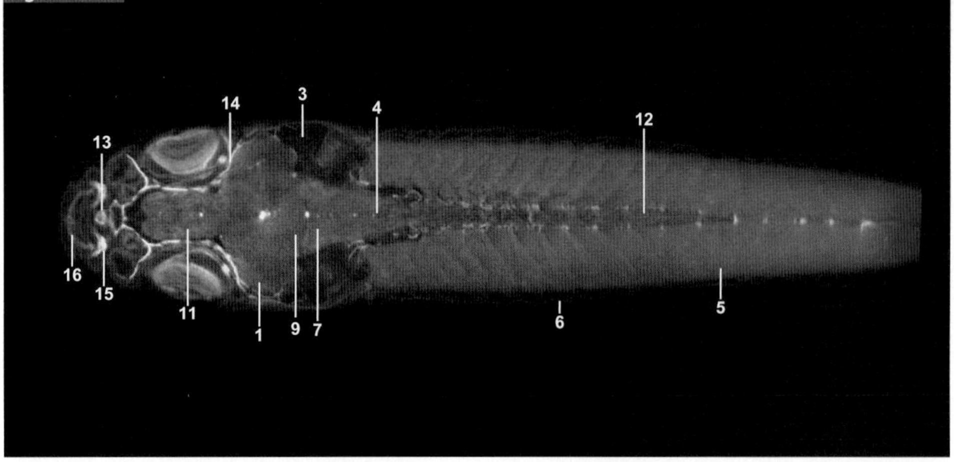

1. optic tectum	**5.** myotome	**9.** cerebellum	**13.** kinethmoid
2. tectal ventricle	**6.** scales	**10.** telencephalic ventricle	**14.** infraorbital
3. semicircular canal	**7.** crista cerebellaris	**11.** telencephalon	**15.** maxilla
4. vagal lobe	**8.** fourth ventricle	**12.** spinal cord	**16.** Meckel's cartilage

Figure 12.37

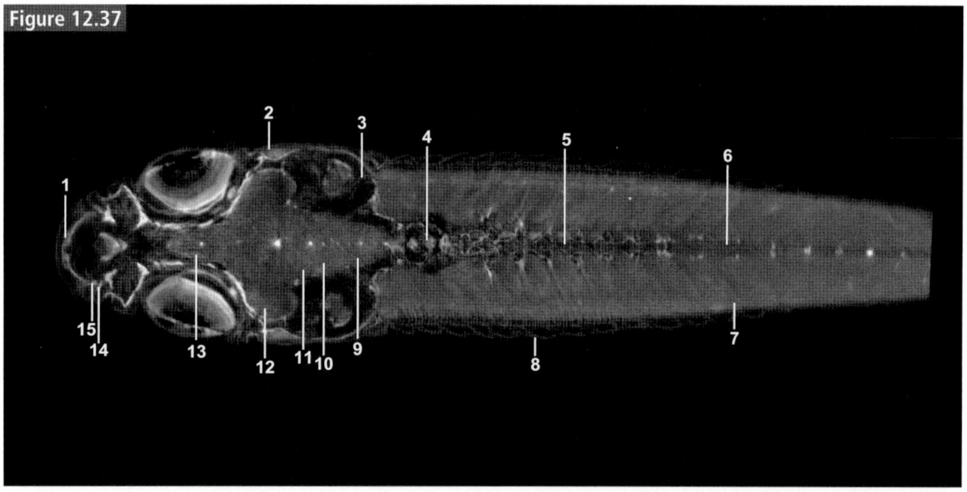

Figure 12.38

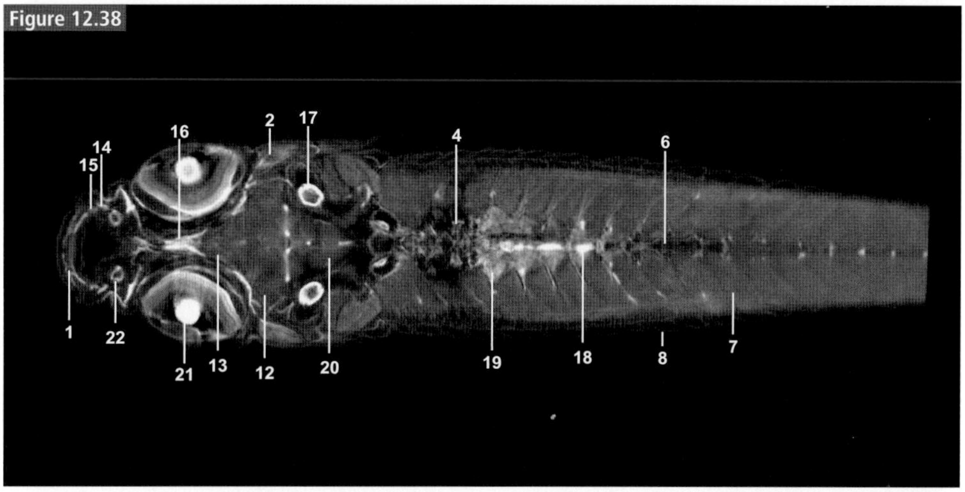

Figure 12.39

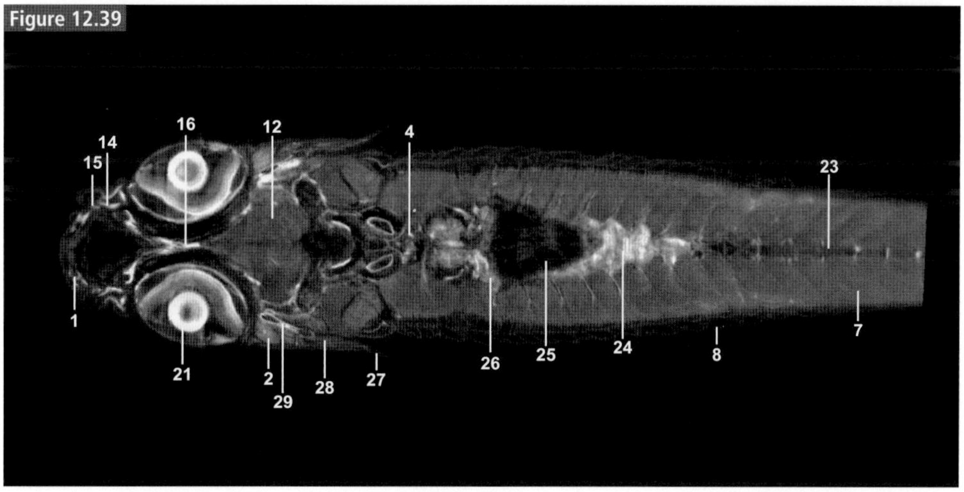

1. Meckel's cartilage
2. dilator operculi
3. semicircular canal
4. Weberian apparatus
5. precaudal vertebra
6. spinal cord
7. myotome
8. scales
9. vagal lobe
10. crista cerebellaris
11. cerebellum
12. optic tectum
13. telencephalon
14. maxilla
15. premaxilla
16. parasphenoid
17. otolith
18. posterior cardinal vein
19. nephric duct
20. medulla oblongata
21. lens
22. nasal bone
23. caudal vertebra
24. mesonephros
25. anterior chamber swim bladder
26. pronephros
27. opercle
28. adductor operculi
29. hyomandibula

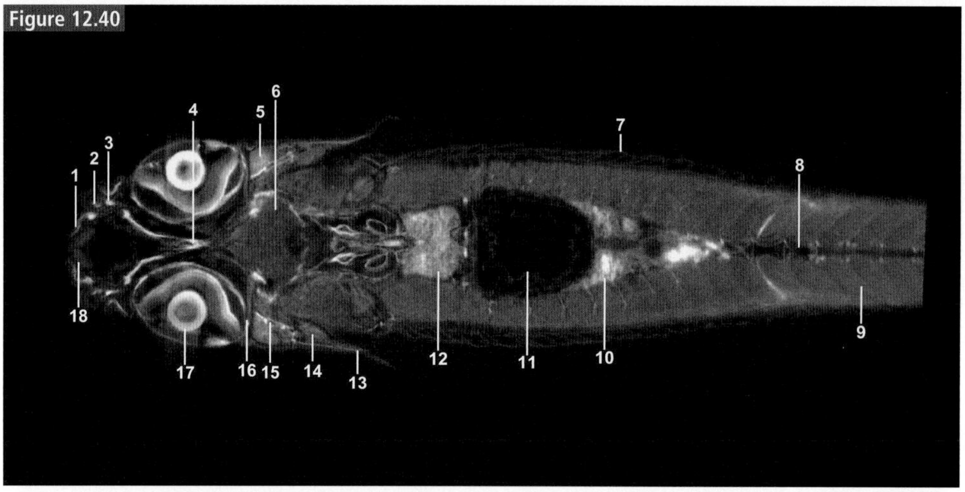

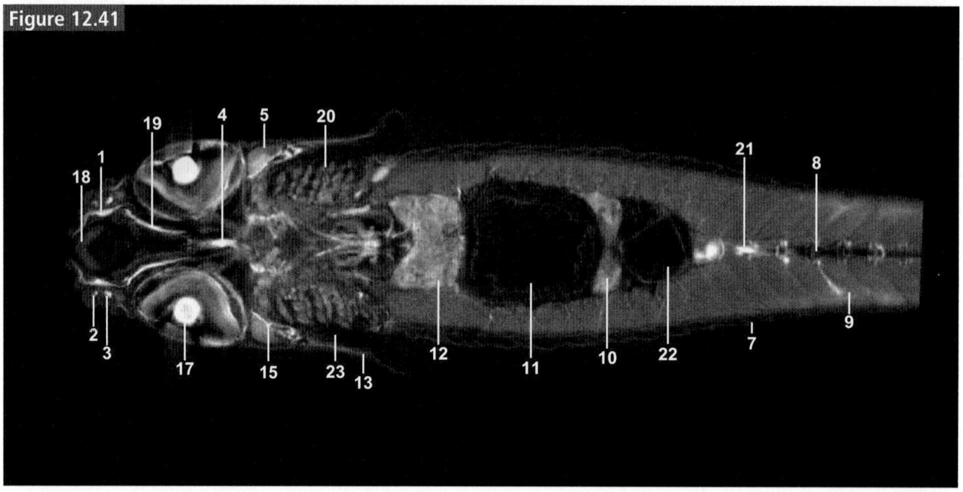

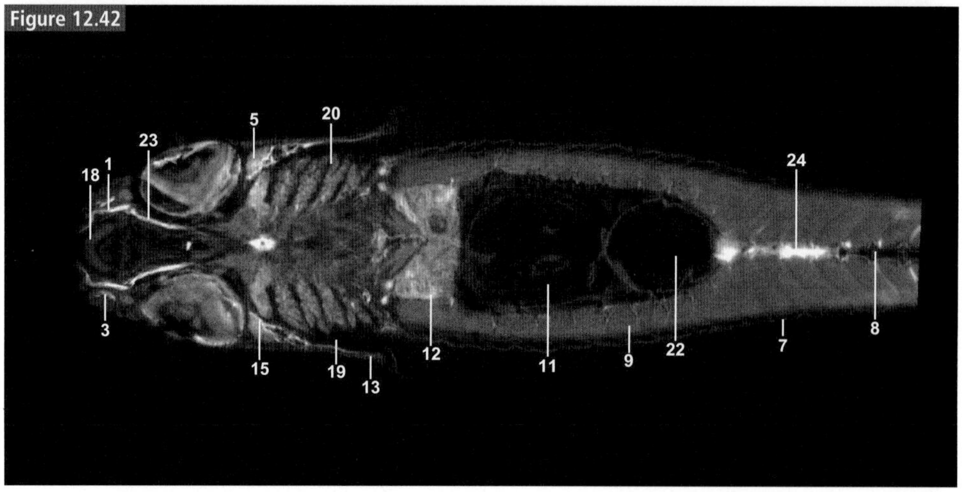

1. dentary
2. premaxilla
3. maxilla
4. parasphenoid
5. dilator operculi
6. optic tectum
7. scales
8. caudal vertebra
9. myotome
10. mesonephros
11. anterior chamber swim bladder
12. pronephros
13. opercle
14. adductor operculi
15. hyomandibula
16. lateral rectus
17. lens
18. intermandibularis
19. palatoquadrate
20. gill filaments
21. caudal artery
22. posterior chamber swim bladder
23. opercular cavity
24. posterior cardinal vein

Figure 12.43

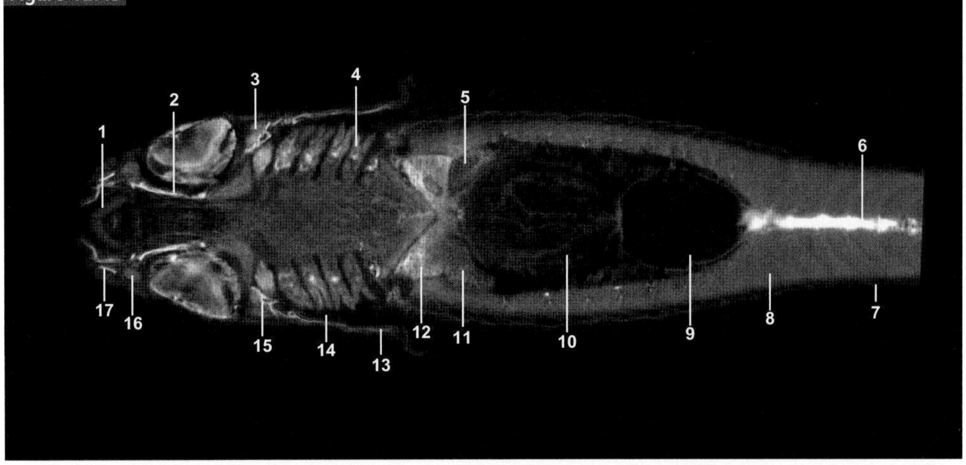

Figure 12.44

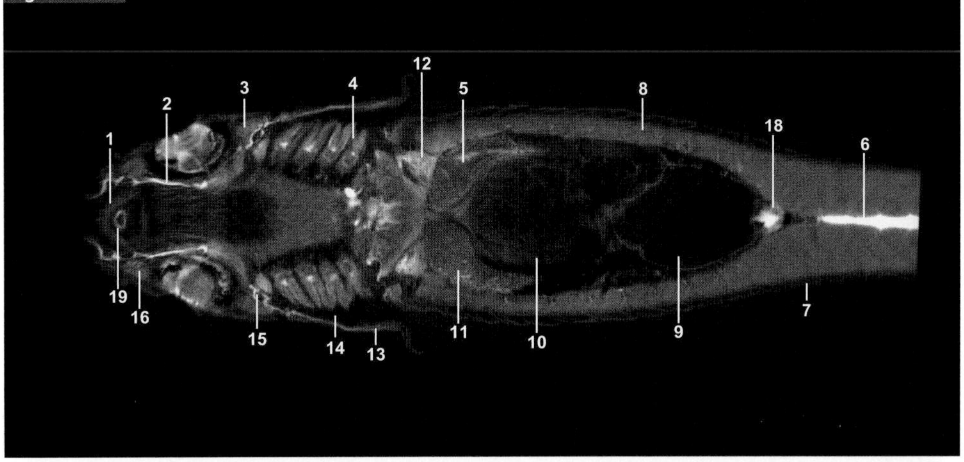

Figure 12.45

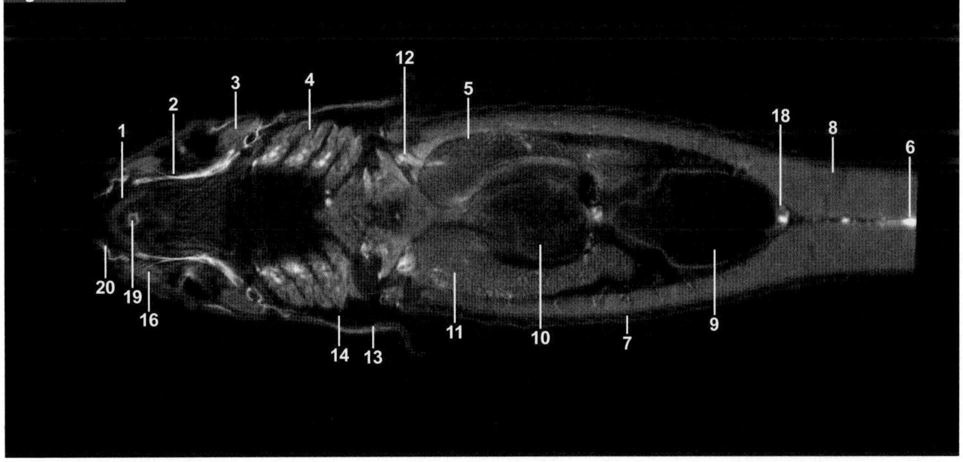

1. protractor hyoidei
2. palatoquadrate
3. dilator operculi
4. gill filaments
5. pancreas
6. caudal vein
7. scales
8. myotome
9. posterior chamber swim bladder
10. anterior chamber swim bladder
11. liver
12. pronephros
13. opercle
14. opercular cavity
15. hyomandibula
16. adductor mandibulae complex
17. maxilla
18. mesonephric duct
19. basihyal
20. dentary

Figure 12.46

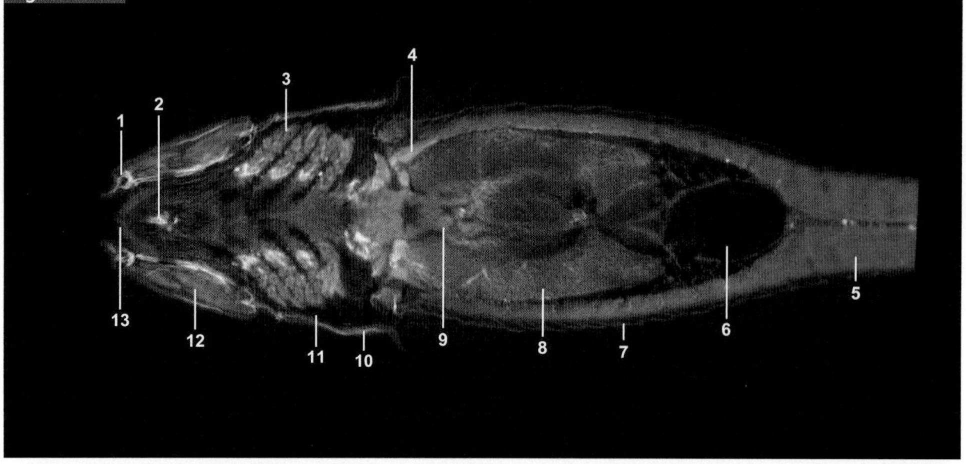

Figure 12.47

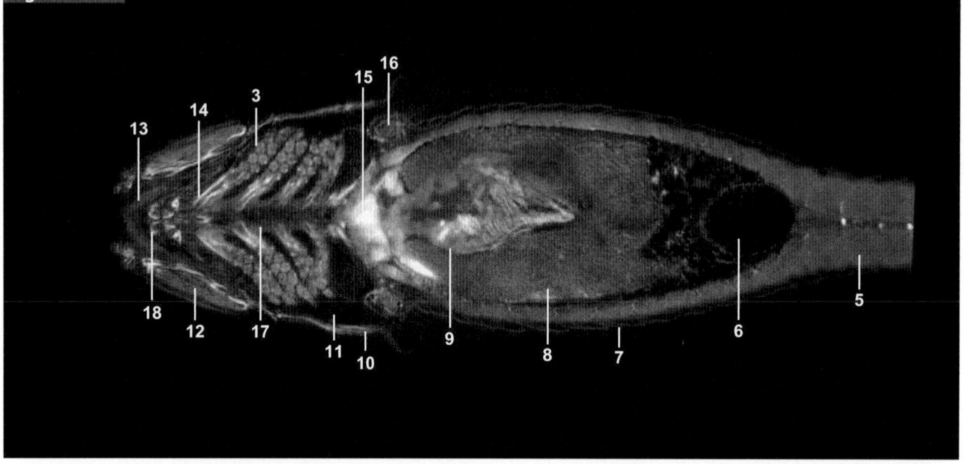

Figure 12.48

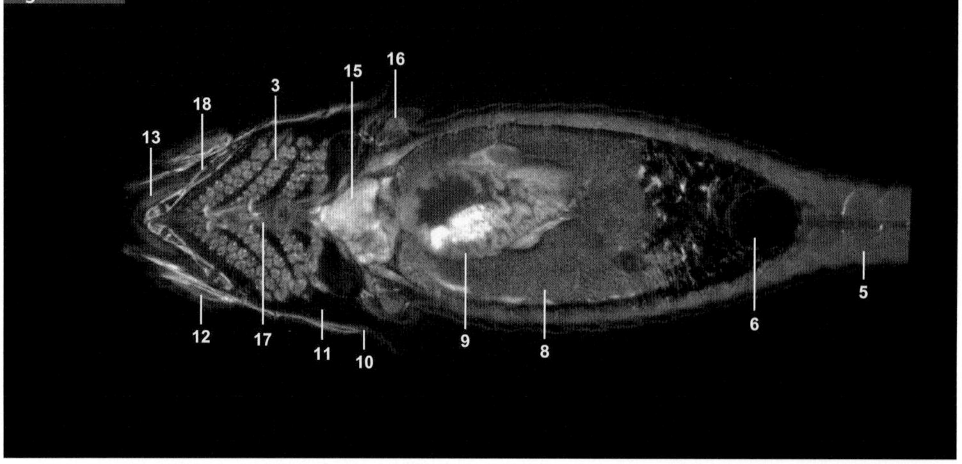

1. palatoquadrate	**5.** myotome	**8.** liver	**12.** adductor mandibulae complex	**15.** atrium
2. basihyal	**6.** posterior chamber swim bladder	**9.** intestinal bulb	**13.** hyohyoideus	**16.** pectoral fin musculature
3. gill filaments	**7.** scales	**10.** opercle	**14.** aortic arches	**17.** basibranchial
4. pronephros		**11.** opercular cavity		**18.** ceratohyal

Figure 12.49

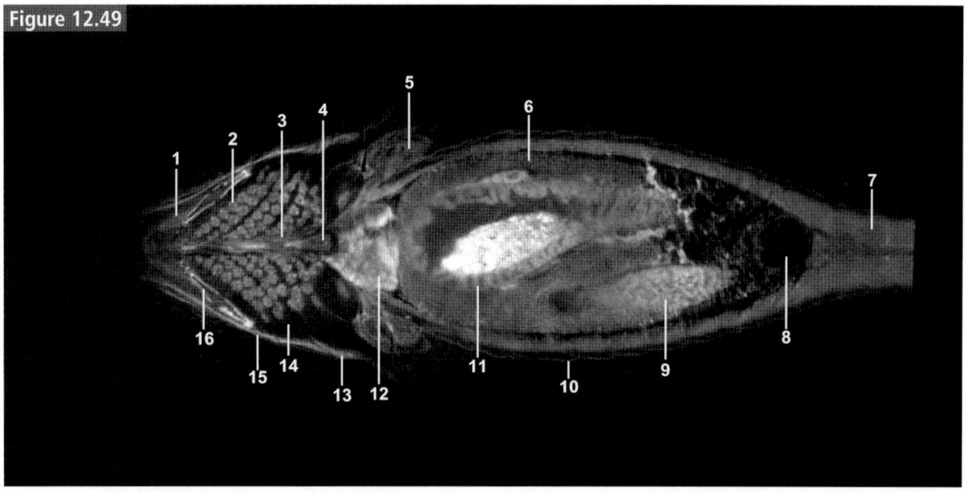

Figure 12.50

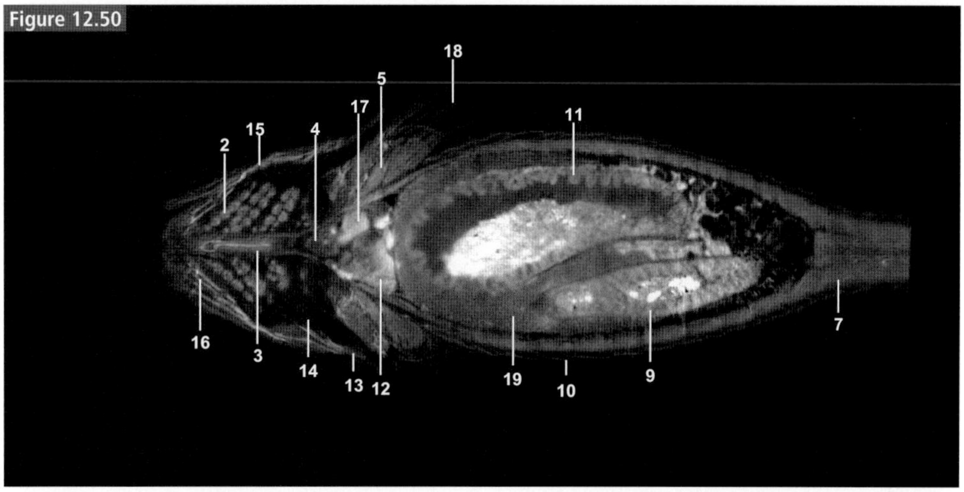

Figure 12.51

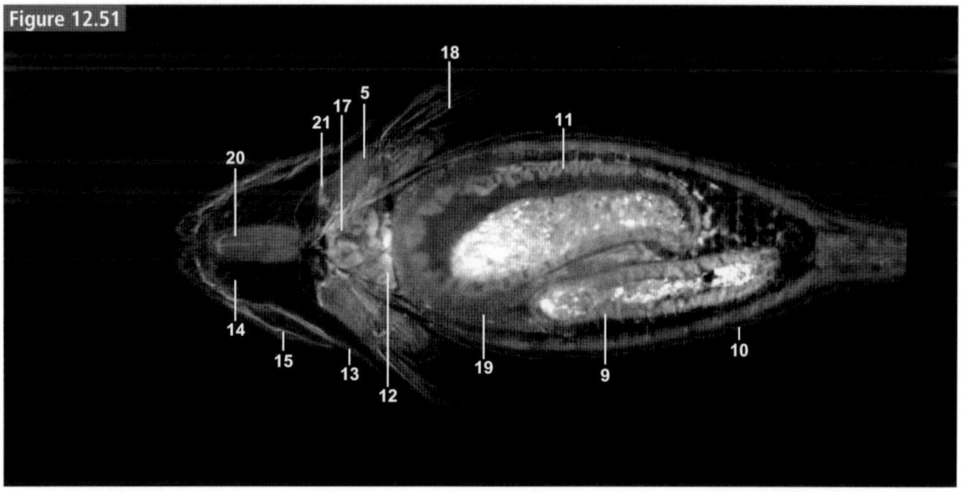

1. hyohyal	**5.** pectoral fin musculature	**9.** posterior intestine	**13.** opercle	**17.** cardiac ventricle
2. gill filaments	**6.** pancreas	**10.** scales	**14.** opercular cavity	**18.** pectoral fin
3. basibranchial	**7.** myotome	**11.** intestinal bulb	**15.** branchiostegal ray	**19.** liver
4. bulbus arteriosus	**8.** posterior chamber swim bladder	**12.** atrium	**16.** ceratohyal	**20.** sternohyoid
				21. cleithrum

Figure 12.52

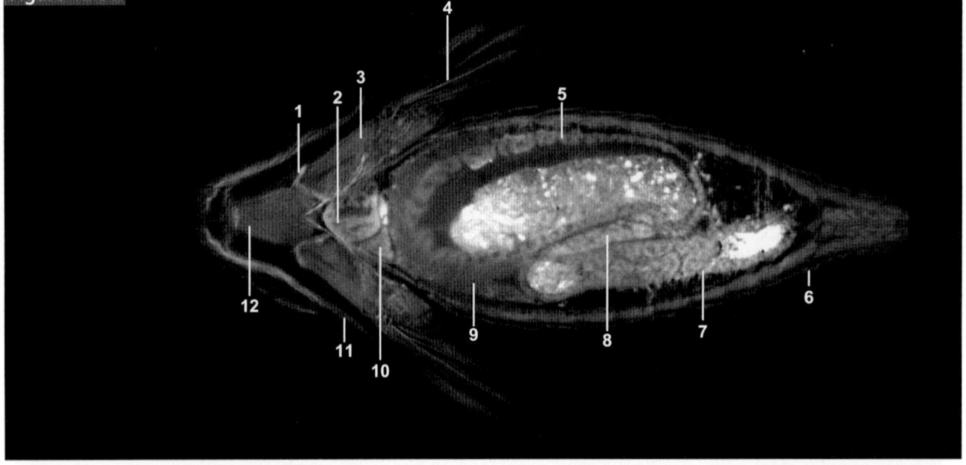

Figure 12.53

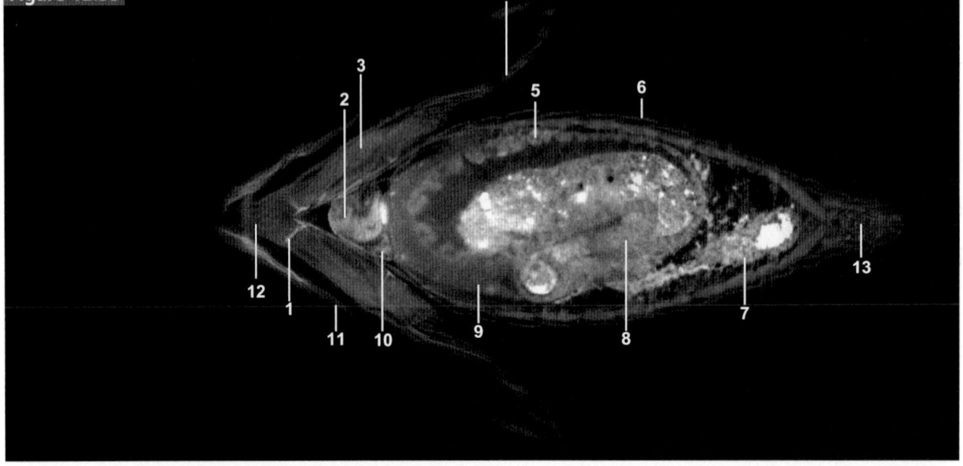

Figure 12.54

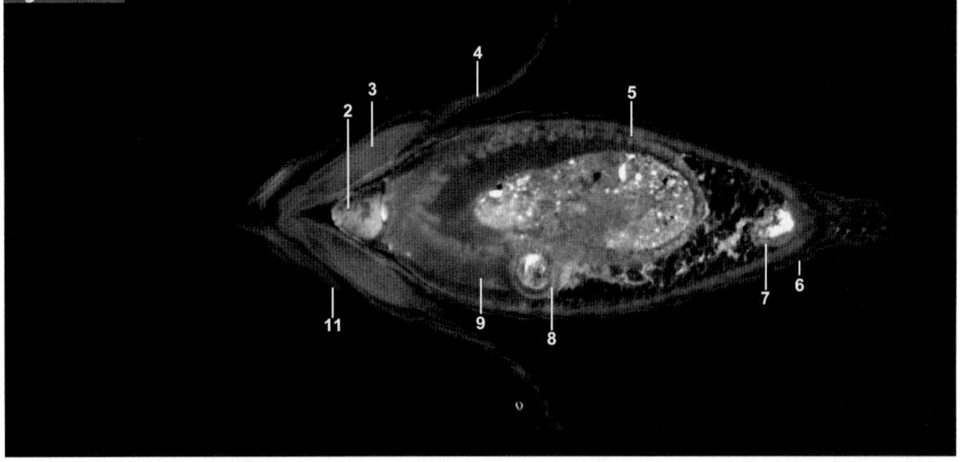

1. cleithrum	**4.** pectoral fin	**7.** posterior intestine	**10.** atrium	**13.** anal fin
2. cardiac ventricle	**5.** intestinal bulb	**8.** mid intestine	**11.** opercle	
3. pectoral fin musculature	**6.** scales	**9.** liver	**12.** sternohyoid	

Figure 12.55

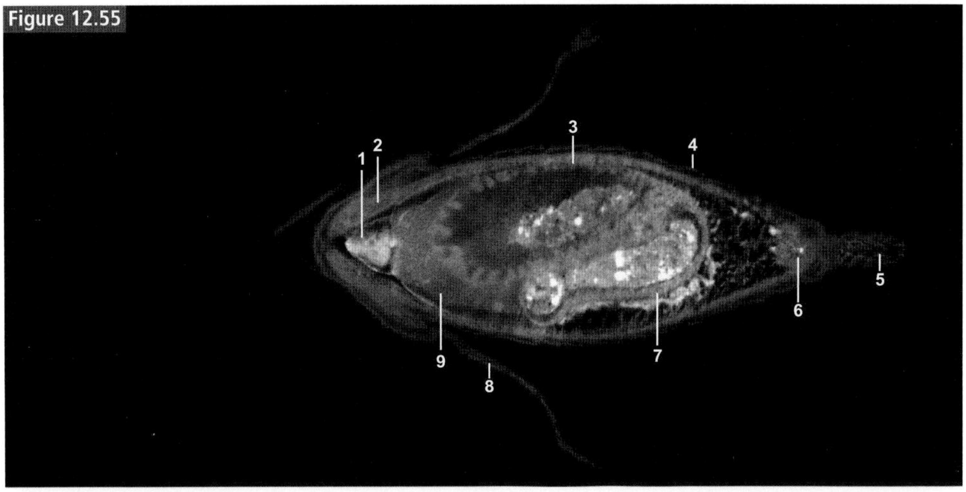

Figure 12.56

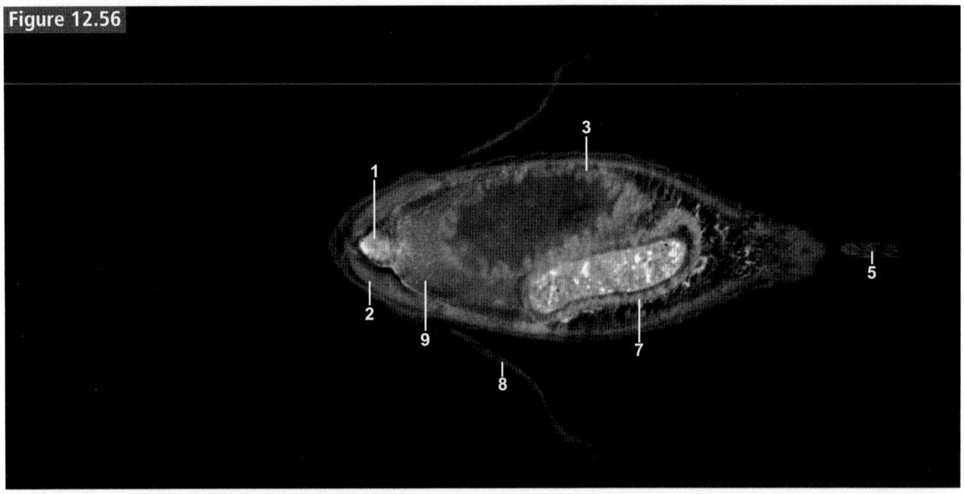

Figure 12.57

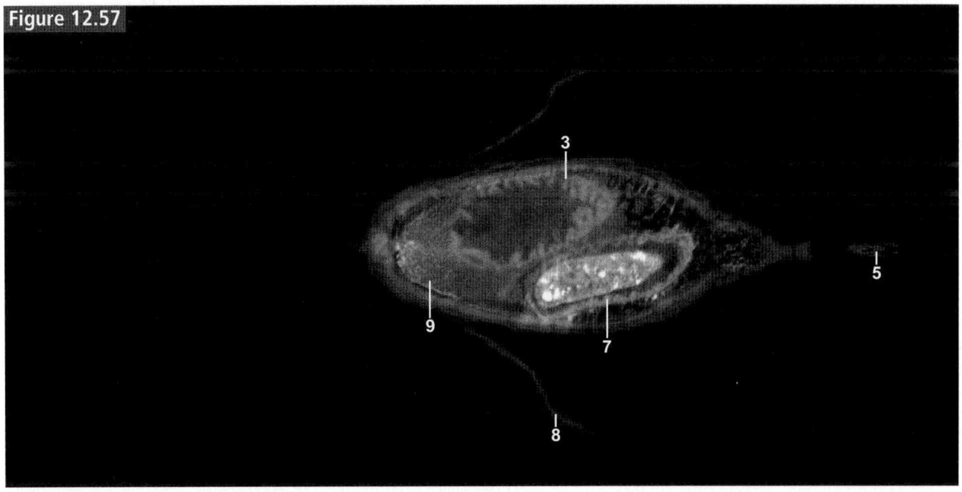

1. cardiac ventricle	**4.** scales	**7.** mid intestine
2. pectoral fin musculature	**5.** anal fin	**8.** pectoral fin
3. intestinal bulb	**6.** urogenital opening	**9.** liver

Figure 12.58

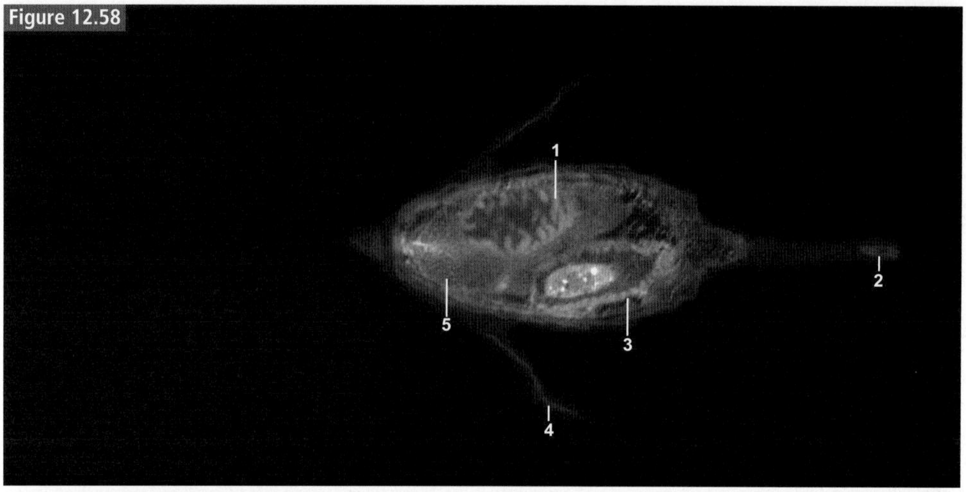

Figure 12.59

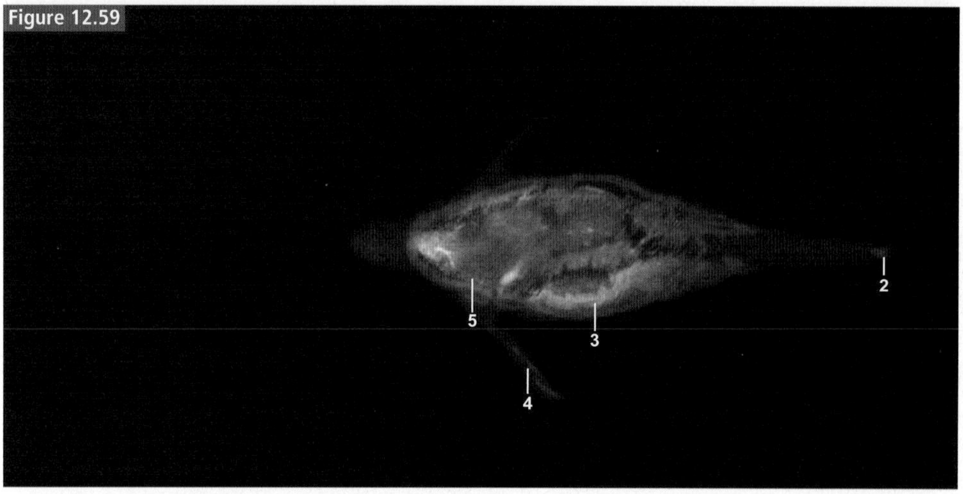

Figure 12.60

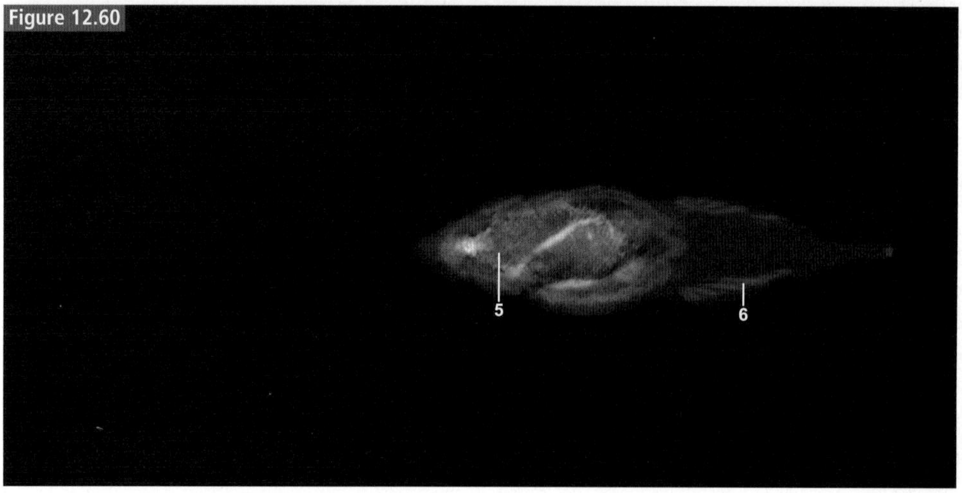

1. intestinal bulb	**2.** anal fin	**3.** mid intestine	**4.** pectoral fin	**5.** liver	**6.** pelvic fin

Transverse Sections

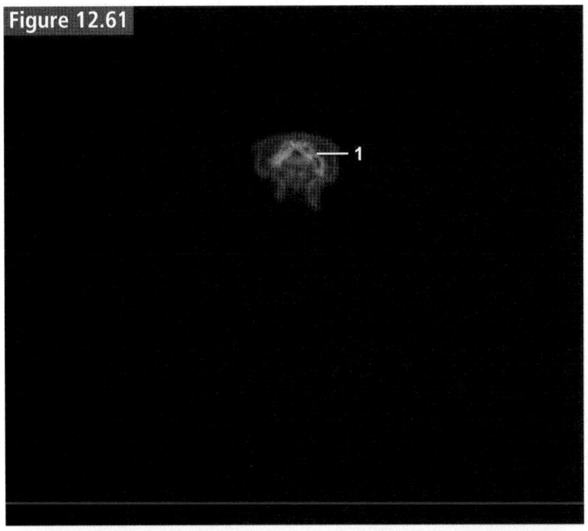

Figure 12.61

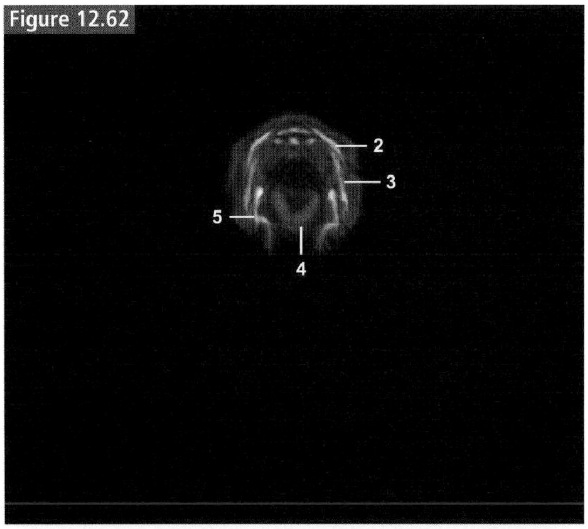

Figure 12.62

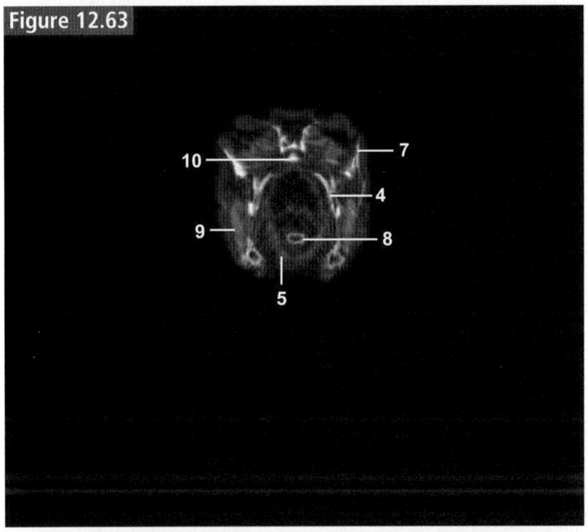

Figure 12.63

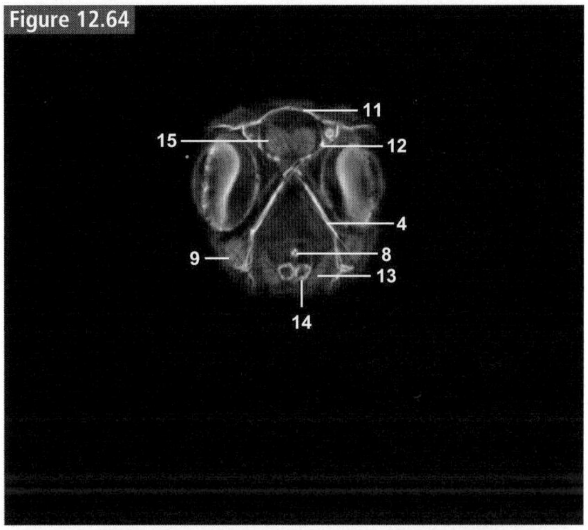

Figure 12.64

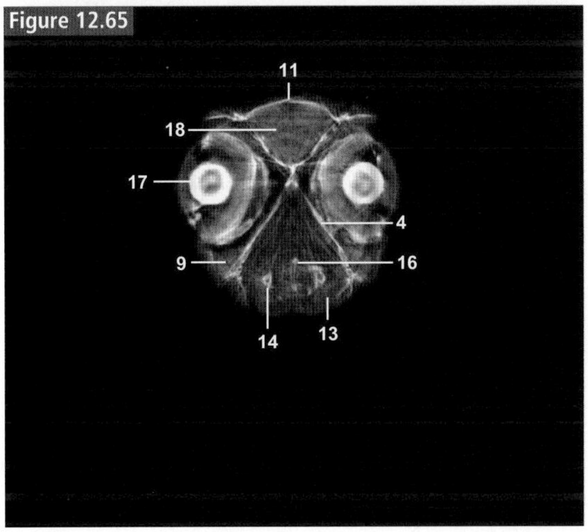

Figure 12.65

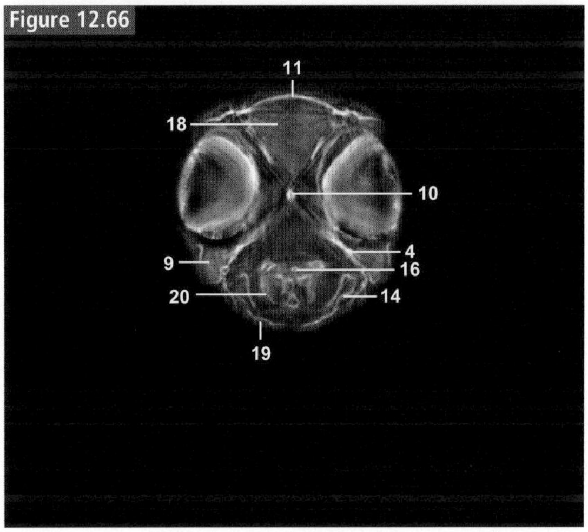

Figure 12.66

1. Meckel's cartilage	**5.** protractor hyoidei	**9.** adductor mandibulae complex	**13.** interhyal	**17.** lens
2. maxilla	**6.** dentary	**10.** parasphenoid	**14.** ceratohyal	**18.** telencephalon
3. premapilla	**7.** infraorbital 1	**11.** frontal bone	**15.** olfactory bulb	**19.** branchiostegal ray
4. palatoquadrate	**8.** basihyal	**12.** basicranium	**16.** basibranchial	**20.** gill filaments

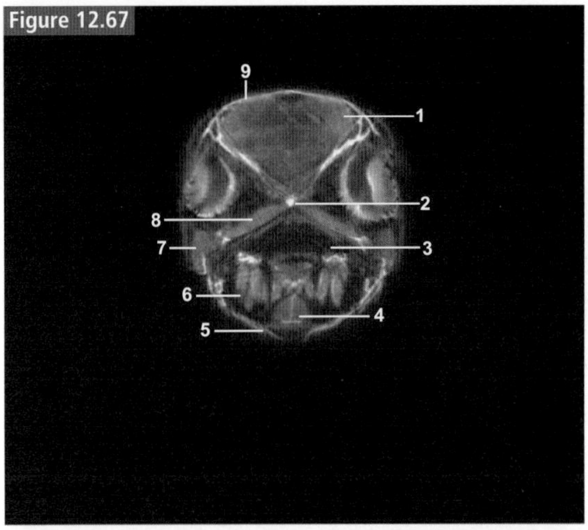

Figure 12.67

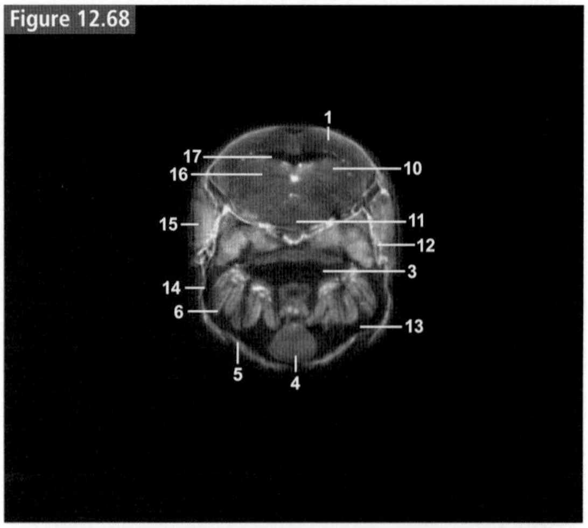

Figure 12.68

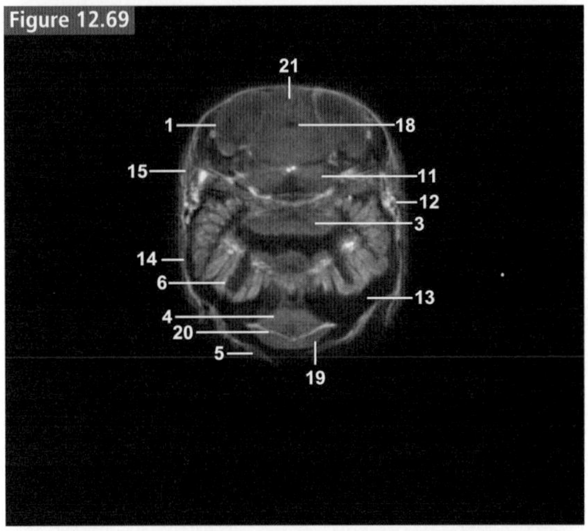

Figure 12.69

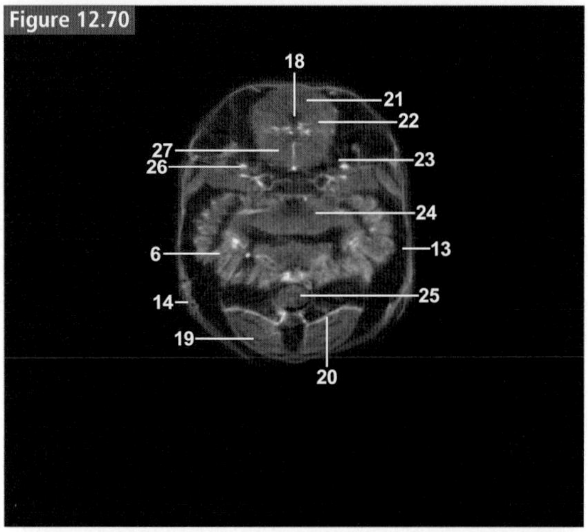

Figure 12.70

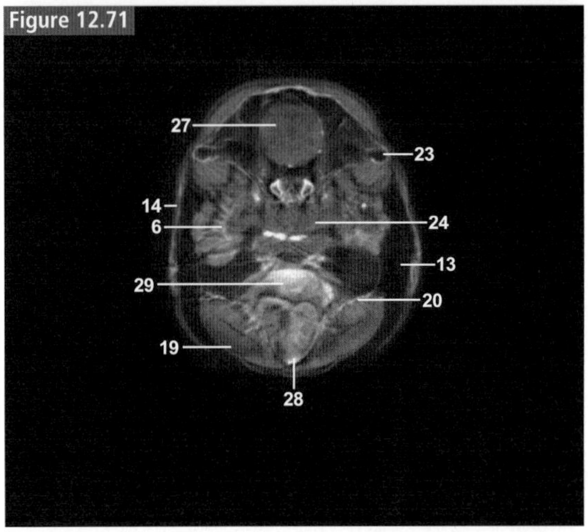

Figure 12.71

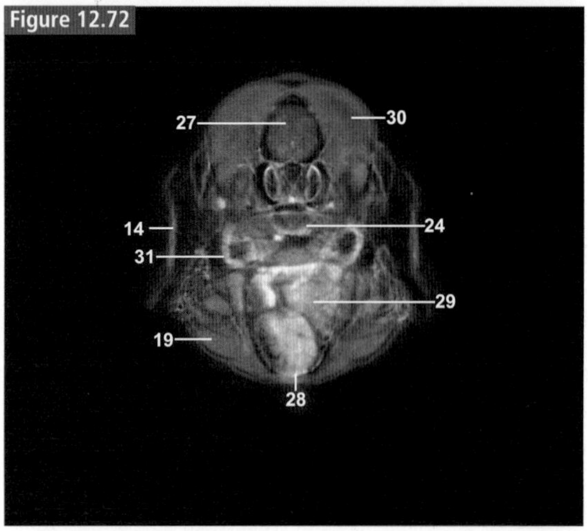

Figure 12.72

1. optic tectum
2. parasphenoid
3. pharynx
4. sternohyoid
5. branchiostegal ray
6. gill filaments
7. adductor mandibulae complex
8. lateral rectus
9. parietal bone
10. torus semicircularis
11. inferior lobe
12. hyomandibula
13. opercular cavity
14. opercle
15. dilator operculi
16. tegmentum
17. tectal ventricle
18. fourth ventricle
19. pectoral fin musculature
20. cleithrum
21. cerebellum
22. crista cerebellaris
23. semicircular canal
24. esophagus
25. bulbus arteriosus
26. otolith
27. medulla oblongata
28. cardiac ventricle
29. atrium
30. myotome
31. ceratobranchial 5

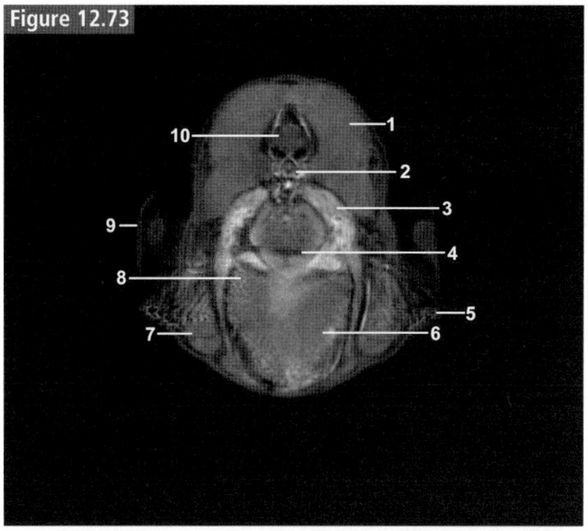

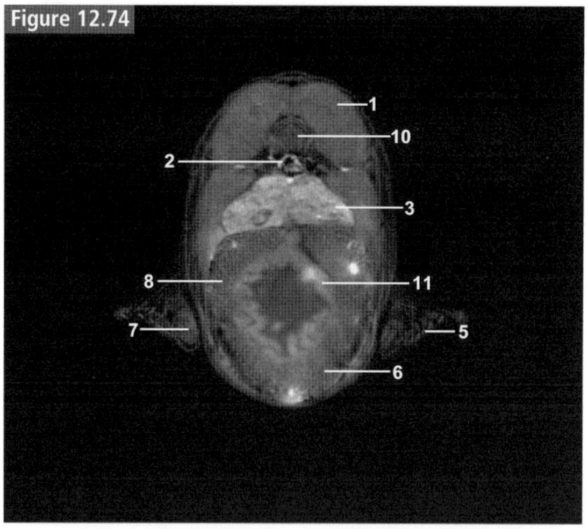

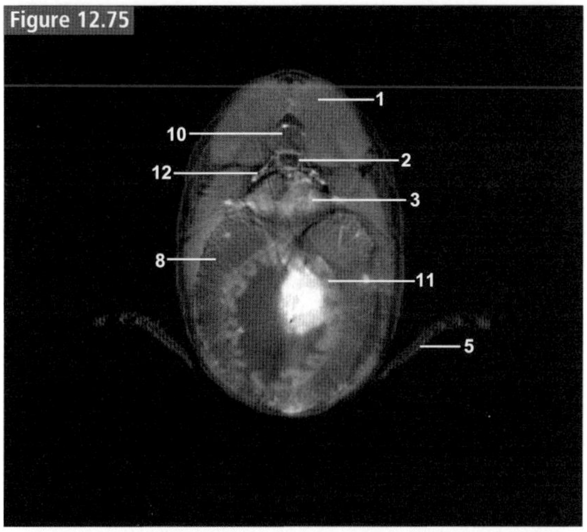

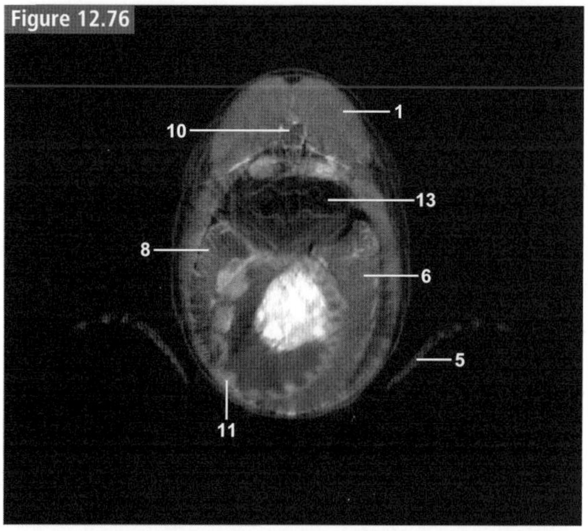

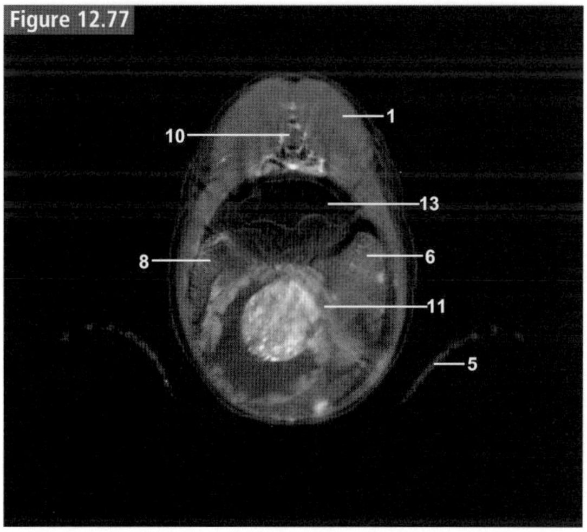

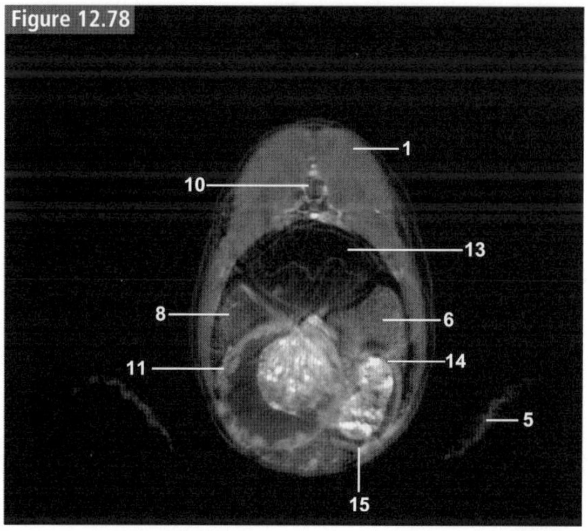

1. myotome	**5.** pectoral fin	**9.** opercle	**13.** anterior chamber swim bladder
2. weberian vertebra	**6.** liver	**10.** spinal cord	**14.** posterior intestine
3. pronephros	**7.** pectoral fin musculature	**11.** intestinal bulb	**15.** mid intestine
4. esophagus	**8.** pancreas	**12.** haemal spine	

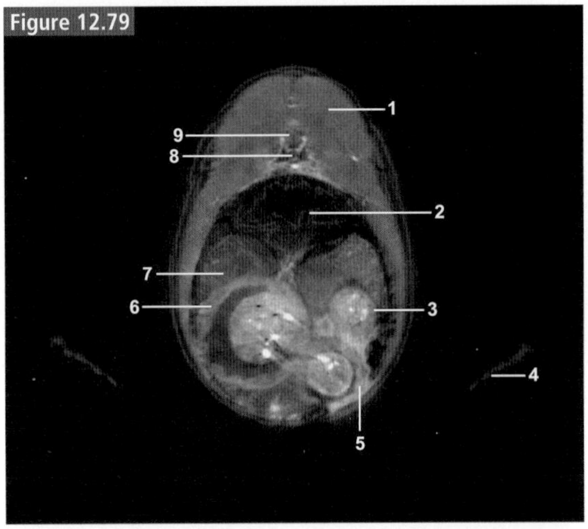

Figure 12.79

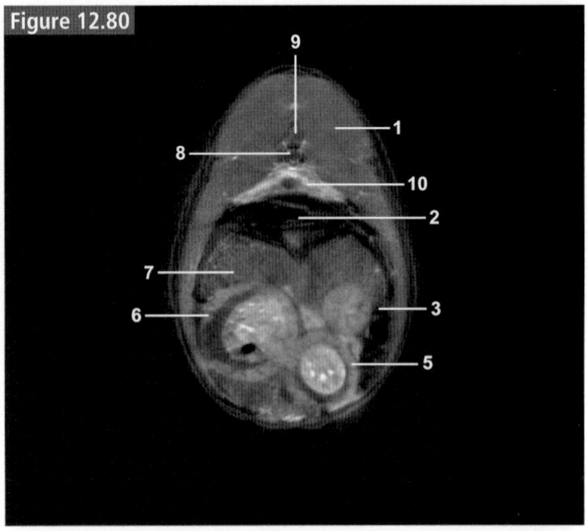

Figure 12.80

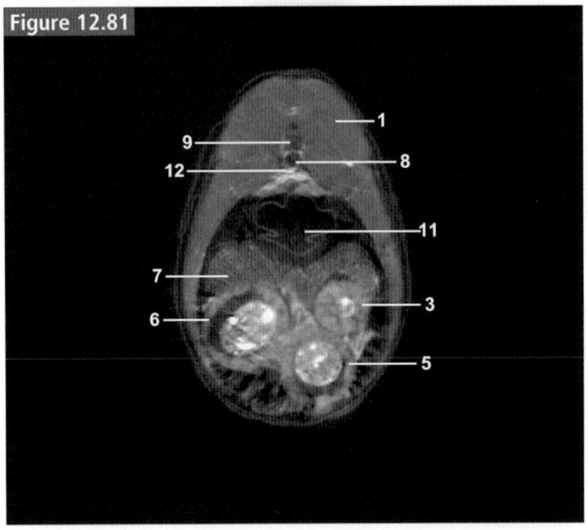

Figure 12.81

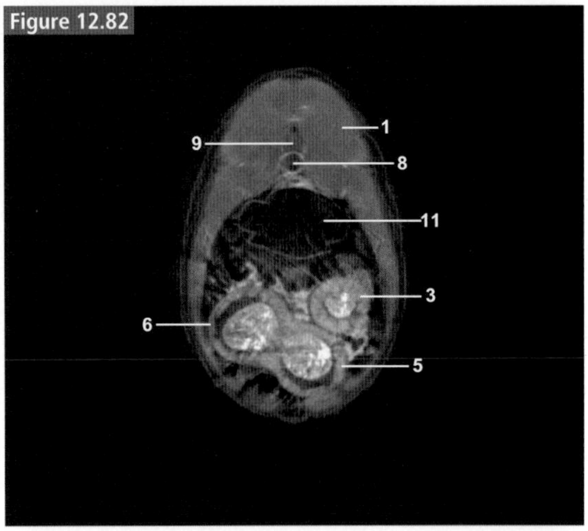

Figure 12.82

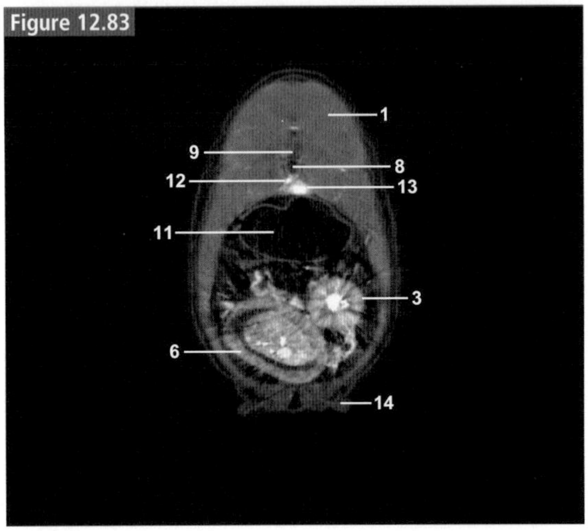

Figure 12.83

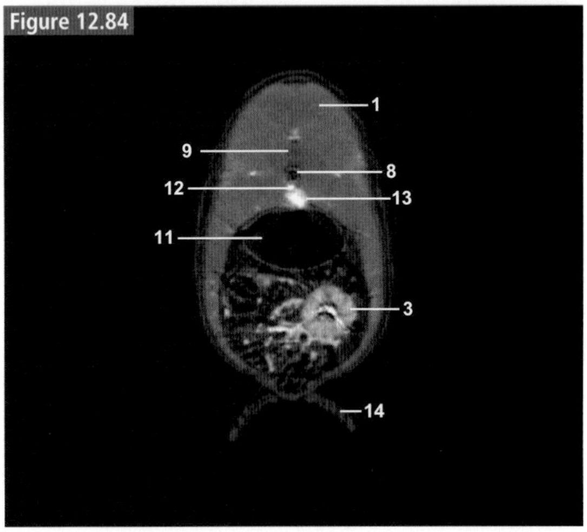

Figure 12.84

1. myotome
2. anterior chamber swim bladder
3. posterior intestine
4. pectoral fin
5. mid intestine
6. intestinal bulb
7. pancreas
8. precaudal vertebra
9. spinal cord
10. mesonephros
11. posterior chamber swim bladder
12. caudal artery
13. posterior cardinal vein
14. pelvic fin

Figure 12.85
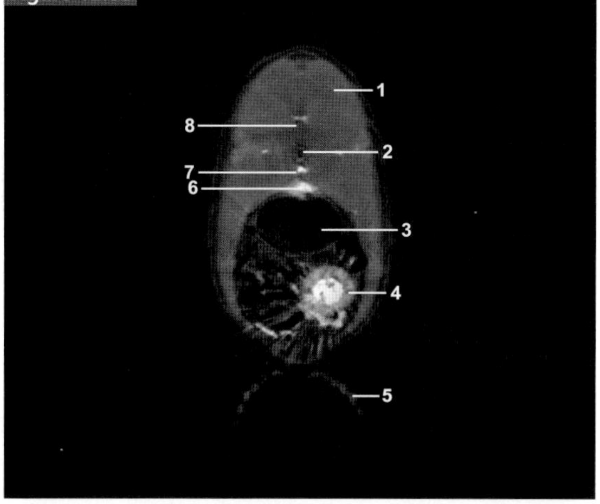

Figure 12.86
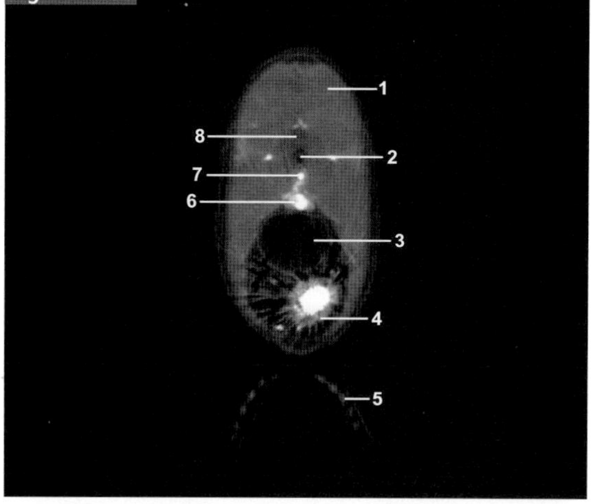

Figure 12.87
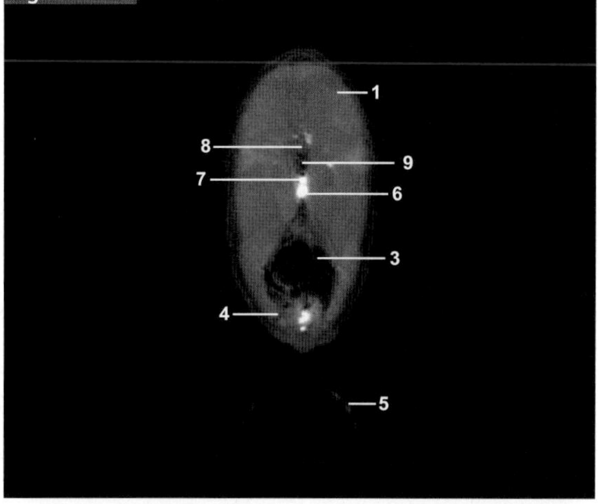

Figure 12.88
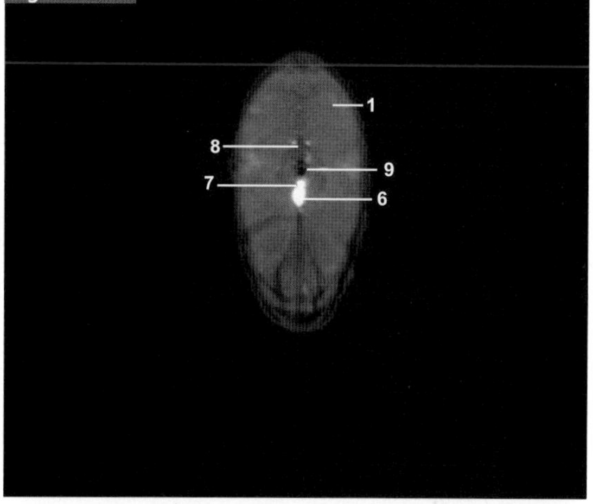

Figure 12.89
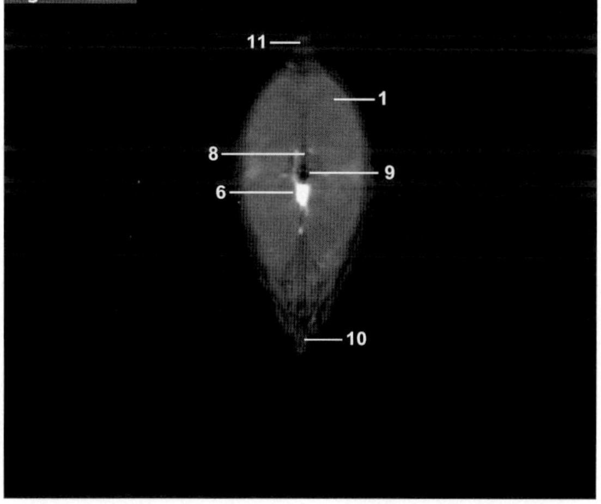

Figure 12.90
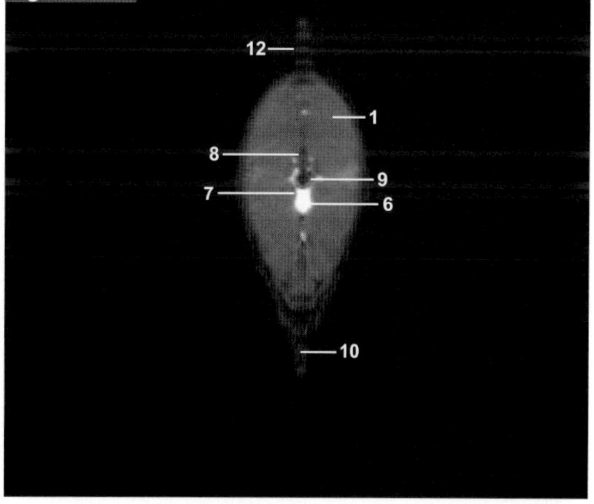

1. myotome	**4.** posterior intestine	**7.** caudal artery	**10.** anal fin
2. precaudal vertebra	**5.** pelvic fin	**8.** spinal cord	**11.** dorsal fin
3. posterior chamber swim bladder	**6.** posterior cardinal vein	**9.** caudal vertebra	**12.** dorsal fin lepidotrichia

REFERENCES

Amsterdam, A., Sadler, K. C., Lai, K., Farrington, S., Bronson, R. T., & Lees, J. A., et al. (2004). Many ribosomal protein genes are cancer genes in zebrafish. *PLoS Biology, 2*, E139.

Berghmans, S., Murphey, R. D., Wienholds, E., Neuberg, D., Kutok, J. L., & Fletcher, C. D. M., et al. (2005). tp53 mutant zebrafish develop malignant peripheral nerve sheath tumors. *Proceedings of the National Academy of Sciences of the United States of America, 102*, 407–412.

Bryson-Richardson, R. J., Berger, S., Schilling, T. F., Hall, T. E., Cole, N. J., & Gibson, A. J., et al. (2007). FishNet: An online database of zebrafish anatomy. *BMC Biology, 5*, 34.

Carvan, M. J., Loucks, E., Weber, D. N., & Williams, F. E. (2004). Ethanol effects on the developing zebrafish: Neurobehavior and skeletal morphogenesis. *Neurotoxicology and Teratology, 26*, 757–768.

Darland, T., & Dowling, J. E. (2001). Behavioral screening for cocaine sensitivity in mutagenized zebrafish. *Proceedings of the National Academy of Sciences of the United States of America, 98*, 11691–11696.

Hurlstone, A. F. L., Haramis, A.-P. G., Wienholds, E., Begthel, H., Korving, J., & van Eeden, F., et al. (2003). The Wnt/beta-catenin pathway regulates cardiac valve formation. *Nature, 425*, 633–637.

Isogai, S., Horiguchi, M., & Weinstein, B. M. (2001). The vascular anatomy of the developing zebrafish: An atlas of embryonic and early larval development. *Developmental Biology, 230*, 278–301.

Kamei, M., Isogai, S., & Weinstein, B. M. (2004). Imaging blood vessels in the zebrafish. *Methods in Cell Biology, 76*, 51–74.

Kimmel, C. B., Ballard, W. W., Kimmel, S. R., Ullmann, B., & Schilling, T. F. (1995). Stages of embryonic development of the zebrafish. *Developmental Dynamics, 203*, 253–310.

Langenau, D. M., Traver, D., Ferrando, A. A., Kutok, J. L., Aster, J. C., & Kanki, J. P., et al. (2003). Myc-induced T cell leukemia in transgenic zebrafish. *Science, 299*, 887–890.

Lieschke, G. J., & Currie, P. D. (2007). Animal models of human disease: Zebrafish swim into view. *Nature Reviews Genetics, 8*, 353–367.

Lockwood, B., Bjerke, S., Kobayashi, K., & Guo, S. (2004). Acute effects of alcohol on larval zebrafish: a genetic system for large-scale screening. *Pharmacology, Biochemistry, and Behavior, 77*, 647–654.

Mueller, T., & Wulliman, M. F. (2005). *Atlas of Early Zebrafish Brain Development*. Amsterdam: Elsevier.

Patton, E. E., Widlund, H. R., Kutok, J. L., Kopani, K. R., Amatruda, J. F., & Murphey, R. D., et al. (2005). BRAF mutations are sufficient to promote nevi formation and cooperate with p53 in the genesis of melanoma. *Current Biology, 15*, 249–254.

Peterson, R. T., Shaw, S. Y., Peterson, T. A., Milan, D. J., Zhong, T. P., & Schreiber, S. L., et al. (2004). Chemical suppression of a genetic mutation in a zebrafish model of aortic coarctation. *Nature Biotechnology, 22*, 595–599.

Schilling, T. (2002). The morphology of larval and adult zebrafish. In R. Dahm & C. Nusslein-Volhard (Eds.), *Zebrafish: A Practical Approach.* Oxford: Oxford University Press.

Sharpe, J., Ahlgren, U., Perry, P., Hill, B., Ross, A., & Hecksher-Sørensen, J., et al. (2002). Optical projection tomography as a tool for 3D microscopy and gene expression studies. *Science, 296*, 541–545.

Shepard, J. L., Amatruda, J. F., Stern, H. M., Subramanian, A., Finkelstein, D., & Ziai, J., et al. (2005). A zebrafish bmyb mutation causes genome instability and increased cancer susceptibility. *Proceedings of the National Academy of Sciences of the United States of America, 102*, 13194–13199.

Wienholds, E., Schulte-Merker, S., Walderich, B., & Plasterk, R. H. A. (2002). Target-selected inactivation of the zebrafish rag1 gene. *Science, 297*, 99–102.

Wienholds, E., van Eeden, F., Kosters, M., Mudde, J., Plasterk, R. H. A., & Cuppen, E. (2003). Efficient target-selected mutagenesis in zebrafish. *Genome Research, 13*, 2700–2707.

Wullimann, M. F., Rupp, B., & Reichart, H. (1996). *Neuroanatomy of the Zebrafish Brain*. Basle: Birkhauser.

INDEX